NUTRITION AND DIET RESEARCH PROGRESS

NUTRACEUTICALS

EVIDENCE BASED USE

NUTRITION AND DIET RESEARCH PROGRESS

Additional books and e-books in this series can be found on Nova's website under the Series tab.

NUTRITION AND DIET RESEARCH PROGRESS

NUTRACEUTICALS

EVIDENCE BASED USE

RALUCA MARIA POP
EDITOR

Medicine & Health
New York

Library of Congress Cataloging-in-Publication Data

Names: Pop, Raluca Maria, editor. Title: Nutraceuticals : evidence based use / Raluca Maria Pop, editor.
Description: New York : Nova Science Publishers, [2020] | Series: Nutrition and diet research progress | Includes bibliographical references and index.
Summary: "Nutraceuticals are commonly defined as a substance that may be considered a food or part of food that provides medical or health benefits, encompassing the prevention and treatment of disease. So far, a lot of emphases were put on nutraceuticals health benefits contrasting the little information available on their efficacy, side-effects or possible drug interactions. Within this context, this book aims to raise the consumers' awareness of nutraceuticals diversity, their potential health benefits, side effects, and drug interactions as well as current legislation. Therefore this publication aims to provide up-to-date information about the safe use of nutraceuticals in everyday life from a critical point of view. Also, their proven health benefits will be approached in relationship with the most common diseases, where they are used to prevent, treat or as a complementary therapy to classical medication. Because, the use of experimental models is a mandatory step for understanding the mechanisms of action and characteristic effects of nutraceuticals, the most relevant in vitro and in vivo studies that proved the beneficial effects of some of the most widespread nutraceuticals were described. Also, because of their high prevalence within the general population, the use of nutraceuticals in cardiovascular diseases, diabetes, respiratory and viral infections will be discussed. Special focus will be given to the use of nutraceuticals in the pediatric population. Also, their effect on dermatological conditions and neurological conditions will be addressed. Nutraceuticals may represent a valuable asset in medicine, but future research is required to establish their full therapeutic potential and to translate current findings into clinical practice"-- Provided by publisher.
Identifiers: LCCN 2020022895 (print) | LCCN 2020022896 (ebook) | ISBN 9781536180527 (hardcover) | ISBN 9781536180534 (adobe pdf)
Subjects: LCSH: Functional foods. Classification: LCC QP144.F85 N8847 2020 (print) | LCC QP144.F85 (ebook) | DDC 613.2--dc23
LC record available at https://lccn.loc.gov/2020022895
LC ebook record available at https://lccn.loc.gov/2020022896

Published by Nova Science Publishers, Inc. † New York

CONTENTS

PREFACE

Nutraceuticals are commonly defined as substances that may be considered a food or part of food that provides medical or health benefits, encompassing the prevention and treatment of disease. The wide spectrum of nutraceuticals can include the following: minerals, vitamins, and other dietary supplements, herbal products, dietary enzymes, dietary fibers, hydrolyzed proteins, phytochemicals like flavonoids, proanthocyanidins, carotenoids, prebiotics, and probiotics. Even though the composition of nutraceuticals can be simple (single compounds like β–carotene, resveratrol, vitamins) or more complex (extracted phytochemicals from vegetable sources, functional foods) the consumers should keep in mind that they are all active substances with pharmacological properties and therefore can induce therapeutic effects. So far, a lot of emphases were put on nutraceuticals health benefits contrasting the little information available on their efficacy, side-effects or possible drug interactions. Within this context, this book aims to raise the consumers' awareness of nutraceuticals diversity, their potential health benefits, side effects, and drug interactions as well as current legislation.

Therefore this publication aims to provide up-to-date information about the safe use of nutraceuticals in everyday life from a critical point of view. Also, their proven health benefits will be approached in relationship with the most common diseases, where they are used to prevent, treat or as a complementary therapy to classical medication.

Because, the use of experimental models is a mandatory step for understanding the mechanisms of action and characteristic effects of nutraceuticals, the most relevant *in vitro* and *in vivo* studies that proved the beneficial effects of some of the most widespread nutraceuticals were described. Also, because of their high prevalence within the general population, the use of nutraceuticals in cardiovascular diseases, diabetes, respiratory and viral infections will be discussed. Special focus will be given to the use of nutraceuticals in the pediatric population. Also, their effect on dermatological conditions and neurological conditions will be addressed.

Nowadays, one of the major health problems worldwide, with a negative impact on the morbidity and mortality or the cost of the patient's health, is cardiovascular disease (CVD). Often, these problems are associated with metabolic disorders (obesity, dyslipidemia, etc). Nutraceuticals exert their cardioprotective effects by lowering lipid plasma levels, improving endothelial function or reducing arterial stiffness. Rapidly increasing interest in nutraceutical consumption may lead to new possibilities for population-based strategies for CVD risk reduction. Lately, great importance has been given to the role of the gut microbiota in many diseases, from gastrointestinal problems to allergies or even cancer. In this regard, the relationship between microbiota and cardiovascular diseases has been studied. Therefore, the role of the microbiota and probiotics in cardiovascular diseases were approached.

Knowing that one from eleven people around the world has diabetes mellitus (DM) the role of nutraceuticals in its prevention was investigated. An update of the information about the potential role of nutraceuticals in the prevention and adjuvant use in the treatment of hyperglycemia and chronic complications of DM was described.

Next, the focus was directed towards a few dietary supplements with antioxidant effects that may play a significant role in respiratory allergic diseases prevention and control. Asthma and allergic rhinitis are chronic respiratory disorders associated with airway inflammation. Environmental factors, including diet in association with genetic susceptibility, play a critical role in their pathophysiology. Systemic inflammation can be a feature of asthma and allergic rhinitis and is associated with worse disease outcomes. Several components found in foods may have antioxidant, anti-allergic and anti-inflammatory properties, which can improve the asthma control or could prevent its occurrence. It has been proposed that oxidative stress which is present in association with chronic inflammation in asthma can be reverted by supplemental or dietary intake of antioxidant vitamins (A, C, D, and E).

Also, various aspects of vitamin D interaction with viral infections, based mainly on *in vivo*, and to a lesser extent on *in vitro* studies, aimed to a better understanding of the clinical benefits of vitamin D. Vitamin D has multiple, important roles in the human organism. It was first recognized for its basic function in bone mineralization and controlling calcium homeostasis. Additionally, through its vitamin D receptor (VDR), it acts as a hormone in the modulation of the innate and adaptive immune responses to various types of viral infection. Observational studies have demonstrated a significant association between a low vitamin D status and an increased risk of certain viral infections, but the findings of randomized controlled trials have been conflicting.

The most common nutraceuticals used in pediatrics were also described. Their role in prophylaxis, frequent pathologies or as a supplement to drug therapy was investigated. The controversial situations of frequently used nutraceuticals in children like probiotics, vitamin D and omega 3 fatty acids together with a comprehensive picture of their use and the available evidence are provided.

An overview of supplements that are truly proved useful for dermatological purposes, whether as adjuvants in dermatological conditions or as aides in maintaining skin, nail and hair health is also provided. Furthermore, dosages, side effects, and specific formulations will be discussed.

It is known that nutrition and dietary components can influence brain functioning. Neurological diseases, Parkinson's disease, Alzheimer's dementia or multiple sclerosis or cerebral atherosclerosis can benefit from nutraceutical use. As the common pathways involved in the neurodegenerative diseases are oxidative stress, inflammation, and mitotoxicity, some nutraceuticals can help in inhibiting those processes. Some of them like, coenzyme Q10 or resveratrol, can be used in several neurodegenerative diseases. However, their use must be closely monitored, as they can have drug-to-drug interaction, especially with cardiovascular or neuropsychiatric medication.

Modern life comes with stress and altered sleep. Illness may arise from this. Nutraceuticals appear as a feasible way to balance this effect or to prevent it. The most important and up-to-date findings regarding our present-day knowledge in the subject, the safety concerns and clinical uses in psychiatry alongside their evidence, and future perspectives will be discussed.

Nutraceuticals may represent a valuable asset in medicine, but future research is required to establish their full therapeutic potential and to translate preclinical findings into clinical practice.

In: Nutraceuticals: Evidence Based Use
Editor: Raluca Maria Pop
ISBN: 978-1-53618-052-7

Chapter 1

INTRODUCTION TO NUTRACEUTICALS

***Raluca Maria Pop*[1,*], *Dan Claudiu Măgureanu*[2], *Ștefan Octavian Macovei*[2] and *Veronica Sanda Chedea*[3]**

[1]Department of Pharmacology, Toxicology and Clinical Pharmacology, "Iuliu Hatieganu" University of Medicine and Pharmacy, Cluj-Napoca, Romania

[2]"Iuliu Hatieganu" University of Medicine and Pharmacy, Cluj-Napoca, Romania

[3]Research Station for Viticulture and Enology, Blaj, Romania

ABSTRACT

Nutraceuticals are commonly defined as a substance that may be considered a food or part of food that provides medical or health benefits, encompassing the prevention and treatment of disease. The wide spectrum of nutraceuticals can include the following: minerals, vitamins, and other

[*] Corresponding Author's E-mail: raluca_parlog@yahoo.com.

dietary supplements, herbal products, dietary enzymes, dietary fibers, hydrolyzed proteins, phytochemicals like flavonoids, proanthocyanidins, carotenoids, prebiotics, and probiotics. Even though the composition of nutraceuticals can be simple (single compounds like β–carotene, resveratrol, vitamins) or more complex (extracted phytochemicals from vegetable sources, functional foods) the consumers should keep in mind that they are all active substances with pharmacological properties and therefore can induce therapeutic effects. So far, a lot of emphases were put on nutraceuticals health benefits contrasting the little information available on their efficacy, side-effects or possible drug interactions. Within this context, this chapter aims to raise the consumers' awareness on nutraceuticals diversity, their potential health benefits, side effects, and drug interactions as well as current legislation.

Keywords: nutraceuticals, nutrients, herbals, phytochemicals, fortified products, drug interactions

INTRODUCTION

Over time there have been several numbers of definitions of the term "nutraceuticals". In 1989 Dr. Stephen DeFelice introduced the term "Nutraceutical" using a combination of the two distinct terms "Nutrition" and "Pharmaceutical" with the meaning of "a food (or part of a food) that provides medical or health benefits, including the prevention and/or treatment of a disease" (DeFelice 1995). The nutraceuticals spectrum can include various and numerous products from natural products to dietary supplements and functional foods. They can be classified either as traditional or non-traditional nutraceuticals. The traditional nutraceuticals are classified in chemical constituents, probiotic microorganisms, and nutraceutical enzymes while non-traditional nutraceuticals are classified in fortified nutraceuticals and recombinant nutraceuticals as depicted in Figure 1.

The most known class of the *traditional nutraceuticals (A)* is the *chemical constituents.* They refer to the nutrients, herbals, and phytochemicals.

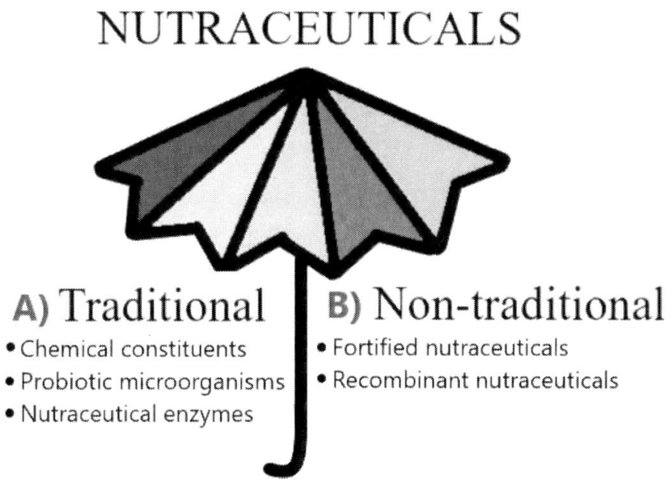

Figure 1. Nutraceuticals classification.

Chemical Constituents - The Nutrients

Nutrients are substances with well-known nutritional functions. The vitamins (A, B complex, C, D, E, K), fatty acids (omega 3 fatty acids, polyunsaturated fatty acids, monosaturated fatty acids), minerals (calcium, copper, selenium, potassium, zinc), amino acids are the most common nutrients (Mohammad 2019). Common nutrients are already known for their health benefits effects like antioxidant action (vitamins A, E, C), their strengthening effect on bone and teeth health (calcium, phosphorus, vitamin D, vitamin C), their influence on energy production and oxygen transportation to tissues (Iron), blood clotting (Vitamin K) and many others. In the next table, the health effects of some of the most used nutrients in the human diet in different experimental models are presented.

Table 1. Nutrients evidenced health benefits

Nutrients	Models	Evidenced Health benefits	References
Folic acid	Tau protein developed in *E. coli* bacteria	-protein aggregation was delayed because of folic acid binding through different sites of tau protein	(Ghasemzadeh and Hossein Riazi 2019)
Magnesium Zinc Calcium Vitamin D	60 GDM (gestational diabetes mellitus) patients**	-reduced FPG (fast plasma glucose), levels of serum insulin, QUIKI (quantitative insulin sensitivity check index), HOMA-IR (homeostatic model assessment; method used to quantify insulin resistance and beta-cell function) -decreased VLDL-cholesterol concentrations, serum triglycerides, AIP (atherogenic index of plasma), CRR (cardiac risk ratio), AP (atherogenic coefficient)	(Karamali et al. 2018)
Glycine	Human umbilical vein endothelial cells (HUVECs)	-HUVECs supported to form networks	(Guo et al. 2017)
	Chorioallantoic angiogenesis membrane (CAM)	-promote angiogenesis	
	Mice femoral artery ligation and section	-improved neovascularization with signs of blood flow from the third day	
Omega 3 fatty acids (n-3 PUFA)	Cirrhosis and hepatectomy in rats	-AST (Aspartate transaminase) and ALT (Alanine transaminase) levels decreased in the group feed with high concentration n-3 PUFA compared to control group -Platelet counts of the groups feed with n-3 PUFA were higher than the control group -IL-6 (pro-inflammatory) was down-regulated in the high concentration PUFA group compared to the control group -IL-10 (anti-inflammatory) was up-regulated in the PUFA fed groups -MMP-9 (proteolysis of collagen fiber) level was increased in the PUFA groups -TIMP-1 (healing process) decreased in high concentration PUFA group compared to control	(Yang et al. 2019)
Amino acids	14 healthy elderly subjects	-increased serum phenylalanine levels from the basal state -decreased the anabolism of amino acid from muscles in both groups	(Volpi et al. 2003)
Monosatu-rated fatty acids (MFA)	4680 subjects in a cross-sectional epidemiological study	-decreased DBP (diastolic blood pressure) -MFA from vegetable sources may prevent and control adverse BP levels	(Miura et al. 2013)

** Double-blind, randomized, placebo-controlled clinical trial.

Chemical Constituents - The Herbals

Table 2. Herbals evidenced health benefits

Herbals	Models	Evidenced Health benefits	References
Garlic	51 healthy adults with obesity**	-decreased IL-6 and TNF-α serum levels -increased innate lymphocytes (γδ-T cells) level -decreased Natural killer T cells (NKT) cells level -decreased LDL serum level	(Xu et al. 2018)
	40 patients with metabolic syndrome	-reduced abdominal obesity -decreased blood pressure -decreased serum fasting blood glucose levels -decreased serum triglycerides levels -increased serum HDL-cholesterol levels	(Choudhary, Jani, and Sharma 2018)
Onion	38 patients with Alopecia areata*	-full hair re-growth in 8 weeks (20 out of 23 patients)	(Sharquie and Al-Obaidi 2002)
	56 breast cancer patients***	-reduced hyperglycemia induced by doxorubicin -reduced insulin resistance induced by doxorubicin	(Jafarpour-Sadegh et al. 2017)
Ginger	20 patients at high risk for colorectal cancer*	-reduced COX-1 expression inducing the decrease of PGE2 synthesis	(Jiang et al. 2013)
	168 patients with primary dysmenorrhea	-reduced pain intensity	(Adib Rad et al. 2018)
	88 patients with type 2 diabetes**	-reduced the levels of HbA1c and FBS (fasting blood sugar)	(Mozaffari-Khosravi et al. 2014)
Turmeric	Human colon cancer cell line SW480	-anti-metastatic effect -inhibitory effect on cancer cells proliferation	(Hosseini, Zand, and Cheraghpour 2019)
	160 patients with knee osteoarthritis***	- decreased oxidative stress - decreased serum IL-1 level	(Srivastava et al. 2016)
Golden seal	Antibacterial activity assay	-antibacterial effect against *Staphylococcus aureus*	(Britton et al. 2018)
Liquorice	Human breast cancer cell line	-anticancer effect	(Wang et al. 2013)
	Human prostate cancer cell line	-anticancer effect	(Seon et al. 2012)
	Human hepatoma cells HepG2	-hepatoprotective activity by up-regulation effect on CYP3A4, CYP3A11, SHP (small heterodimer partner) -down-regulation of CYP7A1	(Wang et al. 2012)
	Murine cell line RAW 264.7	-anti-inflammatory effects by COX-2, iNOS, TNF- α, NO, IL-6, IL-1β, PGE2 and ROS reduced expression	(Wang et al. 2011)

* single-blind, placebo-controlled clinical study.

** double-blind, randomized, placebo-controlled clinical trial.

*** triple-blind, randomized, placebo-controlled clinical trial study.

Herbals refers to the multitude of plants used for centuries in traditional medicine to treat or to prevent different diseases like garlic (anti-inflammatory, hypotensive, antibacterial, antifungal, antithrombotic effects), onion, ginger (carminative, antiemetic, cholagogue effects), Echinacea (immunostimulant effect), St. John's Worth (anxiolytic, antiinflammatory, antidepressant effects), liquorice (expectorant, secretolytic, gastroprotective effects), turmeric (anti-inflammatory, antioxidant, hypolipidemic, hypoglycemic effects), goldenseal (antimicrobial, antihemorrhagic, astringent, anti-inflammatory effects), senna (laxative effect) and others (Gupta et al. 2010; Ruchi et al. 2017). The next table describes the most common herbals used in human diets and their pharmacological effects proved in recent experimental studies.

Chemical Constituents - The Phytochemicals

Phytochemicals refer to carotenoids (lycopene, lutein, α-carotene, β-carotene, zeaxanthin, cryptoxanthin), phenolic compounds (flavonols, flavanones, flavones, anthocyanins, phenolic acids, catechins), tocotrienols and tocopherols, phytosterols, alkaloids (tropane alkaloids, ergot alkaloids, quinine, morphine, coumarin, scopoletin) (Torabally and Rahmanpoor 2019). These are bioactive compounds are extracted from plants and possess distinct biological action. For example carotenoids and phenolic compounds are well known for their cardioprotective, anti-cancer, anti-diabetic, anti-inflammatory, anti-obesity, antioxidant activities and skin protective effects (Sathasivam and Ki 2018; Tungmunnithum et al. 2018). Some of the most representative compounds and their health demonstrated benefits are presented in the next table (Table 3).

Table 3. Phytochemicals evidenced health benefits

Phytochemicals	Models	Evidenced Health benefits	References
Lycopene	HL-1 cardiomyocytes	-improved cells viability (1 μM lycopene concentration)	(Tong et al. 2016)
	I/R (ischemia-reperfusion) induced Myocardial Infarct in mices	-MDA was significantly reduced -AMP activated kinase signaling/AMP activated kinase (AMPK-ACC) pathway was moderately activated	
β-cryptoxanthin (β-CRX)	TSOD (Tsumura Suzuki obese diabetes) and TSNO (Tsumura Suzuki non-obese diabetes) rats	-reduced body weight in TSOD-Experimental (E) mice -reduced β-CRX serum lipids levels in TSOD-E mice -decreased liver weight in obese rats experimental group -reduced adipocyte hypertrophy in TSOD-E -reduced adipose tissue weight - peroxisome proliferator-activated receptors (PPAR-α) up-patients	(Takayanagi 2011)
Tocotrienols	88 patiens with WML (white matter lession)**	-WML progression slowed down	(Gopalan et al. 2014)
Scopoletin	Mice splenic cells and central nervous system (CNS) cells	-reduced dentritic cells activation in mice splenic cells -inhibited Th1 and Th17 cells formation in mice splenic cells - impaired inflamatory response in CNS cells: CD11c+ (DC), CD45+ (leukocyte) -decreased CD11b+ (macrophage/microglia) cell populations -decreased CD4+ and CD8+ T cells	(Zhang et al. 2019)
	Bone-marrow-derived dendritic cells (BM-DCs)	-suppressed biomarkers of DC activation: CD 80 and CD86 -suppressed formation of IL-12 (Th1 cells differentiation), IL-23, IL-6 and TNF-α -inhibited NF-kB signalling pathway	
	EAE (experimental autoimmune encephalomyelitis) induced in mice	-reduced progression of the disease with a reduced number of inflammatory cells in CNS white matter -decreased demyelination areas	
Polyphenols (epigallocatechin gallate – EGCG)	Mouse macrophage cells	-Notch expression was attenuated (role in macrophage-mediated inflammation) -Notch signalling pathway bound with EGCG and suppress the inflammatory response	(Huang et al. 2019)
	Streptozotocin-induced diabetes in mice	-diabetic wound healing was improved -inhibited Notch signalling pathway -decreased IL-6, IL-1β, TNF-α levels	

** Double-blind, randomized, placebo-controlled clinical trial.

Probiotic Microorganisms

The second class of traditional microorganisms is the *probiotic microorganism.* They are considered live microorganisms that offer health benefits to the host when are administered in proper amounts (Martín and Langella 2019). Usually, the traditional probiotics belong to *Lactobacillus spp.*, *Bifidobacterium spp.*, *Bacillus* and *Saccharomyces*, all with a long history of use that proved their safety profile (Martín and Langella 2019). Several clinical studies proved the effect of probiotics in the treatment of different diseases such as obesity, type 2 diabetes, insulin resistance syndrome, fatty liver disease or non-alcoholic fatty liver disease (Markowiak and Ślizewska 2017). Few of the positive effects determined by different probiotics are observed in different experimental settings as described in the next table (Table 4).

Table 4. Probiotics evidenced health benefits

Probiotics	Models	Evidenced Health benefits	References
Lactobacillus spp.	68 elderly nursing home residents**	- reduce the number of days with fever in feverish episodes	(Kushiro et al. 2019)
	high-fat/high-cholesterol diet in rats	- hypocholesterolemic effect -improved cholesterol metabolism	(Jeon, Lee, and Chang 2019)
Bifidobacterium spp.	human gastric epithelial cell line	- inhibitory effects against *Helicobacter pylori*	(Shirasawa et al. 2010)
	40 ageing female mice	- host-immune system modulation by senescence delay	(Fu et al. 2010)
	122 patients**	- improved life quality of patients with irritable bowel syndrome	(Guglielmetti et al. 2011)
Bacillus	HT-29 cells	- protective effects against *Salmonella typhimurium*	(Kawarizadeh et al. 2019)
	36-40 high-fat diet in mice	- decreases body weight - suppresses the chronic inflammation - increases the oxidation of fatty acid in the liver - decreases fat accumulation - effects against induced insulin resistance	(Kim et al. 2018)
Saccharomyces	156 male chicks	- improved intestinal ecosystem, - protective effect against *Campylobacter jejuni* infection	(Massacci et al. 2019)

** Double-blind, randomized, placebo-controlled clinical trial.

Nutraceutical Enzymes

Table 5. Nutraceutical enzymes evidenced health benefits

Nutraceutical enzymes	Models	Evidenced Health benefits	References
Bromelain	Cell lines isolated from human colon adenocarcinomas (CRC)	-CRC survival suppressed -induced autophagy: proteins related to autophagy (ATG5/12, p62, beclin and LC3I/II) -induced apoptosis: biomarkers (apoptosis-inducing factor caspases -8,-9,-3, Endo G)-DNA fragmentation -oxidative stress and superoxide production in CRC	(Chang et al. 2019)
	Zebrafish embryos injected with CRC	-inhibitory effect on CRC proliferation	
	Male BALB mice injected with CRC cells	-tumour growth inhibited	
Pancrelipase	54 patients (with chronic pancreatitis or exocrine pancreatic insufficiency with or without diabetes mellitus**	-Coefficient of Fat Absorption (CFA) was significantly increased in both diabetes mellitus (DM) group/ non-DM group treated with pancreatic enzyme -Coefficient of Nitrogen Absorption (CNA) was significantly increased in DM group with enzyme treatment	(Whitcomb et al. 2016)
Ox bile	*Campylobacter jejuni*, isolated from humans	-inhibition of *C. jejuni* growth (concentration-dependent mode) -down-regulation of GalU enzyme (production of different structure) -up-regulation of Ferritin (in order to survive *C. jejuni* sequesters iron) -up-regulation of Flagellin A and flagellin B -up-regulation of ATP synthase (energy for resistance against bile stress) -increased pump CmeABC activity (bile resistance) -down-regulation of 2 glycolysis enzymes -down-regulation of 4 TCA (tricarboxylic acid cycle) enzymes, of fumarate reductase, of fumarase, of succinyl-CoA synthetase and of malate dehydrogenase	(Fox et al. 2007)
Pancreatic enzyme	31 pediatric patients with persistent diarrhoea**	-reduced the diarrhoea episode with 7 days in the treatment group compared to the placebo group	(Widodo et al. 2018)
Agalsidase Alfa (substitution for α galact-oxidase)	10 patients with Fabry disease (FD) in a 4 years prospective clinical trial	-Treatment tolerated with the improvement of FD symptoms	(Schiffmann et al. 2010)

** Double-blind, randomized, placebo-controlled clinical trial.

The thirds class of traditional nutraceuticals is the *nutraceutical enzymes*. These are derived from microbial, animal and plant sources. The most common nutraceutical enzymes are hemicellulase, catalase, amyloglucosidase, glucoamylase, cellulase, invertase, α-galactosidase, β-amylase, bromelain, OxBile, pancrelipase, trypsin and pepsin (Singh and Sinha 2012). Usually, the nutraceutical enzymes are used in the treatment of gastrointestinal tract related diseases, in ulcerative colitis, in constipation, diarrhea or several rare diseases like Pompe disease, Hunter syndrome, Gaucher disease or Fabry disease (Chanda et al. 2019). The next table summarizes some of the recent studies regarding the use of nutraceutical enzymes in different experimental studies (Table 5).

The second class of nutraceuticals is represented by the *non-traditional nutraceuticals (B)*. Generally, this class refers to the enriched foods with supplements or biotechnologically designed crops to produce specific nutrients known for their bioactive activities and positive health effects for humans. The non-traditional nutraceuticals are classified in two subclasses as follows: the fortified nutraceuticals and the recombinant nutraceuticals.

Fortified Nutraceuticals

Fortified nutraceuticals or *functional foods* are the food containing added ingredients, technologically developed for a specific health benefit (Siró et al. 2008). Among functional food, we can remember the fortified products, the enriched products, the altered products and enhanced products (Tung Bui and Duy Tung 2015).

Fortified Nutraceuticals - The Fortified Products

The fortified products are the products which were added extra nutrients like juices fortified with different vitamins or minerals (C, B, Zn, Ca, Mg), fortified milk with vitamin D or fortified cereals. Although there are a relatively large number of fortified products on the market their impact on human health is poorly studied. Some of the few existing studies in the literature are summarized in the next table (Table 6).

Table 6. Fortified products evidenced health benefits

Fortified Products	Models	Evidenced Health benefits	References
Vitamin D fortified milk	713 children	-increased 25 hydroxyvitamin D serum levels -increased vitamin D serum levels	(Khadgawat et al. 2013)
Folic Acid fortified flour	23.685 women before pregnancy or marriage	-increased folate serum levels -reduced the birth prevalence of Neural Tube Defects	(Wang et al. 2016)
β–carotene fortified synbiotic food	102 patients with Type 2 Diabetes Mellitus**	-improved insulin metabolism -improved TG, cholesterol, HDL, VLDL and magnesium levels	(Asemi et al. 2016)
Zinc fortified water	184 children	-increased serum zinc levels -lowered morbidity	(Kujinga et al. 2018)
Vitamin A fortified rice	50 school children**	-increase vitamin A serum levels and liver stores	(Pinkaew et al. 2014)
Vitamins A, C, zinc, iron and lysine fortified juice	100 anaemic children**	-reduced the anaemia -increased serum zinc levels -increased serum ferritin levels -increased haemoglobin levels	(Angeles-Agdeppa, Magsadia, and Capanzana 2011)

** Double-blind, randomized, placebo-controlled clinical trial.

Fortified Nutraceuticals - The Enriched Products

The enriched products are the products that have been added nutrients to replace other compounds like vitamins or minerals that are degraded during the production process (Tung Bui and Duy Tung 2015; Shamal and Mohan 2017). An example of an enriched product is the white bread with added vitamins. The vitamins are added after the bleaching process since at this step the originally contained vitamins are depleted. These products are specifically designed to offer multiple beneficial health effects as observed in the next table (Table 7).

Table 7. Enriched products evidenced health benefits

Enriched products	Models	Evidenced Health benefits	References
Protein-enriched familiar foods and drinks	147 patients*	-intervention group reached recommended daily dose of protein (1.2 g/kg/d) and higher energy status	(Beelen et al. 2018)
Phytosterols-enriched low-fat milk	180 patients**	-decreased total cholesterol and LDL-C levels -decreased diastolic blood pressure -no significant change in systolic blood pressure, BMI, weight, HDL-C, triglycerides, and fasting glucose levels	(Cheung et al. 2017)
Baked products enriched with n-3 fatty acids, folates, β-glucans, and tocopherol	17 individuals (overweight and adulterated lipid levels)**	-reduced serum triglycerides levels (due to n-3 Fatty Acids) -a postprandial decrease in insulin levels -decreased homocysteine (due to Folates) -satiety feeling was higher after the enriched meal (due to β-glucans) -n-6 fatty acids lower in active diet -chylomicron fraction decreased	(De Natale et al. 2012)
Lactoferrin-enriched fermented milk	36 patients (acne vulgaris)**	-decreased total lesion count, inflammatory lesion count and acne grade in all subjects from lactoferrin group after 12 weeks	(J. Kim et al. 2010)
Plant sterol-enriched fermented milk	83 hyper-cholesterolemic subjects**	-decreased LDL-C and Total cholesterol in plant sterol group	(Plana et al. 2008)

* Single-blind, placebo-controlled clinical study.

** Double-blind, randomized, placebo-controlled clinical trial.

Fortified Nutraceuticals - The Altered Products

The *altered products* are the foods from which a harmful component was either removed either reduced or replaced with a substance that possess health beneficial effect (Table 8). An example is a fat reduction from the ice cream products and its replacement with whey protein isolate, inulin or probiotics (Underdown, Quail, and Smith 2011).

Table 8. Altered products evidenced health benefits

Altered Products	Models	Evidenced Health benefits	References
Products with unsaturated fat: MUFA (monounsaturated fatty acid) and PUFA(ω-6 polyunsaturated fatty acid) instead of saturated fat	195 subjects with moderate CVD risk in a randomized, controlled, single-blind, parallel-group dietary intervention	- reduced the increase in systolic blood pressure during the night - reduced E-selectin levels - reduced total cholesterol, LDL and total cholesterol/HDL ratio serum levels	(Vafeiadou et al. 2015)

Table 9. Enhanced products evidenced health benefits

Enhanced Products	Models	Evidenced Health benefits	References
Vitamin D–enhanced eggs	51 healthy adults*	-increased 25(OH)D concentrations in the Vitamin groups -no difference in BMI, PTH (parathormone), total cholesterol and weight	(Hayes et al. 2016)
Omega-3 polyunsaturated fatty acids-enriched hen eggs	36 young and healthy people**	-post occlusive reactive hyperemia improved in OMEGA group (enhanced eNOS gene or inhibited ROS production) -blood pressure, TGs and hsCRP decreased in OMEGA group	(Stupin et al. 2018)
Enhanced homestead food production and aquaculture	850 women and children in a parallel cluster randomized controlled trial	-enhanced homestead food production (EHFP) group: higher Zinc and Vitamin A -higher iron, vitamin A and riboflavin in EHFP and aquaculture groups	(Verbowski et al. 2018)
Enhanced nutrient supply to very low birth weight (VLBW) birth weight < 1500 g	44 included VLBW in a randomized controlled multi-intervention, nutritional study	-higher protein intake - stimulated growth velocity	(Strømmen et al. 2017)
Liquid milk with an enhanced content of whey protein	47 subjects (>65 years)	-increased albumin and prealbumim serum levels -increased vitamin D serum levels -increased HDL-cholesterol serum levels - improved weight	(Rambousková et al., n.d.)

* Single-blind, placebo-controlled clinical study.

** Double-blind, randomized, placebo-controlled clinical trial.

Fortified Nutraceuticals - The Enhanced Products

Finally, the *enhanced products* are the food products in which some of the compounds are naturally enhanced using special feed and growing conditions or genetic manipulation (Table 9). The most common examples are the genetically modified corn or soy. Also the rice with high content in β-carotene or eggs with high content in omega-3 (Singh and Sinha 2012; Tung Bui and Duy Tung 2015; Hefferon 2015).

Recombinant Nutraceuticals

The *recombinant nutraceuticals* include the food products obtained using biotechnological methods (enzyme/fermentation technologies, genetic engineering technologies) from modified agricultural crops (Torabally and Rahmanpoor 2019; Singh and Sinha 2012). Examples of recombinant nutraceuticals are the yogurts, cheese, fermented starch, vinegar (Table 10).

All these products belonging to the vast spectrum of nutraceuticals have the aim to offer medical or health advantages through their prophylactic use or even to cure different diseases. The increased consumer's interest in nutraceuticals is best observed in its specific market sales. For example, in 2018 in the USA, the nutraceuticals market accounted to US$ 73,986.0 Mn being expected to grow at a CAGR of 7.1% between 2019 – 2027 (approximately to US$ 138,047.1 Mn by 2027) ("USA Nutraceutical Market to 2027: Increasing Spectrum of Nutraceutical in Sport Nutrition & Pet Food"). Regarding the European nutraceutical market, in 2016 it was valued at USD 44,341.7 million being estimated to reach an 83,757.9 million USD by 2025 at a growing CAGR of 7.3% ("Nutraceuticals Market: Dietary Supplements 2/4 | Hi Europe").

With some of the nutraceuticals like nutrients (vitamins, minerals) and herbals (garlic, onion, St. John's Worth) the long use experience over time situates them in the comfort zone since important aspects like side effects, drug-drug interactions, metabolization are already known. The potential of other nutraceuticals, which evolved hand in hand with technological advancement, to induce both health benefits and side effects is still largely unknown. Unlike drugs, which are subject to the strict and clear regulations, the nutraceuticals follow a different and more permissive set of regulations. This poor framing induces the increased occurrence on the market of a large number of new products each year.

Table 10. Recombinant nutraceuticals evidenced health benefits

Recombinant nutraceuticals	Models	Evidenced Health benefits	References
Cheese	*Lactobacillus sakei* A15	- antimicrobial effects of cheese peptides on gram-positive and gram-negative bacteria	(Rizzello et al. 2005)
	30 type 2 induced diabetes in obese mice	- improved glucose tolerance - decreased the expression of hepatic lipogenesis controlling factor, -decreased sterol regulatory element-binding protein 1c - decreased NADPH-oxidase mRNA expression	(Geurts et al. 2012)
Fermented starch	45 diet controlled in rats	-enhanced gastrointestinal tract cells growth, proliferation and differentiation -increased gut hormones	(Keenan et al. 2012)
	40 ovariectomized virgin female rats	-increased the bacterial populations level in the gut	(Keenan et al. 2013)
	20 high glycemic index starch diet in rats	-increased insulin secretion	(Pawlak et al. 2001)
Wood vinegar from *Litchi chinensis*	Antibacterial activity assay	-antibacterial activity against *Staphylococcus aureus* and *Pseudomonas aeruginosa*	(Yang et al. 2016)
Apple cider vinegar	Antibacterial activity assay	- antimicrobial activity against *E-coli, S. aureus* and *C. albicans*	(Yagnik, Serafin, and J Shah 2018)
	E-coli, C. *albicans* and S. *aureus* infected human monocytes	- reduced IL-6 and TNF-α inflammatory cytokine production - upregulated the phagocytic capacity as compared to unstimulated monocytes	
Nipa vinegar	Induced obesity in mice with high-fat-diet	-reduced lipid deposition - improved lipid profile - increased adipokine expression -suppressed inflammation	(Beh et al. 2017)
Vinegar	155 obese subjects**	-reduced body weight, body mass index and waist circumference -reduced serum triglyceride levels in the vinegar groups compared with control	(Kondo et al. 2014)

** Double-blind, randomized, placebo-controlled clinical trial.

More than that most of the products are not standardized and well characterized. Missing data about the botanicals sources, the preparation processes and its' exact formulation contribute to unsafe consumer exposure. Within this context, the aim of this chapter is to raise the consumers' awareness of the potential health benefits of nutraceuticals but also about their risk exposure.

Nutraceuticals Regulatory Aspects

As previously described we can state that nutraceuticals aim to provide medical or health benefits. Because of their varied composition, the technological process used for their production, their commercialization forms (from capsules, plants extracts till food-like products), nutraceuticals are situated at the borderline between food and drugs.

Within the *European Union (EU)* the concept was to harmonize the regulation with respect to nutritional supplements, derived health foods, and herbal medicines (Anadón et al. 2016). In this regard, in 2002 was adopted the Food Supplements Directive (FSD) (Directive 2002/46/ EC (EC, 2002a)) which sets clear rules for food supplements labeling and clear indications about the vitamins and mineral salts that can be used in food supplements and their maximum admitted levels (Anadón et al. 2016) (Toma et al. 2010). On the other hand, the substances not included in the list but used on the market before the mentioned directive and allowed by the European Food Safety Authority could also be used.

Within the *USA*, The Federal Food, Drug, and Cosmetic Act (FDA) include dietary supplement ingredients as substances that are aiming to increase the total dietary intake. The list of the substances having this role includes the vitamins, the minerals, the herbs, the amino acids, metabolites, extracts or combination of these. It is also stipulated that there is a clear difference between these dietary supplements and drugs since their aim is not to treat, cure, prevent or diagnose any disease. The FDA's Center for Food Safety and Applied Nutrition (*CFSAN)* is the responsible authority for dietary supplements.

In *Canada*, the Natural Health Products Regulations (January 1, 2004) refers to nutraceuticals as natural health products. A list including natural health products is given. The vitamins and minerals, herbals, the homoeopathic medicines, the traditional medicines, amino acids, probiotics and fatty acids are included in the list. More important the rules regarding their good manufacturing processes like production, labeling, packaging and storage are well stipulated. Also, aspects referring to their use in the clinical trials with human subjects are also detailed in a specific section. Moreover, nutraceuticals identified as natural health products must be safe and effective for human consumption, their use being supported the evidence obtained in clinical trials or published in the journals ("Natural Health Products Regulations").

In *China*, it is important to mention that there the four big categories of products that could be included in the nutraceuticals spectrum. Among these traditional Chinese medicines, the health or functional foods, the new resource foods and the dietary supplements differ in terms of applied regulatory aspects from medicines. For example, in the case of health or functional foods, their approval on the market will be performed based on aspects like safety, labelling and quality control (Yang 2016). The regulation of new resource foods (NRF) differs from the one of functional food through the lack of scientific proof evidence regarding their efficacy. The efficacy of these new materials (derived or separated from plants, microorganism or animals that can be used in foods) is based solely on the existing extensive literature. Still, their lack of toxicity and harmful effects on humans health must be demonstrated ("China Regulations on New Food Raw Materials | ChemLinked"; Yang 2016). Regarding the dietary supplements, The Food Safety Law of China stipulates that their certification must be similar to the functional foods one. Finally, Traditional Chinese medicines (TCM) can have a more rigorous or not legislation, depending on if the products are manufactured as drugs or as functional foods. The possibility of the manufacturer to market his product as functional food instead of TCM it exempts him of many steps like clinical trials, safety assessment and efficacy tests (Yang 2016).

In *Japan*, there are several health foods categories like nutraceuticals, functional foods, foods with nutrient function claims (FNFC), foods for specified health uses (FoSHU), foods for special dietary uses (FOSDU), nutritional supplements, dietary supplements, health supplements and others (Umegaki 2016). In 1991 The Ministry of Health, Labor, and Welfare created Establishment of foods for specified health uses (FoSHU system) which included functional foods and continued to transform until 2009 when the Consumer Affairs Agency was created to further evaluate and approve the health claims of these products (Umegaki 2016). The products must respect several requirements for FoSHU approval. In summary, their efficacy, scientific evidence of health benefits, safety and characterization must be well detailed. Also, functional components properties must be conserved as possible with the one in the original state, it mustn't be included in the list of ingredients used in drugs and can be consumed on a daily basis (Umegaki 2016).

In *India*, the Chapter IV, Article 22 of Food Safety and Standards Act 2006, approaches the nutraceuticals, dietary supplements and functional foods with regards to their manufacture, commercialization, distribution or import (Keservani, Jarouliya, and Sharma 2019). Indian framework for nutraceuticals requirements largely refers to nutraceuticals active ingredients and additives permissibility, to the vitamins or minerals concentrations, their daily recommended allowance in order to claim their therapeutic or prophylactic effects. Because of the lack of more specific regulations, manufacturers are often unclear whether the products will be categorized as food, food supplements, or drug according to the specific legislation (Keservani, Jarouliya, and Sharma 2019; Prevention of Food Adulteration Act 1954 and Rules 1955, Food Safety and Standards Act 2006, and Drugs and Cosmetics Act 1940 and Rules 1945).

NUTRACEUTICALS SIDE EFFECTS AND INTERACTIONS

The use of nutraceuticals has known continuous growth over the last years despite the many gaps related to their active constituents, their mechanisms of action, safety and efficacy.

Table 11. Nutraceuticals side effects

Nutrace-uticals	Dose	Models	Reported side effects	References
Phytoestro-gens (isoflavones)	5, 50, 500, 1000 ppm (parts per million)	Pregnant Long-Evans dams (n = 9/group) Dams were feed from gestational day (GD) 12 to postnatal day (PND) 21 with isoflavones Male offspring rats were fed with soy-free nutrients from PND 21 to 90	- testis weight higher in 50, 500 and 1000 ppm group - 5 ppm group: drop in testosterone (T) production - 1000 ppm group: increase serum T level and Leydig cell secretion; T production per cell decreased	(Akingbemi et al. 2007)
Homocyste-ine	-	636 black ethnicity subjects: stroke group (n = 457) and control group (n = 179) in a prospective study	- association between high levels of homocysteine and SVD stroke (ischemic or hemorrhagic) and cardioembolic stroke - non-significant in LVD SVD-small vessel disease LVD- large vessel disease	(Khan et al. 2008)
Micronutri-ent fortified rice	-	1977 school children in a double-blinded, cluster-randomized, placebo-controlled trial	- new hookworm infection increased in every school who received fortified rice - all types of fortified rice increased hookworm infection prevalence in schools (>15%)	(De Gier et al. 2016)
Green tea extract*	115.4 mg total catechins	1021 healthy postmenopausal women in a randomized, double-blinded, placebo-controlled phase II clinical trial GTE group: 513 Placebo group: 508	- ALT and AST significantly increased in the GTE group - alkaline phosphatase (AKP) and total bilirubin: non-significant changes	(Yu et al. 2017)
Ferrous sulphate	-65 mg or 130 mg	162 pregnant patient with haemoglobin between 10 g/dl and 14.5 g/dl. in a clinical trial randomised with two arms	-nausea, vomiting, epigastric pain in 130 mg daily dose group due to the metallic taste of iron tablets	(Adaji et al. 2019)

* Green tea extract: epigallocatechin gallate (EGCG), epicatechin gallate (ECG), epigallocatechin (EGC), epicatechin (EC).

It is well known that all these bioactive constituents can interfere in the drugs pharmacokinetic processes like absorption, transport and metabolization. However, current legislation does not require the nutraceuticals investigation in terms of side effects that may occur when administrated as such or in combination with classical medications. Therefore the experimental studies are remaining the principal source which offers information regarding these aspects. However, these studies are insufficient and spread in too many directions. The heterogeneity of these studies comes mainly from the differences in the forms and dosage of the studied nutraceuticals (lack of standardization of different types of extracts, oils, pills, capsules, gels, etc.) and from the various experimental models or physiological condition of the enrolled patients. Thus, these studies are not able to cover all the critically important aspects related to nutraceuticals adverse effects, accumulation, interaction or toxicity. Some examples of studies presenting various nutraceuticals side effects are presented in the next table (Table 11).

Another important and complex aspect is the possible interactions between nutraceuticals and conventional pharmaceuticals. Taking into account that the global nutraceuticals market size is expected to grow from 267.4 Billion USD to 404.8 Billion USD in the time interval between 2019 to 2025 ("Nutraceutical Products Market by Type (Food, Beverages, Dietary Supplements), Source (Probiotics, Prebiotics, Vitamins, Minerals), Distribution Channel (Conventional Stores, Specialty Stores, Drugstores & Pharmacies), Region-Global Forecast to 2025") the possibility of nutraceuticals-drugs interaction are at high risk. Therefore, the worldwide increasing popularity of nutraceuticals makes this aspect particularly urgent.

If the interaction occurs, nutraceuticals can affect drugs pharmacokinetic processes like absorption, transport (drug carriers and drug-efflux proteins), metabolism (enzymes of cytochrome P450 or UDP-glucuronosyl transferases) or excretion (Sprouse and Van Breemen 2016). For example, if the uptake drug carriers are inhibited, the drugs absorption process can be decreased and by this lowering their efficacy (Sprouse and Van Breemen 2016). Also, nutraceuticals can inhibit or induce the activities of the enzymes involved in drug metabolism causing either a prolonged either a shortened drug half-life. If inhibitions occur, prolonged drugs half-lives can induce prolonged action and toxicity. If induction occurs, the half-life of the shortened drug will induce subtherapeutic drug levels decreasing the pharmacological effects. As observed, the interactions can influence drug treatment success. However, the interactions can also be beneficial to the classical treatments, in some cases nutraceuticals being able to improve drug efficacy and even to minimize the adverse effect as observed in the next table (Table 12).

Table 12. Nutraceuticals interaction

Nutraceu-ticals	Drug	Models	Reported interactions	References
Folic acid	Enalapril	20.702 hypertensive subjects**	- reduced the risk of first stroke	(Qin et al. 2016)
	Fluoxetine	42 patients with depressive**	- improved Hamilton Depression Rating Scale (HDRS) and the Beck Depression Inventory (BDI)	(Venkatasubramanian, Kumar, and Pandey 2013)
Vitamins C + E	Metformin	40 T2DM patients*	- improved glycemic values - significantly decreased insulin, HbA1c, homeostasis model assessment of insulin resistance (HOMA-IR), fasting blood sugar (FBS) - significantly increased Quantitative Insulin Sensitivity Check Index (QISCI) - improved lipidic profile	(El-Aal et al. 2018)
Vitamin D	Simvasta-tin	28 patients**p	- reduced the deterioration of exercise-association cardiorespiratory fitness	(Singla et al. 2017)
L-carnitine	Clomiphe-ne citrate	170 women with clomiphene resistant Polycystic ovary syndrome in a randomized, placebo-controlled clinical trial	- improved ovulation and cumulative pregnancy rates - reduced the number of days to reach an adequate follicular maturation - helped to thicken the endometrium - increased serum progesterone levels in the luteal-phase - reduced serum total cholesterol, triglyceride, LDL-C levels -increased serum HDL-C levels	(Ismail et al. 2014)
Vitamins B	Diclofenac	48 patients with severe osteoarthritis and pain level ≥ 7**	- significantly reduced pain - significantly improved pain relief	(Magaña-Villa et al. 2013)
Bitter Melon (*Momordica charantia*)	Vinblastine	MDR (multidrug resistance) cervical carcinoma cell line (KB-V1)	- inhibitor effect on P-glycoprotein -increased intracellular vinblastine accumulation -decreases vinblastine efflux	(Pitchakarn et al. 2012)
Lipid-based nutrient supplements	Nevirapine (NVP)	282 HIV-positive patiens*	- reduced NVP plasma concentration	(Abdissa et al. 2015)
Panax ginseng	Midazolam	12 participants in a single-centre, single–sequence, crossover, open-label investigation	- Panax ginseng influenced CYP3A activity reducing systemic exposure of drugs metabolized by this enzyme	(Malati et al. 2012)

Table 12. (Continued)

Nutraceu-ticals	Drug	Models	Reported interactions	References
American Ginseng	Warfarin	20 healthy patients**	- interaction with the P450 enzyme system -reduced warfarin's anticoagulant activity	(Yuan et al. 2004)
Ginkgo Biloba extract (GBE)	Warfarin	Mice fed with GBE or bilobalide or ginkgolide B	-anticoagulation effects of warfarin attenuated by GBE and bilobalide	(Taki et al. 2012)
		Mice fed with ginkgolide A or B or C or bilobalide	-ginkgolide A, B, C and bilobalide increased cytochrome P450 enzymes (CYP) levels -attenuated the effect of warfarin	

* Randomized, single-blinded controlled clinical trial.
** Randomized, double-blind controlled clinical trial.
**p Double-blind placebo-controlled clinical trial.

CONCLUSION

The studies existing so far highlight the large varieties of nutraceuticals available on the market. Their increasing dietary intake among the population is based mainly on their health-promoting beneficial effects. Also, they are often used for prevention purposes. In order to regulate their consumption on the market, nutraceuticals are subject to different regulations. Because some of these regulations are permissive a large number of new nutraceuticals appear each year on the market, most of them not being standardized or well characterized. Critical points in nutraceuticals development like great variation of used botanicals sources, preparation processes, formulation, mechanisms of action, safety and efficacy are still unresolved. Therefore, there is a risk for the population to be exposed to an unsafe consumption, their knowledge and awareness being highly limited. Within this context, the clinical trials and the *in vivo* and *in vitro* experimental studies are still considered the gold standard and the references are taken into consideration when answers to questions related to nutraceuticals health benefits, side effects and interactions are required.

References

Abdissa, A., M. .F Olsen, D. Yilma, M. Tesfaye, T. Girma, M. Christiansen, C. M. Hagen, et al. 2015. "Lipid-Based Nutrient Supplements Do Not Affect Efavirenz but Lower Plasma Nevirapine Concentrations in Ethiopian Adult HIV Patients." *HIV Medicine* 16 (7): 403–11. https://doi.org/10.1111/hiv.12225.

Adaji, Ja, Ay Isah, T. Otu, and Hi Abdullahi. 2019. "*Daily Versus Twice Daily Dose of Ferrous Sulphate Supplementation in Pregnant Women: A Randomized Clinical Trial.*" https://doi.org/10.4103/njcp.njcp_211_18.

Adib Rad, Hajar, Zahra Basirat, Fatemeh Bakouei, Ali Akbar Moghadamnia, Soraya Khafri, Zeynab Farhadi Kotenaei, Maryam Nikpour, and Somayeh Kazemi. 2018. "Effect of Ginger and Novafen on Menstrual Pain: A Cross-over Trial." *Taiwanese Journal of Obstetrics and Gynecology* 57 (6): 806–9. https://doi.org/10.1016/J.TJOG.2018.10.006.

Akingbemi, Benson T., Tim D. Braden, Barbara W. Kemppainen, Karen D. Hancock, Jessica D. Sherrill, Sarah J. Cook, Xiaoying He, and Jeffrey G. Supko. 2007. "Exposure to Phytoestrogens in the Perinatal Period Affects Androgen Secretion by Testicular Leydig Cells in the Adult Rat." *Endocrinology* 148 (9): 4475–88. https://doi.org/10.1210/en.2007-0327.

Anadón, Arturo, María Rosa Martínez-Larrañaga, Irma Ares, and María Aránzazu Martínez. 2016. *Evaluation and Regulation of Food Supplements: European Perspective. Nutraceuticals: Efficacy, Safety and Toxicity*. Elsevier Inc. https://doi.org/10.1016/B978-0-12-802147-7.00063-2.

Angeles-Agdeppa, Imelda, Clarita R. Magsadia, and Mario V. Capanzana. 2011. "Fortified Juice Drink Improved Iron and Zinc Status of Schoolchildren." *Asia Pacific Journal of Clinical Nutrition* 20 (4): 535–43.

Asemi, Zatollah, Sabihe Alsadat Alizadeh, Khorshidi Ahmad, Mohammad Goli, and Ahmad Esmaillzadeh. 2016. "Effects of Beta-Carotene

Fortified Synbiotic Food on Metabolic Control of Patients with Type 2 Diabetes Mellitus: A Double-Blind Randomized Cross-over Controlled Clinical Trial." *Clinical Nutrition* 35 (4): 819–25. https://doi.org/10.1016/j.clnu.2015.07.009.

Beelen, Janne, Emmelyne Vasse, Nancy Janssen, André Janse, Nicole M. de Roos, and Lisette C. P. G. M. de Groot. 2018. "Protein-Enriched Familiar Foods and Drinks Improve Protein Intake of Hospitalized Older Patients: A Randomized Controlled Trial." *Clinical Nutrition* 37 (4): 1186–92. https://doi.org/10.1016/j.clnu.2017.05.010.

Beh, Boon Kee, Nurul Elyani Mohamad, Swee Keong Yeap, Huynh Ky, Sook Yee Boo, Joelle Yi Heng Chua, Sheau Wei Tan, et al. 2017. "Anti-Obesity and Anti-Inflammatory Effects of Synthetic Acetic Acid Vinegar and Nipa Vinegar on High-Fat-Diet-Induced Obese Mice." *Scientific Reports* 7 (1): 6664. https://doi.org/10.1038/s41598-017-06235-7.

Britton, Emily R., Joshua J. Kellogg, Olav M. Kvalheim, and Nadja B. Cech. 2018. "Biochemometrics to Identify Synergists and Additives from Botanical Medicines: A Case Study with Hydrastis Canadensis (Goldenseal)." *Journal of Natural Products* 81 (3): 484–93. https://doi.org/10.1021/acs.jnatprod.7b00654.

Chanda, Silpi, Raj Kumar Tiwari, Arun Kumar, and Kuldeep Singh. 2019. "Nutraceuticals Inspiring the Current Therapy for Lifestyle Diseases." *Advances in Pharmacological Sciences*. Hindawi Limited. https://doi.org/10.1155/2019/6908716.

Chang, Tung Cheng, Po Li Wei, Precious Takondwa Makondi, Wei Ting Chen, Chien Yu Huang, and Yu Jia Chang. 2019. "Bromelain Inhibits the Ability of Colorectal Cancer Cells to Proliferate via Activation of ROS Production and Autophagy." *PLoS ONE* 14 (1). https://doi.org/10.1371/journal.pone.0210274.

Cheung, Ching-Lung, Daniel Ka-Chun Ho, Chor-Wing Sing, Man-Fung Tsoi, Vincent Ka-Fai Cheng, Grace Koon-Yee Lee, Yuk-Nam Ho, and Bernard M. Y. Cheung. 2017. "Randomized Controlled Trial of the Effect of Phytosterols-Enriched Low-Fat Milk on Lipid Profile in Chinese." *Nature Publishing Group*. https://doi.org/10.1038/srep41084.

China Regulations on New Food Raw Materials | ChemLinked. n.d. Accessed January 24, 2020. https://food.chemlinked.com/foodpedia/china-regulations-new-food-raw-materials.

Choudhary, Prema Ram, Rameshchandra D. Jani, and Megh Shyam Sharma. 2018. "Effect of Raw Crushed Garlic (*Allium sativum* L.) on Components of Metabolic Syndrome." *Journal of Dietary Supplements* 15 (4): 499–506. https://doi.org/10.1080/19390211.2017.1358233.

DeFelice, Stephen L. 1995. "The Nutraceutical Revolution: Its Impact on Food Industry R&D." *Trends in Food Science and Technology*. Elsevier. https://doi.org/10.1016/S0924-2244(00)88944-X.

El-Aal, Ali Abd, Eman A. Abd El-Ghffar, Asmaa Abu Ghali, Mohammed R. Zughbur, and Mahmoud M. Sirdah. 2018. "The Effect of Vitamin C and/or E Supplementations on Type 2 Diabetic Adult Males under Metformin Treatment: A Single-Blinded Randomized Controlled Clinical Trial." *Diabetes and Metabolic Syndrome: Clinical Research and Reviews* 12 (4): 483–89. https://doi.org/10.1016/j.dsx.2018.03.013.

Fox, Edward M., Mark Raftery, Amber Goodchild, and George L. Mendz. 2007. "Campylobacter Jejuni Response to Ox-Bile Stress." *FEMS Immunology and Medical Microbiology* 49 (1): 165–72. https://doi.org/10.1111/j.1574-695X.2006.00190.x.

Fu, Yu-Rong, Zheng-Jun Yi, Jing-liang Pei, and Shizhen Guan. 2010. "Effects of Bifidobacterium Bifidum on the Adaptive Immune Senescence in Aging Mice." *Microbiology and Immunology* 54 (10): no-no. https://doi.org/10.1111/j.1348-0421.2010.00255.x.

Geurts, Lucie, Amandine Everard, Pascale le Ruyet, Nathalie M. Delzenne, and Patrice D. Cani. 2012. "Ripened Dairy Products Differentially Affect Hepatic Lipid Content and Adipose Tissue Oxidative Stress Markers in Obese and Type 2 Diabetic Mice." *Journal of Agricultural and Food Chemistry* 60 (8): 2063–68. https://doi.org/10.1021/jf204916x.

Ghasemzadeh, Samin, and Gholam Hossein Riazi. 2019. "Inhibition of Tau Amyloid Fibril Formation by Folic Acid: In-Vitro and Theoretical Studies." *International Journal of Biological Macromolecules*, November. https://doi.org/10.1016/J.IJBIOMAC.2019.11.032.

Gier, Brechje De, Maiza Campos Ponce, Marlene Perignon, Marion Fiorentino, Kuong Khov, Chhoun Chamnan, Michiel R. De Boer, et al. 2016. "Micronutrient-Fortified Rice Can Increase Hookworm Infection Risk: A Cluster Randomized Trial." *PLoS ONE* 11 (1). https://doi.org/10.1371/journal.pone.0145351.

Gopalan, Yogheswaran, Ibrahim Lutfi Shuaib, Enrico Magosso, Mukhtar Alam Ansari, Mohd Rizal Abu Bakar, Jia Woei Wong, Nurzalina Abdul Karim Khan, et al. 2014. "Clinical Investigation of the Protective Effects of Palm Vitamin e Tocotrienols on Brain White Matter." *Stroke* 45 (5): 1422–28. https://doi.org/10.1161/STROKEAHA.113.004449.

Guglielmetti, S., D. Mora, M. Gschwender, and K. Popp. 2011. "Randomised Clinical Trial: Bifidobacterium Bifidum MIMBb75 Significantly Alleviates Irritable Bowel Syndrome and Improves Quality of Life -- a Double-Blind, Placebo-Controlled Study." *Alimentary Pharmacology & Therapeutics* 33 (10): 1123–32. https://doi.org/10.1111/j.1365-2036.2011.04633.x.

Guo, Dongqing, Colin E. Murdoch, Hao Xu, Hui Shi, Dayue Darrel Duan, Asif Ahmed, and Yuchun Gu. 2017. "Vascular Endothelial Growth Factor Signaling Requires Glycine to Promote Angiogenesis." *Scientific Reports* 7 (1): 14749. https://doi.org/10.1038/s41598-017-15246-3.

Gupta, Sumeet, Devesh Chauhan, Kritika Mehla, Preeti Sood, and Anroop Nair. 2010. "An Overview of Nutraceuticals: Current Scenario." *Journal of Basic and Clinical Pharmacy* 1 (2): 55–62. www.jbclinpharm.com.

Hayes, Aoife, Sarah Duffy, Michael O'grady, Jette Jakobsen, Karen Galvin, Joanna Teahan-Dillon, Joseph Kerry, et al. 2016. "Vitamin D-Enhanced Eggs Are Protective of Wintertime Serum 25-Hydroxyvitamin D in a Randomized Controlled Trial of Adults." *American Journal of Clinical Nutrition* 104 (3): 629–37. https://doi.org/10.3945/ajcn.116.132530.

Hefferon, Kathleen L. 2015. "Nutritionally Enhanced Food Crops; Progress and Perspectives." *International Journal of Molecular Sciences*. MDPI AG. https://doi.org/10.3390/ijms16023895.

Hosseini, Seyed Ahmad, Hamid Zand, and Makan Cheraghpour. 2019. "The Influence of Curcumin on the Downregulation of MYC, Insulin and

IGF-1 Receptors: A Possible Mechanism Underlying the Anti-Growth and Anti-Migration in Chemoresistant Colorectal Cancer Cells." *Medicina (Kaunas, Lithuania)* 55 (4). https://doi.org/10.3390/medicina55040090.

Huang, Ye Wei, Qiang Qiang Zhu, Xiao Ying Yang, Huan Huan Xu, Bin Sun, Xuan Jun Wang, and Jun Sheng. 2019. "Wound Healing Can Be Improved by (2)-Epigallocatechin Gallate through Targeting Notch in Streptozotocin-Induced Diabetic Mice." *FASEB Journal* 33 (1): 953–64. https://doi.org/10.1096/fj.201800337R.

Ismail, Alaa M., Ali Hassan Hamed, Srdjan Saso, and Hossam H. Thabet. 2014. "Adding L-Carnitine to Clomiphene Resistant PCOS Women Improves the Quality of Ovulation and the Pregnancy Rate. A Randomized Clinical Trial." *European Journal of Obstetrics and Gynecology and Reproductive Biology* 180 (1): 148–52. https://doi.org/10.1016/j.ejogrb.2014.06.008.

Jafarpour-Sadegh, Farnaz, Vahid Montazeri, Ali Adili, Ali Esfehani, Mohammad-Reza Rashidi, and Saeed Pirouzpanah. 2017. "Consumption of Fresh Yellow Onion Ameliorates Hyperglycemia and Insulin Resistance in Breast Cancer Patients During Doxorubicin-Based Chemotherapy: A Randomized Controlled Clinical Trial." *Integrative Cancer Therapies* 16 (3): 276–89. https://doi.org/10.1177/1534735416656915.

Jeon, Yu Bin, Jae-Joon Lee, and Hae Choon Chang. 2019. "Characterization of Juice Fermented with *Lactobacillus Plantarum* EM and Its Cholesterol-lowering Effects on Rats Fed a High-fat and High-cholesterol Diet." *Food Science & Nutrition* 7 (11): 3622–34. https://doi.org/10.1002/fsn3.1217.

Jiang, Yan, Danielle K. Turgeon, Benjamin D. Wright, Elkhansa Sidahmed, Mack T. Ruffin, Dean E. Brenner, Ananda Sen, and Suzanna M. Zick. 2013. "Effect of Ginger Root on Cyclooxygenase-1 and 15-Hydroxyprostaglandin Dehydrogenase Expression in Colonic Mucosa of Humans at Normal and Increased Risk for Colorectal Cancer." *European Journal of Cancer Prevention : The Official Journal of the*

European Cancer Prevention Organisation (ECP) 22 (5): 455–60. https://doi.org/10.1097/CEJ.0b013e32835c829b.

Karamali, Maryam, Shahla Bahramimoghadam, Fateme Sharifzadeh, and Zatollah Asemi. 2018. "Magnesium-Zinc-Calcium-Vitamin D Co-Supplementation Improves Glycemic Control and Markers of Cardio-Metabolic Risk in Gestational Diabetes: A Running Title: Magnesium-Zinc-Calcium-Vitamin D and Gestational Diabetes Number of Words (Text): 2433." *Appl. Physiology, Nutrition, and Metabolism* 43 (6): 565–70.

Kawarizadeh, Amin, Farshad Nojoomi, Mohammad Tabatabaei, Saeid Hosseinzadeh, and Mina Farzaneh. 2019. "The Effect of Bacillus Coagulans on Cytotoxicity and Apoptosis Induced by Salmonella Typhimurium in HT-29 Cell Culture." *Iranian Journal of Microbiology* 11 (4): 305–12.

Keenan, Michael J., Marlene Janes, Julina Robert, Roy J. Martin, Anne M. Raggio, Kathleen L. McCutcheon, Christine Pelkman, et al. 2013. "Resistant Starch from High Amylose Maize (HAM-RS2) Reduces Body Fat and Increases Gut Bacteria in Ovariectomized (OVX) Rats." *Obesity (Silver Spring, Md.)* 21 (5): 981–84. https://doi.org/10.1002/oby.20109.

Keenan, Michael J., Roy J. Martin, Anne M. Raggio, Kathleen L. McCutcheon, Ian L. Brown, Anne Birkett, Susan S. Newman, et al. 2012. "High-Amylose Resistant Starch Increases Hormones and Improves Structure and Function of the Gastrointestinal Tract: A Microarray Study." *Journal of Nutrigenetics and Nutrigenomics* 5 (1): 26–44. https://doi.org/10.1159/000335319.

Keservani, Raj K., Urmila Jarouliya, and Anil K. Sharma. 2019. *Nutraceutical and Functional Food Regulations in India. Nutraceutical and Functional Food Regulations in the United States and around the World.* Third Edit. Elsevier Inc. https://doi.org/10.1016/b978-0-12-816467-9.00028-9.

Khadgawat, R., R. K. Marwaha, M. K. Garg, R. Ramot, A. K. Oberoi, V. Sreenivas, M. Gahlot, N. Mehan, P. Mathur, and N. Gupta. 2013. "Impact of Vitamin D Fortified Milk Supplementation on Vitamin D

Status of Healthy School Children Aged 10-14 Years." *Osteoporosis International* 24 (8): 2335–43. https://doi.org/10.1007/s00198-013-2306-9.

Khan, Usman, Carollyn Crossley, Lalit Kalra, Anthony Rudd, Charles D. A. Wolfe, Paul Collinson, and Hugh S. Markus. 2008. "Homocysteine and Its Relationship to Stroke Subtypes in a UK Black Population: The South London Ethnicity and Stroke Study." *Stroke* 39 (11): 2943–49. https://doi.org/10.1161/STROKEAHA.107.513416.

Kim, Bobae, Jeonghyeon Kwon, Min-Seok Kim, Haryung Park, Yosep Ji, Wilhelm Holzapfel, and Chang-Kee Hyun. 2018. "Protective Effects of Bacillus Probiotics against High-Fat Diet-Induced Metabolic Disorders in Mice." *PloS One* 13 (12): e0210120. https://doi.org/10.1371/journal.pone.0210120.

Kim, Jungmin, Yeonjeong Ko, Yu Kyung Park, Nack In Kim, Woel Kyu Ha, and Yunhi Cho. 2010. "Dietary Effect of Lactoferrin-Enriched Fermented Milk on Skin Surface Lipid and Clinical Improvement of Acne Vulgaris." *Nutrition* 26 (9): 902–9. https://doi.org/10.1016/j.nut.2010.05.011.

Kondo, Tomoo, Mikiya Kishi, Takashi Fushimi, Shinobu Ugajin, and Takayuki Kaga. 2014. "Vinegar Intake Reduces Body Weight, Body Fat Mass, and Serum Triglyceride Levels in Obese Japanese Subjects." *Http://Dx.Doi.Org/10.1271/Bbb.90231.* https://doi.org/10.1271/BBB.90231.

Kujinga, Prosper, Valeria Galetti, Elizabeth Onyango, Viktor Jakab, Simone Buerkli, Pauline Andang'o, Inge D. Brouwer, Michael B. Zimmermann, and Diego Moretti. 2018. "Effectiveness of Zinc-Fortified Water on Zinc Intake, Status and Morbidity in Kenyan Pre-School Children: A Randomised Controlled Trial." *Public Health Nutrition* 21 (15): 2855–65. https://doi.org/10.1017/S1368980018001441.

Kushiro, Akira, Kensuke Shimizu, Toshihiko Takada, Itsumi Kusunoki, and Naomi Aiba. 2019. "Decreased Number of Days of Fever Detection and Duration of Fever with Continuous Intake of a Fermented Milk Drink: A Randomized, Double-Blind, Placebo-Controlled Study of Elderly

Nursing Home Residents." *Bioscience of Microbiota, Food and Health* 38 (4): 151–57. https://doi.org/10.12938/bmfh.18-024.

Magaña-Villa, M., H. Rocha-González, C. Fernández del Valle-Laisequilla, V. Granados-Soto, J. Rodríguez-Silverio, F. Flores-Murrieta, M. Carrasco-Portugal, and J. Reyes-García. 2013. "B-Vitamin Mixture Improves the Analgesic Effect of Diclofenac in Patients with Osteoarthritis: A Double Blind Study." *Drug Research* 63 (06): 289–92. https://doi.org/10.1055/s-0033-1334963.

Malati, Christine Y., Sarah M. Robertson, Jennifer D. Hunt, Cheryl Chairez, Raul M. Alfaro, Joseph A. Kovacs, and Scott R. Penzak. 2012. "Influence of Panax Ginseng on Cytochrome P450 (CYP)3A and p-Glycoprotein (P-Gp) Activity in Healthy Participants." *Journal of Clinical Pharmacology* 52 (6): 932–39. https://doi.org/10.1177/0091270011407194.

Markowiak, Paulina, and Katarzyna Ślizewska. 2017. "Effects of Probiotics, Prebiotics, and Synbiotics on Human Health." *Nutrients* 9 (9). https://doi.org/10.3390/nu9091021.

Martín, Rebeca, and Philippe Langella. 2019. "Emerging Health Concepts in the Probiotics Field: Streamlining the Definitions." *Frontiers in Microbiology*. https://doi.org/10.3389/fmicb.2019.01047.

Massacci, Francesca Romana, Carmela Lovito, Silvia Tofani, Michele Tentellini, Domenica Anna Genovese, Alessia Arcangela Pia De De Leo, Paola Papa, et al. 2019. "Dietary Saccharomyces Cerevisiae Boulardii CNCM I-1079 Positively Affects Performance and Intestinal Ecosystem in Broilers during a Campylobacter Jejuni Infection." *Microorganisms* 7 (12): 596. https://doi.org/10.3390/microorganisms7120596.

Miura, Katsuyuki, Jeremiah Stamler, Ian J. Brown, Hirotsugu Ueshima, Hideaki Nakagawa, Masaru Sakurai, Queenie Chan, et al. 2013. "Relationship of Dietary Monounsaturated Fatty Acids to Blood Pressure: The International Study of Macro/Micronutrients and Blood Pressure." *Journal of Hypertension* 31 (6): 1144–50. https://doi.org/10.1097/HJH.0b013e3283604016.

Mozaffari-Khosravi, Hassan, Behrouz Talaei, Beman-Ali Jalali, Azadeh Najarzadeh, and Mohammad Reza Mozayan. 2014. "The Effect of Ginger Powder Supplementation on Insulin Resistance and Glycemic Indices in Patients with Type 2 Diabetes: A Randomized, Double-Blind, Placebo-Controlled Trial." *Complementary Therapies in Medicine* 22 (1): 9–16. https://doi.org/10.1016/j.ctim.2013.12.017.

Natale, Claudia De, Valentina Minerva, Lidia Patti, Raffaella Mazzarella, Ornella Ciano, Simona Maione, Delia Luongo, et al. 2012. "Effects of Baked Products Enriched with N-3 Fatty Acids, Folates, β-Glucans, and Tocopherol in Patients with Mild Mixed Hyperlipidemia." *Journal of the American College of Nutrition* 31 (5): 311–19. https://doi.org/10.1080/07315724.2012.10720427.

Natural Health Products Regulations. n.d. Accessed January 24, 2020. https://laws-lois.justice.gc.ca/eng/regulations/SOR-2003-196/.

Nutraceutical Products Market by Type (Food, Beverages, Dietary Supplements), Source (Probiotics, Prebiotics, Vitamins, Minerals), Distribution Channel (Conventional Stores, Specialty Stores, Drugstores & Pharmacies), Region- Global Forecast to 2025. n.d. Accessed February 17, 2020. https://www.researchandmarkets.com/reports/4825471/nutraceutical-products-market-by-type-food?utm_source=dynamic&utm_medium=GNOM&utm_code=j9j6k9&utm_campaign=1317585+-+Outlook+on+the+Global+Nutraceutical+Products+Market%2C+2019-2025+-+Consumer+Skepticism+%26+High+Costs+are+Projected+to+Hinder+the+Market&utm_exec=joca220gnomd.

Nutraceuticals Market: Dietary Supplements 2/4|Hi Europe. n.d. Accessed November 22, 2019. https://www.figlobal.com/hieurope/visit/ news-and-updates/nutraceuticals-market-dietary-supplements-24.

Pawlak, Dorota B., Janet M. Bryson, Gareth S. Denyer, and Janette C. Brand-Miller. 2001. "High Glycemic Index Starch Promotes Hypersecretion of Insulin and Higher Body Fat in Rats without Affecting Insulin Sensitivity." *The Journal of Nutrition* 131 (1): 99–104. https://doi.org/10.1093/jn/131.1.99.

Pinkaew, Siwaporn, Rita Wegmuller, Emorn Wasantwisut, Pattanee Winichagoon, Richard F. Hurrell, and Sherry A. Tanumihardjo. 2014. "Triple-Fortified Rice Containing Vitamin A Reduced Marginal Vitamin A Deficiency and Increased Vitamin A Liver Stores in School-Aged Thai Children." *The Journal of Nutrition* 144 (4): 519–24. https://doi.org/10.3945/jn.113.182998.

Pitchakarn, Pornsiri, Shinobu Ohnuma, Komsak Pintha, Wilart Pompimon, Suresh V. Ambudkar, and Pornngarm Limtrakul. 2012. "Kuguacin J Isolated from Momordica Charantia Leaves Inhibits P-Glycoprotein (ABCB1)-Mediated Multidrug Resistance." *Journal of Nutritional Biochemistry* 23 (1): 76–84. https://doi.org/10.1016/j.jnutbio.2010.11.005.

Plana, Nuria, Catherine Nicolle, Raimon Ferre, Jordi Camps, Rosa Cos, Jesus Villoria, and Luis Masana. 2008. "Plant Sterol-Enriched Fermented Milk Enhances the Attainment of LDL-Cholesterol Goal in Hypercholesterolemic Subjects." *European Journal of Nutrition* 47 (1): 32–39. https://doi.org/10.1007/s00394-007-0693-4.

Qin, Xianhui, Jianping Li, J. David Spence, Yan Zhang, Youbao Li, Xiaobin Wang, Binyan Wang, et al. 2016. "Folic Acid Therapy Reduces the First Stroke Risk Associated with Hypercholesterolemia among Hypertensive Patients." *Stroke* 47 (11): 2805–12. https://doi.org/10.1161/STROKEAHA.116.014578.

Rambousková, Jolana, Bohumír Procházka, Michael Binder, and Michal Anděl. n.d. "[Effect of the Liquid Milk Nutritional Supplement with Enhanced Content of Whey Protein on the Nutritional Status of the Elderly]." *Vnitrni Lekarstvi* 60 (7–8): 556–61.

Rizzello, C. G., I Losito, M. Gobbetti, T Carbonara, M. D. De Bari, and P. G. Zambonin. 2005. "Antibacterial Activities of Peptides from the Water-Soluble Extracts of Italian Cheese Varieties." *Journal of Dairy Science* 88 (7): 2348–60. https://doi.org/10.3168/jds.S0022-0302(05)72913-1.

Ruchi, Sharma, Kaur Amanjot, Thakur Sourav, Bhardwaj Keerti, and Bose Sujit. 2017. "Role of Nutraceuticals in Health Care: A Review."

International Journal of Green Pharmacy 11 (3): S385–94. https://doi.org/10.22377/ijgp.v11i03.1146.

Sathasivam, Ramaraj, and Jang Seu Ki. 2018. “A Review of the Biological Activities of Microalgal Carotenoids and Their Potential Use in Healthcare and Cosmetic Industries.” *Marine Drugs* 16 (1). https://doi.org/10.3390/md16010026.

Schiffmann, Raphael, Rick A. Martin, Tyler Reimschisel, Karen Johnson, Victoria Castaneda, Y. Howard Lien, Gregory M. Pastores, Christoph Kampmann, Markus Ries, and Joe T. R. Clarke. 2010. “Four-Year Prospective Clinical Trial of Agalsidase Alfa in Children with Fabry Disease.” *Journal of Pediatrics* 156 (5). https://doi.org/10.1016/j.jpeds.2009.11.007.

Seon, Mi Ra, So Young Park, Soo Jin Kwon, Soon Sung Lim, Hyun Ju Choi, Heesook Park, Do Young Lim, et al. 2012. “Hexane/Ethanol Extract of Glycyrrhiza Uralensis and Its Active Compound Isoangustone A Induce G1 Cycle Arrest in DU145 Human Prostate and 4T1 Murine Mammary Cancer Cells.” *Journal of Nutritional Biochemistry* 23 (1): 85–92. https://doi.org/10.1016/j.jnutbio.2010.11.010.

Shamal, S., and Bijuna C. Mohan. 2017. “Consumer Behaviour in Fortified Food Choice Decisions in India.” *Nutrition and Food Science* 47 (2): 229–39. https://doi.org/10.1108/NFS-05-2016-0065.

Sharquie, Khalifa E., and Hala K. Al-Obaidi. 2002. “Onion Juice (Allium Cepa L.), A New Topical Treatment for Alopecia Areata.” *The Journal of Dermatology* 29 (6): 343–46. https://doi.org/10.1111/j.1346-8138.2002.tb00277.x.

Shirasawa, Y., H. Shibahara-Sone, T. Iino, and F. Ishikawa. 2010. “Bifidobacterium Bifidum BF-1 Suppresses Helicobacter Pylori-Induced Genes in Human Epithelial Cells.” *Journal of Dairy Science* 93 (10): 4526–34. https://doi.org/10.3168/jds.2010-3274.

Singh, Jagtar, and Shweta Sinha. 2012. “Classification, Regulatory Acts and Applications of Nutraceuticals for Health.” *International Journal of Pharmacy and Biological Sciences* 2 (1). www.paulingblog.wordpress.com].

Singla, Mandeep, Ashu Rastogi, Ashutosh N. Aggarwal, Owais M. Bhat, Darshan Badal, and Anil Bhansali. 2017. "Vitamin D Supplementation Improves Simvastatin-Mediated Decline in Exercise Performance: A Randomized Double-Blind Placebo-Controlled Study." *Journal of Diabetes* 9 (12): 1100–1106. https://doi.org/10.1111/1753-0407.12541.

Siró, István, Emese Kápolna, Beáta Kápolna, and Andrea Lugasi. 2008. "Functional Food. Product Development, Marketing and Consumer Acceptance-A Review." *Appetite*. https://doi.org/10.1016/j.appet.2008.05.060.

Sprouse, Alyssa A., and Richard B. Van Breemen. 2016. "Pharmacokinetic Interactions between Drugs and Botanical Dietary Supplements." *Drug Metabolism and Disposition*. American Society for Pharmacology and Experimental Therapy. https://doi.org/10.1124/dmd.115.066902.

Srivastava, Shobhit, Anil K. Saksena, Sanjay Khattri, Santosh Kumar, and Raghubendra Singh Dagur. 2016. "Curcuma Longa Extract Reduces Inflammatory and Oxidative Stress Biomarkers in Osteoarthritis of Knee: A Four-Month, Double-Blind, Randomized, Placebo-Controlled Trial." *Inflammopharmacology* 24 (6): 377–88. https://doi.org/10.1007/s10787-016-0289-9.

Strømmen, Kenneth, Alexander Haag, Sissel J. Moltu, Marit B. Veierød, Elin W. Blakstad, Britt Nakstad, Astrid N. Almaas, et al. 2017. "Enhanced Nutrient Supply to Very Low Birth Weight Infants Is Associated with Higher Blood Amino Acid Concentrations and Improved Growth." *Clinical Nutrition ESPEN* 18 (April): 16–22. https://doi.org/10.1016/j.clnesp.2017.01.003.

Stupin, Ana, Lidija Rasic, Anita Matic, Marko Stupin, Zlata Kralik, Gordana Kralik, Manuela Grcevic, and Ines Drenjancevic. 2018. "Omega-3 Polyunsaturated Fatty Acids-Enriched Hen Eggs Consumption Enhances Microvascular Reactivity in Young Healthy Individuals." *Applied Physiology, Nutrition and Metabolism* 43 (10): 988 95. https://doi.org/10.1139/apnm-2017-0735.

Takayanagi, Katsuhiko. 2011. "Prevention of Adiposity by the Oral Administration of β-Cryptoxanthin." *Frontiers in Neurology* 2: 67. https://doi.org/10.3389/fneur.2011.00067.

Taki, Yuko, Kaori Yokotani, Shizuo Yamada, Kazumasa Shinozuka, Yoko Kubota, Yasuo Watanabe, and Keizo Umegaki. 2012. "Ginkgo Biloba Extract Attenuates Warfarin-Mediated Anticoagulation through Induction of Hepatic Cytochrome P450 Enzymes by Bilobalide in Mice." *Phytomedicine* 19 (2): 177–82. https://doi.org/10.1016/j.phymed.2011.06.020.

Toma, Claudia Crina, Georgeta Maria Simu, Daniela Hanganu, Neli Olah, Florina Maria Georgiana Vata, Cyrine Hammami, and Mohamed Hammami. 2010. "Chemical Composition of the Tunisian Nigella Sativa. Note I. Profile on Essential Oil." *Farmacia* 58 (4): 458–64.

Tong, Chao, Chuan Peng, Lianlian Wang, Li Zhang, Xiaotao Yang, Ping Xu, Jinjin Li, Thibaut Delplancke, Hua Zhang, and Hongbo Qi. 2016. "Intravenous Administration of Lycopene, a Tomato Extract, Protects against Myocardial Ischemia-Reperfusion Injury." *Nutrients* 8 (3): 138. https://doi.org/10.3390/nu8030138.

Torabally, Nawsheen Bee, and Hananeh Abdolrahman Rahmanpoor. 2019. "Nutraceuticals: Nutritionally Functional Foods – an Overview." *Biomedical Journal of Scientific & Technical Research* 15 (4): 11480–82. https://doi.org/10.26717/bjstr.2019.15.002728.

Tung Bui, Duy, and Bui Duy Tung. 2015. "Consumer Acceptance of Functional Foods in Ho Chi Minh City." *Eurasian Journal of Business and Economics* 8 (16): 19–34. https://doi.org/10.17015/ejbe.2015.016.02.

Tungmunnithum, Duangjai, Areeya Thongboonyou, Apinan Pholboon, and Aujana Yangsabai. 2018. "Flavonoids and Other Phenolic Compounds from Medicinal Plants for Pharmaceutical and Medical Aspects: An Overview." *Medicines* 5 (3): 93. https://doi.org/10.3390/medicines5030093.

Umegaki, Keizo. 2016. *Regulatory Aspects of Nutraceuticals: Japanese Perspective. Nutraceuticals: Efficacy, Safety and Toxicity*. Elsevier Inc. https://doi.org/10.1016/B978-0-12-802147-7.00065-6.

Underdown, J., P. J. Quail, and K. W. Smith. 2011. "Saturated Fat Reduction in Ice Cream." In *Reducing Saturated Fats in Foods*, 350–69. Elsevier. https://doi.org/10.1533/9780857092472.2.350.

USA Nutraceutical Market to 2027: Increasing Spectrum of Nutraceutical in Sport Nutrition & Pet Food. n.d. Accessed November 22, 2019. https://www.prnewswire.com/news-releases/usa-nutraceutical-market-to-2027-increasing-spectrum-of-nutraceutical-in-sport-nutrition--pet-food-300905970.html.

Vafeiadou, Katerina, Michelle Weech, Hana Altowaijri, Susan Todd, Parveen Yaqoob, Kim G. Jackson, and Julie A. Lovegrove. 2015. "Replacement of Saturated with Unsaturated Fats Had No Impact on Vascular Function but Beneficial Effects on Lipid Biomarkers, E-Selectin, and Blood Pressure: Results from the Randomized, Controlled Dietary Intervention and VAScular Function (DIVAS) Study." *The American Journal of Clinical Nutrition* 102 (1): 40–48. https://doi.org/10.3945/ajcn.114.097089.

Venkatasubramanian, Ramakrishnan, Channaveerachari Naveen Kumar, and Ravi Shankar Pandey. 2013. "A Randomized Double-Blind Comparison of Fluoxetine Augmentation by High and Low Dosage Folic Acid in Patients with Depressive Episodes." *Journal of Affective Disorders* 150 (2): 644–48. https://doi.org/10.1016/j.jad.2013.02.029.

Verbowski, Vashti, Zaman Talukder, Kroeun Hou, Ly Sok Hoing, Kristina Michaux, Victoria Anderson, Rosalind Gibson, et al. 2018. "Effect of Enhanced Homestead Food Production and Aquaculture on Dietary Intakes of Women and Children in Rural Cambodia: A Cluster Randomized Controlled Trial." *Maternal and Child Nutrition* 14 (3): 1–9. https://doi.org/10.1111/mcn.12581.

Volpi, Elena, Hisamine Kobayashi, Melinda Sheffield-Moore, Bettina Mittendorfer, and Robert R. Wolfe. 2003. "Essential Amino Acids Are Primarily Responsible for the Amino Acid Stimulation of Muscle Protein Anabolism in Healthy Elderly Adults." *The American Journal of Clinical Nutrition* 78 (2): 250. https://doi.org/10.1093/AJCN/78.2.250.

Wang, Chung Yi, Tzu Chien Kao, Wen Hsieh Lo, and Gow Chin Yen. 2011. "Glycyrrhizic Acid and 18β-Glycyrrhetinic Acid Modulate Lipopolysaccharide-Induced Inflammatory Response by Suppression of NF-KB through PI3K P110δ and P110γ Inhibitions." *Journal of*

Agricultural and Food Chemistry 59 (14): 7726–33. https://doi.org/10.1021/jf2013265.

Wang, Haochen, Hans De Steur, Gong Chen, Xiaotian Zhang, Lijun Pei, Xavier Gellynck, and Xiaoying Zheng. 2016. "Effectiveness of Folic Acid Fortified Flour for Prevention of Neural Tube Defects in a High Risk Region." *Nutrients* 8 (3): 152. https://doi.org/10.3390/nu8030152.

Wang, Kai Lee, Shih Min Hsia, Chia Jung Chan, Full Young Chang, Chih Yang Huang, Da Tian Bau, and Paulus S Wang. 2013. "Inhibitory Effects of Isoliquiritigenin on the Migration and Invasion of Human Breast Cancer Cells." *Expert Opinion on Therapeutic Targets* 17 (4): 337–49. https://doi.org/10.1517/14728222.2013.756869.

Wang, Yu-Guang, Jian-Ming Zhou, Zeng-Chun Ma, Hua Li, Qian-De Liang, Hong-Ling Tan, Cheng-Rong Xiao, Bo-Li Zhang, and Yue Gao. 2012. "Pregnane X Receptor Mediated-Transcription Regulation of CYP3A by Glycyrrhizin: A Possible Mechanism for Its Hepatoprotective Property against Lithocholic Acid-Induced Injury." *Chemico-Biological Interactions* 200 (1): 11–20. https://doi.org/10.1016/j.cbi.2012.08.023.

Whitcomb, David C., Amit Bodhani, Katrin Beckmann, Suntje Sander-Struckmeier, Shufang Liu, Mahesh Fuldeore, Paul F. Pollack, and Rupal P. Khurmi. 2016. "Efficacy and Safety of Pancrelipase/ Pancreatin in Patients with Exocrine Pancreatic Insufficiency and a Medical History of Diabetes Mellitus." *Pancreas* 45 (5): 679–86. https://doi.org/10.1097/MPA.0000000000000514.

Widodo, Ariani Dewi, Rianto Setiabudy, Ina S. Timan, Saptawati Bardosono, Widdy Winarta, and Agus Firmansyah. 2018. "Pancreatic Enzyme Replacement Therapy (PERT) in Children with Persistent Diarrhea: Avoidance of Elemental Diet Need, Accessibility and Costs." *Asia Pacific Journal of Clinical Nutrition* 27 (3): 512–18. https://doi.org/10.6133/apjcn.082017.05.

Xu, Changjie, Anne E. Mathews, Camila Rodrigues, Brandon J. Eudy, Cheryl A. Rowe, Alison O'Donoughue, and Susan S. Percival. 2018. "Aged Garlic Extract Supplementation Modifies Inflammation and Immunity of Adults with Obesity: A Randomized, Double-Blind, Placebo-Controlled Clinical Trial." *Clinical Nutrition ESPEN* 24 (April): 148–55. https://doi.org/10.1016/j.clnesp.2017.11.010.

Yagnik, Darshna, Vlad Serafin, and Ajit J. Shah. 2018. "Antimicrobial Activity of Apple Cider Vinegar against Escherichia Coli, Staphylococcus Aureus and Candida Albicans; Downregulating Cytokine and Microbial Protein Expression." *Scientific Reports* 8 (1): 1732. https://doi.org/10.1038/s41598-017-18618-x.

Yang, Jyh-Ferng, Cheng-Hong Yang, Ming-Tsai Liang, Zi-Jie Gao, Yuh-Wern Wu, and Li-Yeh Chuang. 2016. "Chemical Composition, Antioxidant, and Antibacterial Activity of Wood Vinegar from Litchi Chinensis." *Molecules (Basel, Switzerland)* 21 (9). https://doi.org/10.3390/molecules21091150.

Yang, Mildred S. 2016. *Regulatory Aspects of Nutraceuticals: Chinese Perspective. Nutraceuticals: Efficacy, Safety and Toxicity.* Elsevier Inc. https://doi.org/10.1016/B978-0-12-802147-7.00067-X.

Yang, Y., C. Shao, W. Zhang, G. Wang, D.-C. Lu, W Han, Z.-S. Wu, and C.-B. Chen. 2019. "Omega-3 Polyunsaturated Fatty Acids Prevent Progression of Liver Fibrosis and Promote Liver Regeneration after Partial Hepatectomy in Cirrhotic Rats." *European Review for Medical and Pharmacological Sciences* 23 (22): 10151–60. https://doi.org/10.26355/eurrev_201911_19585.

Yu, Zheming, Hamed Samavat, Allison M. Dostal, Renwei Wang, Carolyn J. Torkelson, Chung S. Yang, Lesley M. Butler, et al. 2017. "Effect of Green Tea Supplements on Liver Enzyme Elevation: Results from a Randomized Intervention Study in the United States." *Cancer Prevention Research* 10 (10): 571–79. https://doi.org/10.1158/1940-6207.CAPR-17-0160.

Yuan, Chun Su, Gang Wei, Lucy Dey, Theodore Karrison, Linda Nahlik, Spring Maleckar, Kristen Kasza, Michael Ang-Lee, and Jonathan Moss. 2004. "Brief Communication: American Ginseng Reduces Warfarin's Effect in Healthy Patients. A Randomized, Controlled Trial." *Annals of Internal Medicine* 141 (1). https://doi.org/10.7326/0003-4819-141-1-200407060-00011.

Zhang, Fei, Yuan Zhang, Ting Yang, Ze-Qing Ye, Jing Tian, Hai-Rong Fang, Juan-Juan Han, Zhe-Zhi Wang, and Xing Li. 2019. "Scopoletin Suppresses Activation of Dendritic Cells and Pathogenesis of Experimental Autoimmune Encephalomyelitis by Inhibiting NF-KB Signaling." *Frontiers in Pharmacology* 10: 863. https://doi.org/10.3389/fphar.2019.00863.

In: Nutraceuticals: Evidence Based Use
Editor: Raluca Maria Pop
ISBN: 978-1-53618-052-7

Chapter 2

EVIDENCE OF NUTRACEUTICALS IN THE *IN VITRO* AND *IN VIVO* STUDIES

Paul-Mihai Boarescu*

Department of Pathophysiology, "Iuliu Haţieganu," University of Medicine and Pharmacy, Cluj-Napoca, Romania

ABSTRACT

The use of experimental models is a mandatory step for understanding the mechanisms of action and characteristic effects of nutraceuticals. This chapter presents *in vitro* and *in vivo* studies that proved the beneficial effects of some of the most widespread nutraceuticals. *In vivo* and *in vitro* studies are useful to validate the beneficial effects of different nutraceuticals. To have more accurate results, most of the studies' design include an *in vitro* and an *in vivo* part. In this chapter, an attempt has been made to briefly describe *in vitro* and *in vivo* experimental models for evaluation of beneficial effects of some of the most widespread nutraceuticals.

Keywords: nutraceuticals, *in vitro*, *in vivo*

---

* Corresponding Author's E-mail: boarescu.paul@umfcluj.ro.

INTRODUCTION

The *in vivo* and *in vitro* systems play an important role in the safety assessment of nutraceuticals. *In vitro* experimental systems including isolated organs or cells, microorganisms, sub-cellular organelles, and molecular entities (e.g., receptors, transport proteins) are commonly used to gain insight into properties of nutraceuticals. *In vivo* studies are performed in a whole, living organism. The use of animal models for the *in vivo* studies to determine the beneficial effects of a natural compound is required as part of the validation and safety of a product.

A simple search on PubMed shows a result of 1617423 studies that contain the words "nutraceuticals," "*in vitro*" or "*in vivo*", 555758 of these studies are from the last 10 years. This chapter represents a brief synthesis of the most important studies of the most widespread nutraceuticals.

CHEMICAL CONSTITUENTS - THE NUTRIENTS *IN VIVO* AND *IN VITRO* STUDIES

Ascorbic Acid

Ascorbic acid, also known as Vitamin C, is one of the important water-soluble vitamins that plays an essential role in protecting the body from infection and disease. It is necessary for the synthesis of collagen in the connective tissues, syntetis of steroid hormones, neurotransmitters, carnitine, conversion of cholesterol to bile acid, and enhancement of iron bioavailability (Tareen et al. 2015). Most plants and animals synthesize ascorbic acid for their requirement but apes and humans have to supplement ascorbic acid mainly through fruits, vegetables and tablets, as they can not synthesize it, due to lack of gluconolactone oxidase enzyme (Naidu 2003). Ascorbic acid is the most abundant vitamin in citrus fruits such as orange, lemon and grapefruit (Tareen et al. 2015).

In vitro and *in vivo* findings of a study performed by Bassem et al. (2013) identified ascorbic acid as a novel regulator of neutrophil extracellular trap (NET) formation in sepsis. NET plays a crucial role in host defence during local infection by trapping and killing pathogens (Bassem et al. 2013).

Recent studies suggested that, after intravenous administration, ascorbic acid can have cytotoxic effects to cancer cells (Ma et al. 2014; Kim et al. 2012; Vuyyuri et al. 2013; Chen et al. 2005). Yun et al., (2015) reported that the oxidized form of vitamin C, dehydroascorbate (DHA) plays a major role in apoptosis of colorectal cancers C cells with KRAS or BRAF mutations, when exposed to high levels of vitamin C. These effects are explained by the increased uptake of dehydroascorbic acid (DHA), via the glucose transporter 1 (GLUT1). Increased DHA uptake causes oxidative stress associated with depleting glutathione (GSH), accumulation of reactive oxygen species (ROS) and inactivation of glyceraldehyde 3-phosphate dehydrogenase (GAPDH). Inhibition of GAPDH in highly glycolytic KRAS or BRAF mutant cells leads to an energetic crisis and cell death (Yun et al. 2015).

In vitro study performed by Wang et al., (2016) on murine 4T1 breast cancer cells and CT26 colon cancer cells concluded that administration of high-dose ascorbic acid inhibits the proliferation and induces tumour cell apoptosis of murine colon and breast cancer cells. Even more, synergistic anti-tumoral effects of vitamin C and cisplatin were observed and cytotoxicity of ascorbic acid can be partially reversed by N-acetyl-cysteine at a relatively low dosage (Wang, Yin, and Wang 2016). The *in vivo* part of the same study performed on female BALB/c mice with subcutaneously inoculated 4T1 and CT26 cells revealed that intratumoral delivery of vitamin C delayed the tumour growth in these murine solid tumour models (Wang, Yin, and Wang 2016).

Vitamin D

Vitamin D is a fat-soluble vitamin that is naturally present in very few foods, responsible for increasing intestinal absorption of calcium, magnesium, and phosphate. Fish (not only fatty fish), egg yolk, and offal such as liver are good sources of vitamin D (Lamberg-Allardt 2006).

The results of an *in vivo* study performed by Gül and Aygün, (2019) revealed that administration of vitamin D had a cardioprotective effect on doxorubicin-induced cardiotoxicity in rat model, as it decreased the levels of cardiac injury markers such as creatine phosphokinase (CPK) and cardiac troponin T, reduced electrocardiography changes and diminished the level of 99mTechnetium pyrophosphate (99mTc PYP) uptake. They suggested that vitamin D administration may be effective in preventing doxorubicin-induced cardiotoxicity and that 99mTc PYP scintigraphy could be used in the follow-up of chemotherapy patients using doxorubicin (Gül and Aygün 2019).

Vitamin D may protect the airways from pathological pollution-induced inflammation as well. One *in vitro* study showed that it enhanced human bronchial epithelial cells antioxidant responses and modulated the immune response to particulate matter pollutant exposure (Pfeffer et al. 2018).

Zinc

Zinc is an essential micronutrient for human metabolism that catalyzes enzyme activity, regulates gene expression, and contributes to protein structure. Zinc is found in a variety of foods, such as beef, seafood, poultry, and grains (Saper and Rash 2009).

Maywald et al., (2018) studied the impact of zinc supplementation on T helper type 9 cells (Th9) differentiation in two in vitro models: 1) on non-activated resting T cells in peripheral blood mononuclear cells (PBMC) and 2) in mixed lymphocyte cultures (MLC) displaying allogeneic activated T cells in graft-versus-host disease. Their findings suggested that zinc supplementation can be an auspicious tool to ameliorate allogeneic immunoreactions as it significantly diminished IL-4-induced Th9 differentiation in MLC. Even more, Th9 cells were induced in resting T cells in PBMC, hence triggering the immunological defence (Maywald, Wang, and Rink 2018).

Yanqiang Huang et al., (2019) synthesized a zinc linolenate compound by reacting zinc chloride with the ligand linolenic acid and investigated its therapeutic potential for the treatment of *Helicobacter pylori* (*H. pylori*) infection *in vitro* and *in vivo*. The *in vitro* part of the experiment showed that zinc linolenate had effective antibacterial activity against standard strains and drug-resistant clinical isolates of *H. pylori*, with no development of resistance during continuous serial passaging. Their explanation for these results was that zinc linolenate disrupted the bacterial cell membranes and generated reactive oxygen species. The *in vivo* part of the experiment that involved mouse models of multidrug-resistant *H. pylori* infection showed that zinc linolenate had killing efficacy comparable and superior to the triple therapy approach, when used as a monotherapy. Moreover, zinc linolenate induced negligible toxicity against normal tissues and caused minimal effects on both the diversity and composition of the murine gut microbiota (Yanqiang Huang et al. 2019).

Selenium

Selenium is an essential mineral that supports the immune system, cognitive function, and fertility. It has a central role in the conservation of biological compounds, especially deoxyribonucleic acid (DNA), proteins and lipids, exerting protection against free radicals resulting from normal metabolism. Selenium is also essential for the functioning of selenoproteins, especially for selenium-dependent glutathione peroxidase (GPx), an enzyme involved in protecting cells against oxidative damage (Mechlaoui et al. 2019). It can be found in high amounts in seafood and organ meats but also in muscle meats, cereals and other grains, and dairy products (Office and Supplements 2003).

Aribi et al., (2015) observed the effect of selenium on macrophage bactericidal activities against *Staphylococcus aureus* using golden Syrian hamsters and peritoneal macrophages cultured with *Staphylococcus aureus*. They concluded that selenium acts in a dose-dependent manner on macrophage activation, phagocytosis and bacterial killing, suggesting that inadequate doses may cause a loss of macrophage bactericidal activities and that selenium supplementation could enhance the *in vivo* control of the immune response to *Staphylococcus aureus* (Aribi et al. 2015).

Viezeliene el al., (2011) designed an *in vivo* study on BALB/c mice that evaluated possible protective effects of selenium against systemic aluminium toxicity. The redox status of mouse liver after short-term (16 h) exposure to aluminium was also investigated. They found out that selenium had an antioxidant effect as it ameliorated intracellular oxidative stress processes mediated by GSH (Viezeliene et al. 2011).

Alpha-Linolenic Acid

Alpha-linolenic acid a is a plant-based essential omega-3 18-carbon polyunsaturated fatty acid with three double bonds that can be mainly found in rapeseed and walnuts. Seeds (chia and flaxseed) and many common vegetable oils are also sources of alpha-linolenic acid (Sokoła-Wysoczańska et al. 2018).

An *in vitro* study performed by Erdinest et al., (2012) on human corneal epithelial cells revealed that alpha-linolenic acid may serve as a potent anti-inflammatory agent in ocular surface inflammation as it decreased protein contents and messenger RNA (mRNA) expression levels of interleukins (IL) such as IL-6, IL-8, IL-1β, and tumor necrosis factor-alpha (TNF-α) after cell stimulation. The cells were stimulated with either polyinosinic: polycytidylic acid or lipopolysaccharide complex. The anti-inflammatory effects of alpha-linolenic acid were mediated through nuclear factor-κB (NF-κB) signal transduction and were comparable to those of corticosteroids (Erdinest et al. 2012)

Performing an *in vivo* and *in vitro* study, Chamberland and Moon (2015) found out that administration of alpha-linolenic acid directly controlled the malignant potential (proliferation, adhesion, invasion and colony formation) of human and mouse colon cancer cell lines. They concluded that it may be useful for the development of new therapies for the management or chemoprevention of colon cancer (Chamberland and Moon 2015).

CHEMICAL CONSTITUENTS - THE PHYTOCHEMICALS *IN VIVO* AND *IN VITRO* STUDIES

Gallic Acid

Gallic acid is a low molecular weight triphenolic compound reported to possess strong antioxidant activities with efficient antiradical and anticancer properties. It provides efficient protection against oxidative damage caused by various reactive species including, hydroxyl (HO), peroxyl (ROO) or superoxide (O2), and the non-radicals, hypochlorous acid (HOCl) and hydrogen peroxide (H_2O_2) (Badhani, Sharma, and Kakkar 2015). It is widely distributed across the plant kingdom such as leaves (bearberry), roots and bark (pomegranates and gall nuts), but it can be also found in processed beverages such as red wines and green teas (Choubey et al. 2015).

Nabavi et al., (2016) studied the antidepressive-like effects of gallic acid and two synthetic analogues, methyl 3-O-methylgallat (M3OMG) and propyl 3-O-methyl gallate (P3OMG), in BALB/c mice with post-stroke depression, which is a secondary form of depression that could be linked to oxidative stress occurring during cerebral ischemia and the following reperfusion. They determined the *in vivo* antioxidant activity of these compounds through the evaluation of catalase (CAT) and superoxide dismutase (SOD) activity, reduced GSH levels and thiobarbituric acid-reactive substances (TBARS) in mouse brain and they found out that gallic acid and its synthetic analogues played an active in the reduction of oxidative stress, enhancement of antioxidant endogenous defences, and modulation of depressive symptoms, restoring normal behaviour (Nabavi et al. 2016).

A recent study performed by Li at al., (2017) suggested that gallic acid is a promising antifungal natural compound. *In vitro* assays performed on thirty strains of fungi belonging to nine species: six strains each of *Trichophyton rubrum* and *Trichophyton mentagrophytes*; three strains each of *Trichophyton verrucosum*, *Microsporum canis*, *Trichophyton violaceum*, and *Trichophyton schoenleinii*; two strains each of *Candida Albicans,*

Candida glabrata, and *Candida tropicalis* concluded that gallic acid has broad-spectrum antifungal activity, with the most sensitive fungus being *Candida albicans*. Investigations continued with *in vivo* testing that showed that gallic acid significantly reduced the mortality rate in the mice infection model of *Candida albicans* (Li et al. 2017).

Quercetin

Quercetin belongs to polyphenolic flavonoids which are abundantly found in apples, red grapes, citrus fruits, raspberries, honey, cherries, onions, and green leafy vegetables. It exerts various biological effects, including antioxidant, anticancer, apoptosis-inducing, cell cycle modulatory, angiogenesis inhibitory, protein kinase C-inhibitory, and antiviral effects (Hashemzaei et al. 2017).

The *in vitro* and *in vivo* study performed by Hashemzaei et al., (2017) demonstrated that quercetin has potential for therapeutic application in neoplasia, as it inhibited the growth of a panel of 9 cancer cell lines. Cell growth inhibition was attributed to the induction of apoptosis, evident in the CT-26, PC-12 (which is derived from pheochromocytoma and has been widely used as a model of neuronal differentiation), LNCaP (androgen-sensitive cancer line) and PC-3 (androgen-insensitive cancer line) cancer cell lines. Furthermore, their results demonstrated that quercetin not only reduced CT-26 and MCF-7 (estrogen receptor-positive breast cancer cells) tumor volume in a mouse model but also increased animal survival rate (Hashemzaei et al. 2017).

On a diabetes mellitus type 2 animal model, administration of quercetin decreased blood glucose and glycosuria levels, results explained by the ability of quercetin to promote beta-pancreatic islets regeneration and increased insulin release (Vessal, Hemmati, and Vasei 2003).

Also, a recent study performed on male A/J mice revealed that quercetin represents an entirely novel therapeutic option in the treatment of asthma due to its acute bronchodilatory properties, as it acts as an inhaled (phosphodiesterase type 4) PDE4 and protein-phospholipase C (PLCβ) inhibitor. When used alone or in combination with existing therapies such as short-acting β-agonists it may have beneficial effects of relaxing airway smooth muscle during an acute exacerbation. Additionally, in combination with anti-inflammatory effects, quercetin may be useful in asthma therapy prevent exacerbations, when administered daily (Townsend and Emala 2013).

Resveratrol

Resveratrol belongs to polyphenols' stilbenoids group and has been detected in more than 70 plant species, especially in grapes' skin and seeds, and was found in discrete amounts in red wines. It was reported to have a variety of anti-inflammatory, anti-platelet, and anti-carcinogenic effects (Salehi et al. 2018).

A study performed by Costa et al., (2011) investigated whether resveratrol can regulate sirtuin 1 (SIRT1), forkhead box protein O1 (FOXO1), adiponectin, PPARγ1–3, and PPARβ/δ (proliferator-activated nuclear receptors) in freshly isolated adipocytes prepared from visceral fat tissue samples obtained during bariatric surgery. Their results showed that resveratrol positively modulates SIRT1, adiponectin, and FOXO1 and decreases PPARγ1–3 mRNA expression in isolated adipocytes. Therefore by modulating genes that are related to lipid metabolism, resveratrol may be useful in preventing metabolic disease in human visceral adipose tissue (Costa et al. 2011).

Harikumar et al., (2010) performed an *in vitro* and *in vivo* study to determine whether resveratrol has the antineoplastic potential either alone or in combination with gemcitabine in the treatment of pancreatic cancer. They observed that resveratrol in combination with gemcitabine inhibited the proliferation of 4 different human advanced pancreatic cancer cell lines, synergized the apoptotic effects of gemcitabine, inhibited the constitutive activation of NF-κB and expression of cyclooxygenase-2 (COX-2), B-cell lymphoma (Bcl-2), Bcl-xL, cyclin D1, matrix metalloproteinase 9 (MMP-9) and vascular endothelial growth factor (VEGF). They concluded that resveratrol can potentiate the effects of gemcitabine through suppression of markers of proliferation, invasion, angiogenesis and metastasis (Harikumar et al. 2010).

Lin et al. (2012) suggested that a novel strategy of combining resveratrol and temozolomide may be beneficial in patients with brain tumors. After performing the *in vitro* experiment on human glioblastoma U87 MG and GBM8401 cell lines classified as grade IV and the *in vivo* experiment on 6 week-old female nude mice (BALB/c *nu/nu*), they observed that coadministration of resveratrol and temozolomide reduced tumor volumes by suppressing ROS/ extracellular signal-regulated kinase (ERK-mediated) autophagy and subsequently induced apoptosis (Lin et al. 2012).

Chemopreventive potential of resveratrol against bladder cancer and its mechanism of action were studied by Bai et al., (2010). They used human bladder cancer cell line T24 for the *in vitro* study and male BALB/c-nude mice (4 weeks old) for the *in vivo* study. Their results showed that resveratrol induced accumulation of T24 cells at the G1 phase of the cell cycle and caused dose- and time-dependent cytotoxicity and induction of apoptosis in bladder cancer T24 cells. Resveratrol inhibited the phosphorylation of protein kinase B (Akt), enhanced the phosphorylation of p38 mitogen-activated protein kinase (MAPK) and decreased the expression of VEGF and fibroblast growth factor-2, which might contribute to the inhibition of tumor growth on the bladder cancer xenograft model (Bai et al. 2010).

Beta Carotene

Beta carotene (β-carotene or provitamin A) is one of the lipophilic plant pigments known to possess anti-oxidation, anti-inflammatory, and anti-cancer effects. β-carotene can be found in red, yellow or orange coloured fruits and vegetables (Khoo et al. 2011).

Bai et al., (2005) studied the *in vitro* and *in vivo* regulatory function of β-carotene on the expression of inducible NO synthase (iNOS), COX-2, TNF-α, IL-1β as well as production of nitric oxide (NO) and prostaglandin E2 (PGE2) after lipopolysaccharides (LPS) administration. β-carotene administration inhibited, in a dose-dependent fashion, the production and expression of these inflammatory mediators in both LPS stimulated RAW264.7 cells and primary macrophages as well as in LPS-administrated mice. They suggested that β-carotene may be beneficial in the prevention or treatment of human inflammatory diseases, such as sepsis, atherosclerosis and rheumatoid arthritis (S. K. Bai et al. 2005).

Matos et al., (2006) observed the protective effects of β-carotene against iron-induced oxidative stress. The data they reported, using iron as a model of oxidative stress generation *in vivo*, indicated that β-carotene provides strong protection against the membrane and DNA oxidation in a prostate cancer model induced by iron administration (Matos et al. 2006).

Curcumin

Curcumin, a plant-derived polyphenolic compound, naturally present in turmeric (*Curcuma longa*), has been shown to target multiple signalling molecules, while also demonstrating activity at the cellular level, which has helped to support its multiple health benefits (Boarescu, Chirilă, et al. 2019).

Platania et al., (2018) tested the *in vitro* antioxidant and anti-inflammatory activity of curcumin on human retinal pigmented epithelial cells exposed to oxidative stress and on human retinal endothelial and human retinal pericytes cultured with high glucose. They observed that curcumin significantly decreased ROS concentration and TNF-α release in retinal pigmented epithelial cells and retinal endothelial cells and protected retinal pericytes from high glucose damage as assessed by cell viability and lactate dehydrogenase (LDH) release (Platania et al. 2018).

In vivo studies proved it's antioxidative effects on cardiovascular disease as curcumin pre-administration reduced oxidative stress parameters such as malondialdehyde (MDA), NO, and total oxidative stress (TOS) and enhanced antioxidant parameters such as thiol and total antioxidant capacity (TAC) in rats with acute myocardial infarction induced by isoproterenol (Boarescu, Boarescu, Bocșan, Pop, et al. 2019; Boarescu, Chirilă, et al. 2019) and in diabetic rats with acute myocardial infarction (Boarescu,

Boarescu, Bocşan, Gheban, et al. 2019). The anti-inflammatory effects were also proved in rats with acute myocardial infarction as curcumin pretreatment reduced the serum levels of TNF-α, IL-1α, IL-1β, IL-6 and regulated upon activation, normal T cell expressed and secreted (RANTES) (Boarescu, Chirilă, et al. 2019).

Pretreatment with curcumin offered a cardioprotective effect in doxorubicin-induced myocardial toxicity as well. Pretreatment with curcumin significantly reduced the level of biomarker enzymes like LDH, CPK, aspartate aminotransferase (AST), alanine aminotransferase (ALT) and alkaline phosphatase (ALP) back to normal. It had also antioxidant effects as it increased the level of antioxidant parameters such as GSH, SOD and CAT and decreased the elevated level of MDA in cardiac tissue (Swamy et al. 2012).

Performing an *in vitro* study on L1210 murine lymphocytic leukaemia, 4T1 murine breast, B16 murine melanoma and CT26 murine colon tumor cell lines and *in vivo* experiment on six-week-old female athymic Nude (nu-/nu-) mice Bianchi et al., (2018) observed that curcumin had an anti-neoplastic effect. Curcumin treatment inhibited the activity of adenosine triphosphate (ATP) synthase in isolated mitochondrial membranes leading to a dramatic drop of ATP and a reduction of oxygen consumption and a decreased tumor cell viability (Bianchi et al. 2018).

PROBIOTIC MICROORGANISMS *IN VIVO* AND *IN VITRO* STUDIES

Lactobacillus Acidophilus

Lactobacillus acidophilus is a Gram-positive lactic acid bacterium that has been used in the dairy industry for a long period and more recently is used as a probiotic. *Lactobacillus acidophilus* is one of the most commonly selected *Lactobacillus* species for dietary use and it is added to commercial yoghurts and dairy formulations both for its flavour and for probiotic effects (Bull et al. 2014).

Lee et al., (2015) revealed that probiotic bacteria, including *Lactobacillus acidophilus*, could be considered useful probiotics for the treatment of rotaviral gastroenteritis or as an alternative therapy since it inhibited infection by rotavirus *in vitro* and decreased the duration of diarrhea in pediatric patients (Lee et al. 2015).

The *in vitro* and *in vivo* study performed by Nakazato et al., (2011) suggested that *Lactobacillus acidophilus* could act as a probiotic agent attenuating the severity of salmonellosis. It was observed to reduce the adhesion to *in vitro* cultured cells and to decrease in the colony-forming units of *Salmonella* in the Peyer's patches and spleen of lactobacilli-treated mice. They demonstrated that *Lactobacillus* interfered with the adhesion, invasion and dissemination processes caused by *Salmonella* (Nakazato et al. 2011).

Huang et al., (2010) evaluated the effect of *Lactobacillus acidophilus* on cholesterol metabolism *in vivo*, using an experimental model with rats fed with a cholesterol-enriched experimental diet with or without *Lactobacillus acidophilus*. Rats with *Lactobacillus acidophilus* had significantly lower total serum cholesterol, low-density lipoprotein cholesterol (LDL-cholesterol) and triglycerides concentrations, but there was no change in the serum high-density lipoprotein cholesterol (HDL-cholesterol) concentrations. They also observed that the expression of Niemann-Pick C1-Like 1 (NPC1L1) in the duodenum and jejunum was significantly decreased following *Lactobacillus acidophilus* feeding, so the reduction in cholesterol levels is at least partially mediated by down-regulation of NPC1L1 (Huang et al. 2010).

Lactobacillus Bulgaricus

Lactobacillus bulgaricus belongs to the acidophilus complex, a group of lactobacilli related to *Lactobacillus gasseri*, *Lactobacillus johnsonii* and *Lactobacillus acidophilus* which have been used as probiotic cultures. It is the most important representatives of the heterogeneous group of lactic acid bacteria, with a worldwide application in yoghurt production. Consumption of yoghurt containing live *Lactobacillus bulgaricus* improves lactose digestion and attenuates the lactose intolerance, and due to its immune modulation activity has diarrhoea-alleviating effects (Van De Guchte et al. 2006).

An *in vitro* study performed by Abedi et al., (2013) on human colon adenocarcinoma cell line, Caco-2 revealed that *Lactobacillus bulgaricus* was able to inhibit *Escherichia coli (E. coli)* infection in the gut, by offering a good anti-adhesion activity against *E. coli*. One possible explanation is that the administration of some of the lactobacillus strains can decrease the pH of gut and faeces. Another explanation might be that the inhibitory effect of bacteria may consist of competition for the same adhesion sites and nutrients. For a better understanding of the mechanism of action of *Lactobacillus bulgaricus*, the authors suggested that *in vivo* studies are necessary to be performed (Abedi et al. 2013).

Takamura et al. (2011) observed that *Lactobacillus bulgaricus* had the potential to regulate the inflammatory responses and to ameliorate colitis. An *in vivo* study proved that *Lactobacillus bulgaricus* lead to the activation of the aryl hydrocarbon receptor and inhibited the development of dextran sodium sulfate-induced colitis on mice, which is a model of human ulcerative colitis induced by the administration of prostaglandin E2 in the large intestine (Takamura et al. 2011).

Li et al., (2017) suggested that *Lactobacillus bulgaricus* can be used as a potential probiotic against acute lead toxicity. *In vitro* part of the experiment revealed that *Lactobacillus bulgaricus* had high lead biosorption and lead tolerance while the *in vivo* part of the experiment showed that *Lactobacillus bulgaricus* favorized detoxication by alleviating tissue lead enrichment, increased lead levels in the faeces, improved the antioxidant index, and relieved renal pathological damages (Li et al. 2017).

OTHER TRADITIONAL NUTRACEUTICALS *IN VIVO* AND *IN VITRO* STUDIES

Beta-Glucans

Beta-glucans (β-Glucans) are groups of dietary fibres or polysaccharides naturally found in barley and oat grain but also in the cell wall of bacteria, fungi and algae (Jacob and Pescatore 2014). Among soluble fibres, β-glucans are the most commonly-consumed immunomodulator with strong anticancer effects. β-Glucans are believed to stimulate the immune system, to modulate cellular and humoral immunity, and thereby to have beneficial effects in fighting infectious diseases, such as bacterial, viral, fungal, and parasitic diseases (Bashir and Choi 2017). Even more, their beneficial role in insulin resistance, dyslipidemia, hypertension, and obesity are being continuously documented (El Khoury et al. 2012).

Javmen et al., (2015) observed that orally administered β-glucan from *Saccharomyces cerevisiae* (*S. cerevisiae*) enhanced interferon gamma (IFNγ) production in BALB/c mice *in vivo*, but did not influence IFNγ synthesis *in vitro*, on isolated leukocytes from BALB/c mice. Another *in vivo* study revealed that β-glucan from *S. cerevisiae* was a potent activator of anti-inflammatory cytokines and had better antioxidant effects compared to β-glucan from *Avena sativa* or *Hericium erinaceus* in BALB/cMlac mice (Chaiyasut et al. 2018). In this study β-glucan from *S. cerevisiae* induced the maximum level of the anti-inflammatory cytokines such as IL-10 and transforming growth factor β (TGF-β) and stimulated the expression of pro-inflammatory cytokines as IL-6, IL-17, and IFN-γ (Chaiyasut et al. 2018). Concerning oxidative stress, it increased the concentration of major antioxidant enzymes such as CAT, SOD, GPx and antioxidant capacity, on the one hand, and reduced the MDA formation by inhibiting lipid peroxidation, on the other hand (Chaiyasut et al. 2018).

An *in vitro* study performed by Jafaar et al., (2014) on human breast cancer cells revealed that β-glucan regulated breast cancer-relevant gene expression and may be useful to inhibit the cell proliferation in endocrine-resistant breast cancer.

A study performed by Andrade et al., (2016) on adult male Wistar rats with diabetes mellitus type 2 (induced by high-fat diet associated with a low dose of streptozotocin, 35 mg/kg) concluded that isolated administration of β-glucan from *S. cerevisiae* decreased the serum levels of glucose, glycated haemoglobin, triglycerides, total cholesterol, low-density lipoprotein and also reduced the atherogenic index of plasma. When it was combined with physical training, an additive effect for reducing the atherogenic index of plasma and renal lesions was observed (Andrade et al. 2016).

Coenzyme Q-10

Coenzyme Q-10 is a member of the ubiquinone family of compounds found in most aerobic organisms from bacteria to mammals. The primary biochemical action of coenzyme Q-10 is essential in the series of redox reactions that are involved in the synthesis of ATP, where it acts as a cofactor in the electron-transport chain. Therefore, coenzyme Q-10 is essential for the health of virtually of all human tissues and organs, as most cellular functions are dependent on an adequate supply of ATP. Moreover, coenzyme Q10 is one of the most significant lipid antioxidants, which prevents the generation of free radicals and modifications of DNA, proteins and lipids (Saini 2011).

Coenzyme Q-10 is widely distributed in nature, as it is found in both plants and animals. Animal products, especially tissues with the highest energy demands including the heart, liver, and kidney tend to have the greatest amount of Coenzyme Q-10 compared with other sources. Nonanimal products such as broccoli, spinach, soybean/canola/palm oils and nuts also contain fair amounts of coenzyme Q10 (Rodick et al. 2018).

Burić et al., (2019) highlighted the benefits of Coenzyme Q-10 supplementation in glioblastoma therapy. They used an RC6 rat glioma model and have demonstrated that Coenzyme Q-10 sensitizes RC6 cells to temozolomide and enhances temozolomide-induced cell death in a 3D model. Coenzyme Q-10 applied in combination with temozolomide modified mitochondrial antioxidant capacity and expression of matrix metalloproteinase 9 and epithelial-mesenchymal transition markers and eventually inhibited invasion of RC6 cells *in vitro* and *in vivo* (Burić et al. 2019).

Coenzyme Q-10 could be a potential skin-whitening agent without side effects. *In vitro* experiment conducted by Hseu et al., (2019) concluded that Coenzyme Q-10 can be characterized as a potent tyrosinase inhibitor by downregulating melanogenic regulatory genes (pro-opiomelanocortin; α-melanocyte-stimulating hormone; tyrosinase, microphthalmia-associated transcription factor) and inducing antioxidant genes (heme oxygenase-1, γ-glutamyl cysteine ligand, GSH) in human skin keratinocyte HaCaT cells when exposed to ultraviolet A. Moreover, *in vivo* part of the experiment proved that Coenzyme Q-10 reversed ultraviolet induced hyperpigmentation in zebrafish body pigmentation (Hseu et al. 2019).

Lulli et al., (2012) found out that Coenzyme Q-10 prevented ultraviolet and γ radiation-induced apoptosis of retinal cells by inhibiting mitochondrial permeability transition. In the in *vitro* experiment performed on cultured retinal pigment epithelial cells (ARPE-19) and in *vivo* experiment performed on adult Wistar albino rats, they observed that that Coenzyme Q-10 is effective in inhibiting the progression of retinopathies, for which solar ultraviolet radiation is recognized as one of the main causes. They suggested that therapeutic application of Coenzyme Q-10 eye drops can be efficient in the treatment of apoptosis-sustained retinopathies, as Coenzyme Q-10 extends its anti-apoptotic property also to retinal cells committed to apoptosis by serum deprivation or chemical hypoxia (Lulli et al. 2012).

Conclusion

Nutraceuticals not only they are widespread but they also have massive chances for growth and extension in the term of health benefit.

In vivo and *in vitro* studies are useful to validate the beneficial effects of different nutraceuticals. To have more accurate results, most of the studies' design includes an *in vitro* and an *in vivo* part.

Although, several nutraceuticlas are tested in both *in vitro* and *in vivo*, further clinical trials are required to study the complete effect of nutraceuticals in the body and on diferent pathologies. Even if *in vitro* and *in vivo* studies suggest a protective effect of nutraceuticals against different diseases, the results must be supported fully by clinical evidence.

REFERENCES

Abedi, D., S. Feizizadeh, V. Akbari, and A. Jafarin-Dehkordi. 2013. "In Vitro Anti-Bacterial and Anti-Adherence Effects of *Lactobacillus Delbrueckii* Subsp Bulgaricus on *Escherichia Coli*." *Research in Pharmaceutical Sciences* 8 (4): 261–68.

Andrade, Eric Francelino, Andressa Ribeiro Veiga Lima, Ingrid Edwiges Nunes, Débora Ribeiro Orlando, Paula Novato Gondim, Márcio Gilberto Zangeronimo, Fernando Henrique Ferrari Alves, and Luciano José Pereira. 2016. "Exercise and Beta-Glucan Consumption (*Saccharomyces Cerevisiae*) Improve the Metabolic Profile and Reduce the Atherogenic Index in Type 2 Diabetic Rats (HFD/STZ)." *Nutrients* 8 (12). https://doi.org/10.3390/nu8120792.

Aribi, Mourad, Warda Meziane, Salim Habi, Yasser Boulatika, Hélène Marchandin, and Jean Luc Aymeric. 2015. "Macrophage Bactericidal Activities against Staphylococcus Aureus Are Enhanced in Vivo by Selenium Supplementation in a Dose-Dependent Manner." *PLoS ONE* 10 (9): 1–18. https://doi.org/10.1371/journal.pone.0135515.

Badhani, Bharti, Neha Sharma, and Rita Kakkar. 2015. "Gallic Acid: A Versatile Antioxidant with Promising Therapeutic and Industrial Applications." *RSC Advances* 5 (35): 27540–57. https://doi.org/10.1039/c5ra01911g.

Bai, Se Kyung, Seon Jin Lee, Hee Jun Na, Kwon Soo Ha, Jeong A. Han, Hansoo Lee, Young Guen Kwon, Cha Kwon Chung, and Young

Myeong Kim. 2005. "β-Carotene Inhibits Inflammatory Gene Expression in Lipopolysaccharide-Stimulated Macrophages by Suppressing Redox-Based NF-KB Activation." *Experimental and Molecular Medicine* 37 (4): 323–34. https://doi.org/10.1038/emm.2005.42.

Bai, Yu, Qi-Qi Mao, Jie Qin, Xiang-Yi Zheng, Yun-Bin Wang, Kai Yang, Hua-Feng Shen, and Li-Ping Xie. 2010. "Resveratrol Induces Apoptosis and Cell Cycle Arrest of Human T24 Bladder Cancer Cells *in Vitro* and Inhibits Tumor Growth *in Vivo*." *Cancer Science* 101 (2): 488–93. https://doi.org/10.1111/j.1349-7006.2009.01415.x.

Bassem, Mohammed, M., Bernard J. Fisher, Donatas Kraskauskas, Daniela Farkas, Donald F. Brophy, Alpha A. Fowler, and Ramesh Natarajan. 2013. "Vitamin C: A Novel Regulator of Neutrophil Extracellular Trap Formation." *Nutrients* 5 (8): 3131–51. https://doi.org/10.3390/nu5083131.

Bashir, Khawaja Muhammad Imran, and Jae Suk Choi. 2017. "Clinical and Physiological Perspectives of β-Glucans: The Past, Present, and Future." *International Journal of Molecular Sciences*. MDPI AG. https://doi.org/10.3390/ijms18091906.

Bianchi, Giovanna, Silvia Ravera, Chiara Traverso, Adriana Amaro, Francesca Piaggio, Laura Emionite, Tiziana Bachetti, Ulrich Pfeffer, and Lizzia Raffaghello. 2018. "Curcumin Induces a Fatal Energetic Impairment in Tumor Cells *in vitro* and *in vivo* by Inhibiting ATP-Synthase Activity." *Carcinogenesis* 39 (9): 1141–50. https://doi.org/10.1093/carcin/bgy076.

Boarescu, Boarescu, Bocșan, Gheban, Bulboacă, Nicula, Pop, Râjnoveanu, and Bolboacă. 2019. "Antioxidant and Anti-Inflammatory Effects of Curcumin Nanoparticles on Drug-Induced Acute Myocardial Infarction in Diabetic Rats." *Antioxidants* 8 (10): 504. https://doi.org/10.3390/antiox8100504.

Boarescu, Paul Mihai, Ioana Boarescu, Ioana Corina Bocșan, Raluca Maria Pop, Dan Gheban, Adriana Elena Bulboacă, Cristina Nicula, Ruxandra Mioara Râjnoveanu, and Sorana D. Bolboacă. 2019. "Curcumin Nanoparticles Protect against Isoproterenol Induced Myocardial

Infarction by Alleviating Myocardial Tissue Oxidative Stress, Electrocardiogram, and Biological Changes." *Molecules (Basel, Switzerland)* 24 (15). https://doi.org/10.3390/molecules24152802.

Boarescu, Paul Mihai, Ioana Chirilă, Adriana E. Bulboacă, Ioana Corina Bocşan, Raluca Maria Pop, Dan Gheban, and Sorana D. Bolboacă. 2019. "Effects of Curcumin Nanoparticles in Isoproterenol-Induced Myocardial Infarction." *Oxidative Medicine and Cellular Longevity* 2019. https://doi.org/10.1155/2019/7847142.

Bull, Matthew J., Keith A. Jolley, James E. Bray, Maarten Aerts, Peter Vandamme, Martin C. J. Maiden, Julian R. Marchesi, and Eshwar Mahenthiralingam. 2014. "The Domestication of the Probiotic Bacterium *Lactobacillus Acidophilus*." *Scientific Reports* 4 (November). https://doi.org/10.1038/srep07202.

Burić, Sonja Stojković, Ana Podolski-Renić, Jelena Dinić, Tijana Stanković, Mirna Jovanović, Stefan Hadžić, Jose M. Ayuso, et al. 2019. "Modulation of Antioxidant Potential with Coenzyme Q10 Suppressed Invasion of Temozolomide-Resistant Rat Glioma in Vitro and in Vivo." *Oxidative Medicine and Cellular Longevity* 2019. https://doi.org/10.1155/2019/3061607.

Chaiyasut, Chaiyavat, Noppawat Pengkumsri, Bhagavathi Sundaram Sivamaruthi, Sasithorn Sirilun, Periyanaina Kesika, Manee Saelee, Khontaros Chaiyasut, and Sartjin Peerajan. 2018. "Extraction of Β-Glucan of *Hericium Erinaceus*, *Avena Sativa* L., and *Saccharomyces Cerevisiae* and *in vivo* Evaluation of Their Immunomodulatory Effects." *Food Science and Technology* 38 (December): 138–46. https://doi.org/10.1590/fst.18217.

Chamberland, John P., and Hyun Seuk Moon. 2015. "Down-Regulation of Malignant Potential by Alpha Linolenic Acid in Human and Mouse Colon Cancer Cells." *Familial Cancer* 14 (1): 25–30. https://doi.org/10.1007/s10689-014-9762-z.

Chen, Qi, Michael Graham Espey, Murali C. Krishna, James B. Mitchell, Christopher P. Corpe, Garry R. Buettner, Emily Shaded, and Mark Levine. 2005. "Pharamacologic Ascorbic Acid Concentrations Selectively Kill Cancer Cells: Action as a pro-Drug to Deliver Hydrogen

Peroxide to Tissuse." *Proceedings of the National Academy of Sciences of the United States of America* 102 (38): 13604–9. https://doi.org/10.1073/pnas.0506390102.

Choubey, Sneha, Lesley R.achel Varughese, Vinod Kumar, and Vikas Beniwal. 2015. "Medicinal Importance of Gallic Acid and Its Ester Derivatives: A Patent Review." *Pharmaceutical Patent Analyst*. https://doi.org/10.4155/ppa.15.14.

Costa, Címtia Dos Santos, Francieli Rohden, Thais Ortiz Hammes, Rogério Margis, Josiane Woutheres Bortolotto, Alexandre Vontobel Padoin, Cláudio Cora Mottin, and Regina Maria Guaragna. 2011. "Resveratrol Upregulated SIRT1, FOXO1, and Adiponectin and Downregulated PPARγ1-3 MRNA Expression in Human Visceral Adipocytes." *Obesity Surgery* 21 (3): 356–61. https://doi.org/10.1007/s11695-010-0251-7.

Erdinest, Nir, Or Shmueli, Yoni Grossman, Haim Ovadia, and Abraham Solomon. 2012. "Anti-Inflammatory Effects of Alpha Linolenic Acid on Human Corneal Epithelial Cells." *Investigative Ophthalmology and Visual Science* 53 (8): 4396–4406. https://doi.org/10.1167/iovs.12-9724.

Fazel Nabavi, Seyed, Solomon Habtemariam, Arianna Di Lorenzo, Antoni Sureda, Sedigheh Khanjani, Seyed Mohammad Nabavi, and Maria Daglia. 2016. "Post-Stroke Depression Modulation and in Vivo Antioxidant Activity of Gallic Acid and Its Synthetic Derivatives in a Murine Model System." Accessed December 5, 2019. https://doi.org/10.3390/nu8050248.

Guchte, M. Van De, S. Penaud, C. Grimaldi, V. Barbe, K. Bryson, P. Nicolas, C. Robert, et al. 2006. "The Complete Genome Sequence of Lactobacillus Bulgaricus Reveals Extensive and Ongoing Reductive Evolution." *Proceedings of the National Academy of Sciences of the United States of America* 103 (24): 9274–79. https://doi.org/10.1073/pnas.0603024103.

Harikumar, Kuzhuvelil B., Ajaikumar B. Kunnumakkara, Gautam Sethi, Parmeswaran Diagaradjane, Preetha Anand, Manoj K. Pandey, Juri Gelovani, Sunil Krishnan, Sushovan Guha, and Bharat B. Aggarwal. 2010. "Resveratrol, a Multitargeted Agent, Can Enhance Antitumor

Activity of Gemcitabine *in vitro* and in Orthotopic Mouse Model of Human Pancreatic Cancer." *International Journal of Cancer*, NA-NA. https://doi.org/10.1002/ijc.25041.

Hashemzaei, Mahmoud, Amin Delarami Far, Arezoo Yari, Reza Entezari Heravi, Kaveh Tabrizian, Seyed Mohammad Taghdisi, Sarvenaz Ekhtiari Sadegh, et al. 2017. "Anticancer and Apoptosis-Inducing Effects of Quercetin *in vitro* and *in vivo*." *Oncology Reports* 38 (2): 819–28. https://doi.org/10.3892/or.2017.5766.

Hseu, You Cheng, Yi Geng Ho, Dony Chacko Mathew, Hung Rong Yen, Xuan Zao Chen, and Hsin Ling Yang. 2019. "The in Vitro and in Vivo Depigmenting Activity of Coenzyme Q10 through the Down-Regulation of α-MSH Signaling Pathways and Induction of Nrf2/ARE-Mediated Antioxidant Genes in UVA-Irradiated Skin Keratinocytes." *Biochemical Pharmacology* 164 (June): 299–310. https://doi.org/10.1016/j.bcp.2019.04.015.

Huang, Yanqiang, Xudong Hang, Xueqing Jiang, Liping Zeng, Jia Jia, Yong Xie, Fei Li, and Hongkai Bi. 2019. "*In vitro* and *in vivo* Activities of Zinc Linolenate, a Selective Antibacterial Agent against *Helicobacter Pylori*." *Antimicrobial Agents and Chemotherapy* 63 (6): 1–18.

Huang, Ying, Jinfeng Wang, Yi Cheng, and Yongchen Zheng. 2010. "The Hypocholesterolaemic Effects of *Lactobacillus Acidophilus* American Type Culture Collection 4356 in Rats Are Mediated by the Down-Regulation of Niemann-Pick C1-Like 1." *British Journal of Nutrition* 104 (6): 807–12. https://doi.org/10.1017/S0007114510001285.

Jacob, Jacqueline P., and Anthony J. Pescatore. 2014. "Barley β-Glucan in Poultry Diets." *Annals of Translational Medicine*. AME Publishing Company. https://doi.org/10.3978/j.issn.2305-5839.2014.01.02.

Jafaar, Zainab M. T., Lacey M. Litchfield, Margarita M. Ivanova, Brandie N. Radde, Numan Al-Rayyan, and Carolyn M. Klinge. 2014. "β-D-Glucan Inhibits Endocrine-Resistant Breast Cancer Cell Proliferation and Alters Gene Expression." *International Journal of Oncology* 44 (4): 1365–75. https://doi.org/10.3892/ijo.2014.2294.

Javmen, Artur, Aušra Nemeikaitė-Čėnienė, Maksim Bratchikov, Saulius Grigiškis, Fortūnatas Grigas, Irena Jonauskienė, Danguolė Zabulytė, and Mykolas Mauricas. 2015. "β-Glucan from Saccharomyces Cerevisiae Induces IFN-γ Production In Vivo in BALB/c Mice." *In Vivo (Athens, Greece)* 29 (3): 359–63. Accessed December 4, 2019. http://www.ncbi.nlm.nih.gov/pubmed/25977381.

Khoo, Hock Eng, K. Nagendra Prasad, Kin Weng Kong, Yueming Jiang, and Amin Ismail. 2011. "Carotenoids and Their Isomers: Color Pigments in Fruits and Vegetables." *Molecules*. https://doi.org/10.3390/molecules16021710.

Khoury, D. El, C. Cuda, B. L. Luhovyy, and G. H. Anderson. 2012. "Beta Glucan: Health Benefits in Obesity and Metabolic Syndrome." *Journal of Nutrition and Metabolism*. https://doi.org/10.1155/2012/851362.

Kim, Jinsun, Soon Duck Lee, Boogi Chang, Dong Hoon Jin, Sam Il Jung, Mee Young Park, Youngsoo Han, et al. 2012. "Enhanced Antitumor Activity of Vitamin C via P53 in Cancer Cells." *Free Radical Biology and Medicine* 53 (8): 1607–15. https://doi.org/10.1016/j.freeradbiomed.2012.07.079.

Lamberg-Allardt, Christel. 2006. "Vitamin D in Foods and as Supplements." *Progress in Biophysics and Molecular Biology*. https://doi.org/10.1016/j.pbiomolbio.2006.02.017.

Lee, Do Kyung, Jae Eun Park, Min Ji Kim, Jae Goo Seo, Ji Hyuk Lee, and Nam Joo Ha. 2015. "Probiotic Bacteria, B.Longum and L.Acidophilus Inhibit Infection by Rotavirus in Vitro and Decrease the Duration of Diarrhea in Pediatric Patients." *Clinics and Research in Hepatology and Gastroenterology* 39 (2): 237–44. https://doi.org/10.1016/j.clinre.2014.09.006.

Li, Bailiang, Da Jin, Shangfu Yu, Smith Etareri Evivie, Zafarullah Muhammad, Guicheng Huo, and Fei Liu. 2017. "*In Vitro* and *In Vivo* Evaluation of *Lactobacillus Delbrueckii* Subsp. Bulgaricus KLDS1.0207 for the Alleviative Effect on Lead Toxicity." *Nutrients* 9 (8): 845. https://doi.org/10.3390/nu9080845.

Li, Zhi Jian, Meng Liu, Gulina Dawuti, Qin Dou, Yu Ma, Heng Ge Liu, and Silafu Aibai. 2017. "Antifungal Activity of Gallic Acid In Vitro and In Vivo." *Phytotherapy Research* 31 (7): 1039–45. https://doi.org/10.1002/ptr.5823.

Lin, Chien Ju, Chin Cheng Lee, Yung Luen Shih, Tsung Yao Lin, Sheng Hao Wang, Yuh Feng Lin, and Chwen Ming Shih. 2012. "Resveratrol Enhances the Therapeutic Effect of Temozolomide against Malignant Glioma *in Vitro* and *in Vivo* by Inhibiting Autophagy." *Free Radical Biology and Medicine* 52 (2): 377–91. https://doi.org/10.1016/j.freeradbiomed.2011.10.487.

Lulli, M., E. Witort, L. Papucci, E. Torre, N. Schiavone, M. Dal Monte, and S. Capaccioli. 2012. "Coenzyme Q10 Protects Retinal Cells from Apoptosis Induced by Radiation in Vitro and in Vivo." *Journal of Radiation Research* 53 (5): 695–703. https://doi.org/10.1093/jrr/rrs025.

Ma, Yan, Julia Chapman, Mark Levine, Kishore Polireddy, Jeanne Drisko, and Qi Chen. 2014. "Cancer: High-Dose Parenteral Ascorbate Enhanced Chemosensitivity of Ovarian Cancer and Reduced Toxicity of Chemotherapy." *Science Translational Medicine* 6 (222). https://doi.org/10.1126/scitranslmed.3007154.

Matos, H. R., S. A. Marques, O. F. Gomes, A. A. Silva, J. C. Heimann, P. Di Mascio, and M. H.G. Medeiros. 2006. "Lycopene and β-Carotene Protect in Vivo Iron-Induced Oxidative Stress Damage in Rat Prostate." *Brazilian Journal of Medical and Biological Research* 39 (2): 203–10. https://doi.org/10.1590/S0100-879X2006000200006.

Maywald, Martina, Fudi Wang, and Lothar Rink. 2018. "Zinc Supplementation Plays a Crucial Role in T Helper 9 Differentiation in Allogeneic Immune Reactions and Non-Activated T Cells." *Journal of Trace Elements in Medicine and Biology* 50 (February): 482–88. https://doi.org/10.1016/j.jtemb.2018.02.004.

Mechlaoui, Marwa, David Dominguez, Lidia Robaina, Pierre Andre Geraert, Sadasivam Kaushik, Reda Saleh, Mickael Briens, Daniel Montero, and Marisol Izquierdo. 2019. "Effects of Different Dietary Selenium Sources on Growth Performance, Liver and Muscle Composition, Antioxidant Status, Stress Response and Expression of Related Genes in Gilthead Seabream (*Sparus Aurata*)." *Aquaculture* 507 (December 2018): 251–59. https://doi.org/10.1016/j.aquaculture.2019.04.037.

Naidu, K. Akhilender. 2003. "Vitamin C in Human Health and Disease Is Still a Mystery? An Overview." *Nutrition Journal* 2 (1): 7. https://doi.org/10.1186/1475-2891-2-7.

Nakazato, G., F. L. Paganelli, J. C. Lago, F. H. Aoki, C. Mobilon, M. Brocchi, E. G. Stehling, and W. D. Silveira. 2011. "*Lactobacillus Acidophilus* Decreases *Salmonella Typhimurium* Invasion *in Vivo*." *Journal of Food Safety* 31 (2): 284–89. https://doi.org/10.1111/j.1745-4565.2011.00299.x.

Office, N. I. H., and Dietary Supplements. 2003. "NIH Office of Dietary Supplements (Http://Ods.Od.Nih.Gov)." *Office* 03: 1–18.

Pfeffer, Paul E., Haw Lu, Elizabeth H. Mann, Yin-Huai Chen, Tzer-Ren Ho, David J. Cousins, Chris Corrigan, Frank J. Kelly, Ian S. Mudway, and Catherine M. Hawrylowicz. 2018. "Effects of Vitamin D on Inflammatory and Oxidative Stress Responses of Human Bronchial Epithelial Cells Exposed to Particulate Matter." Edited by Stelios Loukides. *PLOS ONE* 13 (8): e0200040. https://doi.org/10.1371/journal.pone.0200040.

Platania, Chiara B. M., Annamaria Fidilio, Francesca Lazzara, Cateno Piazza, Federica Geraci, Giovanni Giurdanella, Gian Marco Leggio, Salvatore Salomone, Filippo Drago, and Claudio Bucolo. 2018. "Retinal Protection and Distribution of Curcumin *in Vitro* and *in Vivo*." *Frontiers in Pharmacology* 9 (JUN). https://doi.org/10.3389/fphar.2018.00670.

Rodick, Taylor C., Donna R. Seibels, Jeganathan Ramesh Babu, Kevin W. Huggins, Guang Ren, and Suresh T. Mathews. 2018. "Potential Role of Coenzyme Q_{10} in Health and Disease Conditions." *Nutrition and Dietary Supplements* Volume 10 (February): 1–11. https://doi.org/10.2147/nds.s112119.

Saini, Rajiv. 2011. "Coenzyme Q10: The Essential Nutrient." *Journal of Pharmacy and Bioallied Sciences*. https://doi.org/10.4103/0975-7406.84471.

Salehi, Bahare, Abhay Prakash Mishra, Manisha Nigam, Bilge Sener, Mehtap Kilic, Mehdi Sharifi-Rad, Patrick Valere Tsouh Fokou, Natália Martins, and Javad Sharifi-Rad. 2018. "Resveratrol: A Double-Edged

Sword in Health Benefits." *Biomedicines*. MDPI AG. https://doi.org/10.3390/biomedicines6030091.

Saper, Robert B., and Rebecca Rash. 2009. "Zinc: An Essential Micronutrient." *American Family Physician*. American Academy of Family Physicians.

Savaş Gül, Serdar, and Hatice Aygün. 2019. "Cardioprotective Effect of Vitamin D and Melatonin on Doxorubicin-Induced Cardiotoxicity in Rat Model: An Electrocardiographic, Scintigraphic and Biochemical Study." *The European Research Journal* 5 (4): 649–57. https://doi.org/10.18621/eurj.410029.

Sokoła-Wysoczańska, Ewa, Tomasz Wysoczański, Jolanta Wagner, Katarzyna Czyż, Robert Bodkowski, Stanisław Lochyński, and Bożena Patkowska-Sokoła. 2018. "Polyunsaturated Fatty Acids and Their Potential Therapeutic Role in Cardiovascular System Disorders—a Review." *Nutrients*. MDPI AG. https://doi.org/10.3390/nu10101561.

Swamy, AgadihiremathViswanatha, Sumeet Gulliaya, Agadihiremath Thippeswamy, BasavarajC Koti, and DonnahalliV Manjula. 2012. "Cardioprotective Effect of Curcumin against Doxorubicin-Induced Myocardial Toxicity in Albino Rats." *Indian Journal of Pharmacology* 44 (1): 73. https://doi.org/10.4103/0253-7613.91871.

Takamura, Takeyuki, Daisuke Harama, Suguru Fukumoto, Yuki Nakamura, Naomi Shimokawa, Kayoko Ishimaru, Shuji Ikegami, Seiya Makino, Masanori Kitamura, and Atsuhito Nakao. 2011. "*Lactobacillus Bulgaricus* OLL1181 Activates the Aryl Hydrocarbon Receptor Pathway and Inhibits Colitis." *Immunology and Cell Biology* 89 (7): 817–22. https://doi.org/10.1038/icb.2010.165.

Tareen, Huma, Fariha Mengal, Zubia Masood, Rabia Mengal, Sana Ahmed, Sherino Bibi, Sara Shoaib, et al. 2015. *Determination of Vitamin C Content in Citrus Fruits and in Non-Citrus Fruits by Titrimetric Method, with Special Reference to Their Nutritional Importance in Human Diet.* Accessed December 4, 2019. www.researchtrend.net.

Townsend, Elizabeth A., and Charles W. Emala. 2013. "Quercetin Acutely Relaxes Airway Smooth Muscle and Potentiates β-Agonist-Induced Relaxation via Dual Phosphodiesterase Inhibition of PLCβ and PDE4."

American Journal of Physiology - Lung Cellular and Molecular Physiology 305 (5). https://doi.org/10.1152/ajplung.00125.2013.

Vessal, Mahmood, Mina Hemmati, and Mohammad Vasei. 2003. "Antidiabetic Effects of Quercetin in Streptozocin-Induced Diabetic Rats." *Comparative Biochemistry and Physiology. Toxicology & Pharmacology : CBP* 135C (3): 357–64. https://doi.org/10.1016/s1532-0456(03)00140-6.

Viezeliene, Dale, Eugene Jansen, Hiliaras Rodovicius, Arturas Kasauskas, and Leonid Ivanov. 2011. "Protective Effect of Selenium on Aluminium-Induced Oxidative Stress in Mouse Liver in Vivo." *Environmental Toxicology and Pharmacology* 31 (2): 302–6. https://doi.org/10.1016/j.etap.2010.11.008.

Vuyyuri, Saleha B., Jacob Rinkinen, Erin Worden, Hyekyung Shim, Sukchan Lee, and Keith R. Davis. 2013. "Ascorbic Acid and a Cytostatic Inhibitor of Glycolysis Synergistically Induce Apoptosis in Non-Small Cell Lung Cancer Cells." Edited by Srikumar P. Chellappan. *PLoS ONE* 8 (6): e67081. https://doi.org/10.1371/journal.pone.0067081.

Wang, Guoping, Tao Yin, and Yongsheng Wang. 2016. "In Vitro and in Vivo Assessment of High-Dose Vitamin C against Murine Tumors." *Experimental and Therapeutic Medicine* 12 (5): 3058–62. https://doi.org/10.3892/etm.2016.3707.

Yun, Jihye, Edouard Mullarky, Changyuan Lu, Kaitlyn N. Bosch, Adam Kavalier, Keith Rivera, Jatin Roper, et al. 2015. "Vitamin C Selectively Kills KRAS and BRAF Mutant Colorectal Cancer Cells by Targeting GAPDH." *Science* 350 (6266): 1391–96. https://doi.org/10.1126/science.aaa5004.

In: Nutraceuticals: Evidence Based Use ISBN: 978-1-53618-052-7
Editor: Raluca Maria Pop

Chapter 3

NUTRACEUTICALS IN CARDIOVASCULAR DISEASES

Sabina Istratoaie[1,*], Diana Pepine[2], Adela Nicoleta Roșianu[3] and Ștefan Horia Roșianu[4]

[1]Department of Pharmacology, Toxicology and Clinical Pharmacology, "Iuliu Hatieganu" University of Medicine and Pharmacy, Cluj-Napoca, Romania

[2]Rehabilitation Hospital, Cluj-Napoca, Romania

[3]Heart Institute "Niculae Stancioiu", Cluj-Napoca, Romania

[4]Department of Cardiology, Heart Institute "Niculae Stancioiu", "Iuliu Hatieganu" University of Medicine and Pharmacy, Cluj-Napoca, Romania

[*] Corresponding Author's E-mail:sabina.istratoaie@gmail.com.

Abstract

Cardiovascular disease (CVD) remains the leading cause of death worldwide. The modifiable risk factors include hypertension, dyslipidemia, diabetes mellitus, obesity and lifestyle risk factors such as smoking, physical inactivity, and dietary factors. The nutritional factors play a critical role in the development and progression of CVD, either directly, or through their effects on various cardiovascular risk factors. Nutraceuticals exert their cardioprotective effects by lowering lipid plasma levels, improving endothelial function or reducing arterial stiffness. Rapidly increasing interest in nutraceutical consumption may lead to new possibilities for population-based strategies for CVD risk reduction. This chapter aims to update the evidence relating to the beneficial effects of nutraceuticals in the prevention and treatment of CVD.

Keywords: nutraceuticals, dietary, cardiovascular disease (CVD)

Introduction

Cardiovascular diseases (CVDs) are the leading global cause of death. By 2030, the World Health Organization estimated CVDs to be responsible for almost 23.6 million deaths worldwide.

Risk factors that play a major role in the etiology and development of CVD are divided into those modifiable by taking a pharmacologic treatment or through lifestyle changes (e.g., hyperlipidemia, hypertension, diabetes mellitus, obesity) and those non-modifiable (advancing age, male gender, family history). Moreover, dietary factors are of major significance in cardiovascular risk, either directly or through their influence on other CVD risk factors such as dyslipidemia or hypertension. The reduction of these risk factors is important to reduce CVD mortality (O'Keeffe et al. 2013).

It has been extensively investigated the cardiovascular benefit of a diet rich in fruits, vegetables, fish, nuts, vegetable oils, low-fat dairy products as well as low in sodium and saturated fats (Slavin and Lloyd 2012; Wall et al. 2010; Mozaffarian 2016). Moreover, epidemiological studies indicate that the replacement of animal fats with vegetable sources of fats and polyunsaturated fatty acids (PUFAs) is associated with a lower incidence of CVD (Chen et al. 2016). The major role the diet plays in CVD prevention is mediated through multiple biologic pathways such as systemic inflammation, oxidative stress, endothelial dysfunction or insulin sensitivity (Giugliano, Ceriello, and Esposito 2006; Alissa and Ferns 2012).

In this context, the role of nutraceuticals and functional foods to cardiovascular health has acquired greater relevance and it might be a key factor for the prevention and treatment of CVD. The intake of nutraceuticals has been shown to have many advantages. Apart from the lipid-lowering effect they also present pleiotropic effects such as reducing oxidative stress and inflammation, improving the function of vascular endothelial cells and arterial stiffness (Houston 2012).

Nutraceuticals That Improve Plasma LDL Cholesterol Levels

Currently, the treatment of dyslipidemia is based on lifestyle changes including dietary habits and physical activity, or pharmacological therapy (Booth et al. 2016). Nevertheless, a healthy lifestyle should always accompany the lipid-lowering therapy (Banach et al. 2017). Numerous epidemiological studies and RCTs (randomized controlled trial) have shown a strong association between a decrease in the LDL-cholesterol (LDL-C) level and a reduction in the relative risk of atherosclerotic cardiovascular disease (Hobbs et al. 2016).

A meta-analysis of the Cholesterol Treatment Trialists' (CTT) Collaboration that included 14 randomized controlled trials (RCTs) and involved 90 056 individuals, showed a significant reduction in coronary and vascular events related to a greater decrease in absolute levels of LDL-C (Baigent et al. 2005).

Statin therapy has proven to reduce the incidence of major coronary events, coronary revascularization, and stroke (Baigent et al. 2005). However, a significant number of patients did not reach their lipid targets and consequently are still at high cardiovascular risk. Moreover, patients on statins may experience side effects such as myalgia, liver disease or rhabdomyolysis. If patients are intolerant to statin or if the statins do not achieve the guideline atherogenic cholesterol targets, an alternative approach is recommended. This includes either other lipid-lowering drugs (such as ezetimibe) that can be added to statin therapy, and/or natural compounds. According to 2019 EAS/ESC Guidelines for the management of dyslipidaemias, in addition to a decrease of trans-unsaturated and saturated fatty acids consumption and increase of fiber consumption, the use of functional foods rich in phytosterols or nutraceuticals containing red yeast rice may be considered (Mach et al. 2019).

Lipid-lowering nutraceuticals may, therefore, be used alone or in combination with pharmacological therapy, especially for patients with cholesterol level moderately elevated but not high enough to require treatment, or for patients intolerant to statins (Patti et al. 2017). Currently, there are not enough studies to recommend nutraceuticals instead of lipid-lowering therapy in patients at cardiovascular disease risk.

Nutraceuticals with lipid-lowering effects can be divided into 3 categories according to their mechanisms of action: natural inhibitors of intestinal cholesterol absorption, inhibitors of hepatic cholesterol synthesis, and enhancers of the excretion of LDL-C (Cicero et al. 2017).

There is a significant number of nutraceuticals that show different levels of effectiveness and evidence of a lipid-lowering effect, mainly on LDL Cholesterol. A summary of the major nutraceuticals with the greatest evidence related to the lipid-modifying effects is outlined in Table 1. For each nutraceutical, a brief explanation of the main mechanism of action, active principles and effective dosages is provided.

Table 1. Nutraceutical influence on LDL-C levels

Agent	Mechanism of action	Active daily doses *	Expected effects on LDL-c **	References
Artichoke (*Cynara scolymus*)	Luteolin interacts with the HMG-CoA reductase enzyme and the pathways of regulation in the liver of sterol regulatory element-binding proteins (SREBPs) and acetyl-CoA C-acetyltransferase	1-3 g	5% to 15%	(Rangboo et al. 2016)
Berberine	Inhibits PCSK9; causes up-regulation of the LDL receptors; decreases intestinal absorption of cholesterol	500-1500 mg	16% to 24%	(Dong et al. 2015)
Bergamot (*Citrus bergamia*)	Inhibits HMG-CoA reductase and ACAT; reduces the formation of cholesterol esters and limits cholesterol transport in the blood	500-1000 mg	15% to 0%	(Miceli et al. 2007)
Curcumin	Inhibits the expression of the NPC1L1 transporter via the SREBP2 transcription factor; increases the efflux of cholesterol enhances the number of LDL-receptor and promotes LDL particle uptake through downregulation of the expression of PCSK9	1-3 g	5%	(Kumar et al. 2011) (Tai et al. 2014)
Garlic (*Allium sativum*)	Inhibits HMG-CoA reductase, squalene monooxygenase, and acetyl-CoA synthetase enzymes	5-6 g (extract)	5% to 10%	(Ried, Toben, and Fakler 2013)
Green tea	Inhibits the expression of inducible nitric oxide synthase; stimulates lipogenesis; inhibits HMG-CoA reductase	25–100 g	5%	(Zheng et al. 2011).
Plant sterols and stanols	Inhibits cholesterol absorption via the formation of mixed micelles; modulates of the inflammatory cascade	400-3000 mg	8% to 12%	(Ferguson et al. 2016) (Ras et al. 2013)
Polyunsaturated omega-3 fatty acids	Reduces VLDL hepatic synthesis; reduces the substrate for the synthesis of new TG; increases beta-oxidation of fatty acids; increases the synthesis of phospholipids	1-4 g	Not applicable	(Harris and Bulchandani 2006)
Red yeast rice	Inhibits HMG-CoA reductase	3–10 mg (monacolin K)	15% to 25%	(Y. Li et al. 2014).

Table 1. (Continued)

Agent	Mechanism of action	Active daily doses *	Expected effects on LDL-c **	References
Soluble fibers (β-glucan, glucomannn , psyllium)	Prolonged gastric emptying time; increases satiety; inhibits of hepatic cholesterol synthesis; increases fecal excretion of cholesterol and bile salts	5-15g	5% to 15%	(Brownet al. 1999)
Soy and lupin proteins	Downregulation of the expression of the hepatic transcription factor of SREBP-1; regulation of SREBP-2; decreases cholesterol synthesis	25-100g	3% to 10%	(Lammi et al. 2014)

*, ** - Based on the Lipid-lowering nutraceuticals in clinical practice: position paper from an International Lipid Expert Panel (Cicero et al. 2017).

Artichoke

Artichoke (*Cynara scolymus*) via its antioxidant action has potential hypolipidemic and hepatoprotective effects. An RCT of 75 patients treated for 12 weeks with 1.28 g/day of artichoke leaf extract versus placebo showed a mean reduction of total cholesterol by 4.2% compared to baseline in the group who took artichoke but no difference as far as LDL-C, HDL-C and TG levels were concerned (Bundy et al. 2008). Another RCT showed a mean reduction of 18.5% compared to 8.6% in the placebo group (Englisch et al. 2000). Also, a beneficial effect on the levels of glucose and the systolic blood pressure has been reported (Rangboo et al. 2016).

Berberine

Berberine is an alkaloid extracted from the root of *Berberis aristate*. Current evidence suggests that berberine has a beneficial effect on LDL-c by lowering it from 20 to 50 mg/dl. Moreover, it can also reduce TG from 25 to 55 mg/dl (Koppen et al. 2017).

The mechanisms by which berberine regulates plasma cholesterol levels are as follows: it inhibits the proprotein convertase subtilisin/kexin type 9 (PCSK9) leading to an increased level of hepatic LDL receptor and it can also cause up-regulation of the LDL receptor through activation of extracellular signal-regulated kinases (ERK) and Jun amino-terminal kinase (JNK)-dependent pathways (H. Li et al. 2009; Abidi et al. 2005). It is also postulated that combined therapy with berberine and red yeast rice is pharmacologically similar to a PCSK inhibitor in combination with statins.

Bergamot

The bergamot (*Citrus bergamia*) is a citrus fruit that contains an especially high amount of flavonoids. Multiple clinical trials using bergamot in different forms suggested that bergamot can reduce total cholesterol and LDL cholesterol levels (Mollace et al. 2018; Toth et al. 2016; Babish et al. 2016). Several possible mechanisms of action have been proposed such as activation of monophosphate-activated protein kinase (AMPK), inhibition of pancreatic cholesterol ester hydrolase, reduction of the intestinal absorption and improving the excretion of bile acids. Activation of AMPK plays a significant role in improving glucose homeostasis, lipid profile and blood pressure control (Nauman and Johnson 2019).

Curcumin

Curcumin is the key component of the spice turmeric, an Indian spice used worldwide either in cooking, cosmetic or medicines. Curcumin has been correlated to antioxidative effects (Pulido-Moran et al. 2016) and it plays a cardioprotective role through its lipid-lowering properties (Maithilikarpagaselvi et al. 2016). An analysis of 7 studies that included 649 patients found that turmeric and curcumin significantly reduced LDL-c and TG levels (Qin et al. 2017). However current evidence is inconsistent and contradictory results were reported. A meta-analysis of 5 RCT with limited sample sizes, showed no significant lipid-lowering effect of curcumin (Sahebkar 2014).

Garlic

Garlic (*Allium sativum*) has multiple beneficial effects. The main molecule of the garlic in called allicin (diallyl thiosulfinate) that is responsible for garlic's action on LDL cholesterol by inhibiting HMG-CoA reductase. Put forward for consideration are other mechanisms such as blocking the absorption of dietary cholesterol and fatty acids and increasing the excretion of bile acids.

According to a recent meta-analysis of RCTs, at a dose of 6 g/day garlic could be useful in patients with mild hypercholesterolemia. It was also demonstrated that garlic has an anti-hypertensive and antiplatelet effect (Ried, Toben, and Fakler 2013). However further studies need to be conducted to attest the efficacy of garlic lowering lipids.

Green tea Extracts

Green tea is a widely consumed beverage worldwide and is made from fresh leaves of *Camellia sinensis*. It is rich in antioxidants and it contains a significant amount of vitamins, minerals, and polyphenols. Several studies have suggested that daily green tea consumption could be protective against CVD risk (Kuriyama et al. 2006). It has been found that green tea catechins exert a cardioprotective effect through multiple mechanisms such as improvement of lipid profile and the inhibition of oxidation, inflammation, and thrombogenesis (Stangl et al. 2007; Bursill, Abbey, and Roach 2007).

A meta-analysis of 14 RCT that included 1136 patients found that green tea consumption leads to a significant reduction of total cholesterol level by 7.20 mg/dl and a lowering of LDL-c by 2.19 mg/dl, with no change on HDL-c level (Zheng et al. 2011).

Phytosterols

Plant sterols and stanols, known as phytosterols, can be found in almost all plant-based products (particularly in nuts, seeds, vegetable oils, legumes) and have a similar structure to cholesterol. It was shown that phytosterols can inhibit both the dietary and biliary cholesterol intestinal absorption in a dose-dependent manner. The meta-analysis of Ras et al., (2013) included 41 clinical trials with 2084 individuals who were provided with a mean dose of phytosterols of 1.6 g/day. After a median duration of 28 days, a significant reduction of 8.5% in LDL-c was observed (Ras et al. 2013). The lipid-lowering effect of phytosterols was shown to be proportional up to reaching a plateau (attained at about 3 g phytosterol/day that determines a 12% in LDL-c decrease). The effect on triglycerides is limited only for patients with high TG levels at baseline. They have no effect on plasma HDL cholesterol and triglycerides. Another important phytosterols property is the improvement of vascular endothelial function and also the anti-inflammatory effect (Cabral and Klein 2017).

Polyunsaturated ω-3 Fatty Acids

Polyunsaturated fatty acids (PUFAs) can be found in animal products (e.g., fish, egg, squid) and in plant sources (algae, seeds, walnut, clary sage).

The daily doses necessary to maintain a normal blood TG level are 2-3 grams as shown by a meta-analysis 47 RCTs which included 16 511 patients. This meta-analysis concluded that after a daily intake of Omega-3 for 24 weeks the blood lever of TG was significantly reduced. Also, a small reduction in LDL-C was observed with no differences in HDL-C (Eslick et al. 2009).

A note worth mentioning is that a dose-dependent effect was observed with higher benefits in reducing TG levels when the dose was raised to 4.9 g/day (Di Stasi et al. 2004). At this dose, an anti-inflammatory effect was observed.

Another important source of PUFAs is sesame. A meta-analysis of 10 RCTs found that regular intake of sesame significantly reduced TG with 21.3 mg/dl, but no changes were observed as far as LDL-C and HDL-C were concerned.

Red Yeast Rice

Red yeast rice (RYR) is derived from rice fermented with *Monascus purpureus* yeast and is well known for the property of lowering cholesterol level. The major active component in RYR is monacolin K(MK), a molecule that inhibits endogenous cholesterol synthesis.

At doses of 3-10 mg/day, monacolin K reduces LDL cholesterol by about 20-25%. Monacolin K also decreases triglycerides in patients with hypertriglyceridemia, where its action on HDL is minimal (Y. Li et al. 2014). Although monacolin K is structurally identical to the statin lovastatin, the RYR monacolin is considered to be more effective on decreasing cholesterol concentration than lovastatin because its bioavailability is higher than that of pure lovastatin (Chen et al. 2013). The first RCT that investigated the RYR effect in cardiovascular prevention (China Coronary Secondary Prevention Study) was performed on 5000 patients with a history of myocardial infarction. It was demonstrated that RYR extracts (xuezhikang) reduced LDL cholesterol level by 20%. Moreover, this decrease in cholesterol level was associated with a significant decrease of stroke, coronary deaths and all-cause mortality (with 44%, 31% and 32%, respectively) (Ye et al. 2007).

RYR might be considered as an acceptable option for patients intolerant to statins. Stefanutti et al., (2017) studied 55 patients with familial hypercholesterolemia who discontinued statin therapy due to myalgia. They were placed on a cholesterol-lowering diet with the addition of 300 mg of RYR (providing 10 mg monacolin K) daily. After 12 months of follow up, LDL cholesterol levels had decreased by 24% for men and 27% for women. No patient presented elevated levels of serum aminotransferase or C-reactive protein level (Stefanutti et al. 2017).

Gerards et al., (2015) conducted a meta-analysis of 20 trials and found that compared with a placebo, RYR extract could significantly decrease LDL concentration, as effective as statin therapy. They also highlighted that RYR might be an effective and safe alternative for statin-intolerant patients in dyslipidemia treatment. Researchers have collected clinical data from a large number of patients with hypercholesterolemia, including some who have used RYR as a therapy, for a follow-up survey (Gerards et al. 2015).

Due to the involvement of cytochrome P 450 and in particular the isoenzyme 3A4 in monacolin K metabolism, caution should be exerted when coadministering other drugs (e.g., erythromycin, cyclosporine, ketoconazole) or grapefruit juice that inhibits the same enzyme since it can lead to adverse reactions to monacolin K (Younes et al. 2018).

Soluble Fibers

Dietary fibers are substances of vegetable origin that are resistant to enzymatic digestion from the gastrointestinal tract. According to their solubility, dietary fibers can be soluble and insoluble fibers. In the last years, soluble fibers have been studied for their lipid-lowering properties, decreasing both total cholesterol and LDL-C level.

The 2019 ESC/EAS guidelines for the management of dyslipidaemias recommends a fat-modified diet that provides 7-13 g of soluble fibers per day (25 to 40 g of total dietary fiber), to be well tolerated and effective for plasma lipid control (Mach et al. 2019).

β-glucan is a soluble fiber that is found in the cell walls of bacteria, fungi, yeasts, and plants. Its lipid-lowering action was studied in a meta-analysis that included 17 RCTs with 916 patients and demonstrated that β-glucan consumption significantly lowered the LDL-c level in a hypercholesterolemic population. However, no effect on HDL-c and TG were noticed (Zhu et al. 2015).

Psyllium (as husk or powder) is a soluble fiber derived from the plant *Plantago psyllium*. A meta-analysis of 21 studies concluded that consumption of psyllium lowered serum LDL-C by 0.278 mmol/l, compared to placebo. They also found that an intake of 10g/day of psyllium is associated with an average reduction of LDL-C of 7% (Wei et al. 2009). Current evidence confirmed that psyllium has a good safety profile, with only a few reported cases of minor gastrointestinal side effects.

Glucomannan, commonly known as konjac root is a soluble fiber derived from *Amorphophallus konjac*. A meta-analysis of 14 RCTs showed that glucomannan intake significantly lowers LDL-c by 0.41mmol/l while reducing TG level by 0.13mmol/l. While other soluble fibers have a low action on triglyceride levels, glucomannan is of high viscosity and can interact with lipoprotein metabolism, thus with a significant impact on lowering TG level (Sood, Baker, and Coleman 2008).

Soy and Lupin Proteins

Soy and lupin proteins contain bioactive peptides that are responsible for the lipid-lowering effect (Cicero, Fogacci, and Colletti 2017). A meta-analysis of 35 RCTs that included 2670 patients showed that soy proteins significantly reduced the TC by 2%, LDL-c by 3%, TG by 4% while increasing HDL-c by 3% (Tokede et al. 2015). An RCT that studied the cholesterol-lowering properties of lupin revealed a reduction in LDL-c by 12% (Fechner, Kiehntopf, and Jahreis 2014).

Concerning side effects, the chronic use of a high quantity of soy proteins was associated with thyroid dysfunction and infertility. Moreover, due to the phytic acid contained in soybean, absorption of minerals such as magnesium, calcium, iron, copper, and zinc may be reduced.

Nutraceuticals That Improve Endothelial Function

Table 2. Nutraceutical influence on endothelial function

Agent	Mechanism of action	Dose	Duration	References
Vitamin C	Improvement in flow-mediated dilation and on endothelium-dependent vasodilatation	1-2 g	2 h	(Heitzer, Just and Munzel 1996)
Vitamin E	Improvement on endothelium-dependent vasodilatation only in hypercholesterolemic smokers	544 UI/day	4 months	(Heitzer et al. 1999)
Vitamin D2	Improvement in flow-mediated dilation; Reduced systolic blood pressure	100,000 IU per os	8 weeks	(Sugden et al. 2008)
Vitamin D3	Improvement in flow-mediated dilation; Reduced lipid peroxidation	300,000 IU per os	3 months	(Tarcin et al. 2009)
Folic acid, Vitamin B12, Vitamin B6	No difference in flow-mediated dilation and carotid intima-media thickness between groups	2 mg, 25 mg and 0.5 mg/day,respectively	2 years	(Potter et al. 2008)
Niacin	Improvement in flow-mediated dilation	1.5 g/day	12 weeks	(Sorrentino et al. 2010)
Omega-3 fatty acids	Improvement in flow-mediated dilation	4 g/diet	4 months	(Goodfellow et al. 2000)
Eicosapentaenoic acid	Improvement in endothelium-dependent and endothelium independent vasodilatation in the forearm	1.8 g/day	6 weeks	(Tagawa et al. 1999)
Black tea	Improvement in flow-mediated dilation	900 ml/day	4 weeks	(Duffy et al. 2001)
Green tea	Improvement in flow-mediated dilation (Nagaya et al. 2004)	400 ml	2 h	(Nagaya et al. 2004)
Cocoa drink	An acute increase in flow-mediated dilation	176–185 mg of flavanol	2 h	(Heiss et al. 2005)
Flavonoid-rich dark chocolate	Improvement in endothelium-dependent coronary vasomotion and decreased platelet adhesion	40 g of dark (70% cocoa) chocolate	2 h	(Flammer et al. 2007)
Red wine with or without alcohol	Improvement in flow-mediated dilation; Reduces the risk of cardiovascular events	250 ml	30 and 60 min	(Agewall et al. 2000)

Endothelial dysfunction and injury are known to play an important role both in the development and in the progression of atherosclerosis. Several clinical studies have observed the association between endothelial dysfunction and cardiovascular events (Lerman and Zeiher 2005). One of the most important molecules involved in the development of atherosclerosis is endothelial nitric oxide. This molecule action acts via inhibition of chemokine expression, vascular remodeling, and endothelial platelet aggregation.

Several nutraceuticals with a beneficial effect on endothelial function are listed in Table 2.

Vitamin C and Vitamin E – The Antioxidant Vitamins

There is a large number of clinical studies that show the relationship between increased vascular oxidative stress and reduced endothelial nitric oxide availability, thus enhancing cardiovascular risk and promoting the progression of atherosclerosis.

Large scale epidemiological studies have connected low plasma levels of Vitamin C with increased risk of cardiovascular events (Khaw et al. 2001). However, large studies did not show a significant reduction in cardiovascular events in people who supplemented their diets with vitamins C and E.

Vitamin E reduces the levels of oxidants in the body that may lead to unwanted effects. It has been studied the benefit of antioxidants on preventing LDL-C oxidation limiting atherosclerosis but at the price of blunting the vasoprotective action of preconditioning making the consequences of ischemia severe. In supporting this theory come two large trials which conclude that supplementation with vitamin E in patients with coronary artery disease showed fewer ischemic events but with more fatal infarcts (Stephens et al. 1996; Rapola et al. 1997).

Vitamin D

Vitamin D is concentrated in fish, egg yolk, mushrooms and is synthesized in the skin after exposure to UV light from the sun. Vitamin D receptors have been identified on endothelial cells and influence cell growth. Studies on mice have shown that a reduction of vitamin D receptors leads to an increase of activity in the renin-angiotensin system which leads to hypertension and left ventricular hypertrophy (Li et al. 2002).

Vitamin D deficiency is associated with a higher incidence of cardiovascular events as vitamin D deficiency is connected to endothelial dysfunction and increased lipid peroxidation (Tarcin et al. 2009).

Omega-3 Fatty Acid (Marine and Plant Derived)

α-linoleic acid (ALA) is a plant-derived omega-3 fatty acid, while eicosapentaenoic acid (EPA) docosahexaenoic acid (DHA) are found especially in fish. The fatty acids cannot be synthesized in the human body and therefore must be procured from the diet. A Mediterranean diet consisting mostly of fish was associated with a lower incidence of coronary artery disease. Moreover, other cardiovascular protective benefits have arisen in omega-3 fatty acids supplements such as hypotriglycer-idemic effects, hypotensive effect, decrease in platelet aggregation, increased endothelial function and antiarrhythmic effects. There seems to be a positive effect of fatty acids on endothelial function, but further randomized trials need to be carried out to certify this action (Zuchi et al. 2010).

Vitamins B6/12 and Folic Acid

A study conducted in 1969 showed that children suffering from homozygous homocystinuria had a higher incidence of coronary artery disease and peripheral artery disease (McCully 1969). Therefore vitamins B12 and folic acid might be an important component of endothelial function and their deficiency can be associated with an increased risk of ischemic heart disease and stroke. A more recent study concluded that elevated levels and homocysteine accompanied by lower levels of vitamin B6 increased the risk of coronary and cerebrovascular events (Vanuzzo et al. 2007).

Flavonoids

Flavonoids are contained in vegetables and fruits, cocoa, tea, and wine and are the most common group of polyphenols in the human diet.

The incidence of coronary artery disease appeared to be lower in those who consumed tea regularly. Moreover, tea consumption, especially green and black tea, in the year after an acute coronary syndrome was associated with lower mortality rates after infarction (Mukamal et al. 2002).

Consumption of rich flavonoid food (e.g., cocoa, black and green tea) in patients suffering from coronary artery disease, or in those after heart transplant or even in healthy individuals improved their endothelial function (Heiss et al. 2005).

Niacin (Vitamin B3)

Vitamin B3 is an endogenous vitamin with an endothelium-dependent vasodilatation effect in individuals with low HDL-C plasma levels. It was also observed that niacin consumption (at a dose of 1.5 g/day) in patients with a diagnosis of diabetes with low HDL-C levels improved endothelial mediated vasodilatation and increased HDL-C plasma levels (Sorrentino et al. 2010).

Other potential cardioprotective effects nutraceuticals from human studies are summarized in Table 3.

Table 3. Nutraceutical cardioprotective effects

Nutraceutical	Observation	References
Butyrate	Reduced the risk of peripheral artery disease Reduced the risk of a cardiovascular event	(Merchant et al. 2003; Pereira et al. 2004)
Coenzyme Q10	Improved endothelial function Together with garlic supplementation reduced serum CRP levels	(Gao et al. 2012; Zeb et al. 2012)
Curcumin	Improved endothelial function Ameliorated arterial stiffness	(Karimian et al. 2017; Fleenor et al. 2013)
Flavanols	Increased vasodilatation Reduced the expression of proinflammatory cytokines Improved endothelial function	(Almeida Rezende et al. 2015; Hsu et al. 2007; Perez-Vizcaino and Duarte 2010)
Hydroxytyrosol	Phenolic content of olive oils was associated with increased serum HDL levels Reduces serum oxLDL levels Reduces the level of inflammatory biomarkers	(Covas et al. 2006; Fitó et al. 2005; Rigacci and Stefani 2016)
Omega-3 PUFAs	Reduced coronary heart disease death Higher plasma eicosapentaenoic acid was associated with reduced risk of heart failure Might prevent arrhythmias	(Rizos et al. 2012; London et al. 2007)
Omega-6 PUFAs	Had a beneficial impact on inflammatory markers Intake from 5 to 20% of the energy intake lowered plasma LDL-C level	(Ferrucci et al. 2006; Czernichow, Thomas, and Bruckert 2010)
Vitamin C	Reduced both systolic and diastolic blood pressure Plasma ascorbic acid was inversely associated with the risk of heart failure Ineffective in improving total cholesterol, LDL, HDL, or triglycerides	(Juraschek et al. 2012; Pfister et al. 2011; Ashor et al. 2016)

CONCLUSION

Many studies have shown that nutraceuticals can exert cardioprotective effects and reduce the risk of cardiovascular disease. Further research is required to fully evaluate the safety and effectiveness of the nutraceuticals, before their widespread use can be recommended. Advances in the field of nutraceuticals cardioprotective properties may lead to new treatments and prevention approaches to reduce the prevalence of cardiovascular diseases.

REFERENCES

Abidi, Parveen, Yue Zhou, Jian Dong Jiang, and Jingwen Liu. 2005. "Extracellular Signal-Regulated Kinase-Dependent Stabilization of Hepatic Low-Density Lipoprotein Receptor MRNA by Herbal Medicine Berberine." *Arteriosclerosis, Thrombosis, and Vascular Biology* 25 (10): 2170–76. https://doi.org/10.1161/01.ATV.0000181761.16341.2b.

Agewall, S, S Wright, R N Doughty, G A Whalley, M Duxbury, and N Sharpe. 2000. "Does a Glass of Red Wine Improve Endothelial Function?" *European Heart Journal* 21 (1): 74–78. https://doi.org/10.1053/euhj.1999.1759.

Alissa, Eman M., and Gordon A. Ferns. 2012. "Functional Foods and Nutraceuticals in the Primary Prevention of Cardiovascular Diseases." *Journal of Nutrition and Metabolism* 2012. https://doi.org/10.1155/2012/569486.

Almeida Rezende, Bruno, Aline Pereira, Steyner Cortes, and Virginia Lemos. 2015. "Vascular Effects of Flavonoids." *Current Medicinal Chemistry* 23 (1): 87–102. https://doi.org/10.2174/0929867323666151111143616.

Ashor, Ammar W., Mario Siervo, Femke van der Velde, Naomi D. Willis, and John C. Mathers. 2016. "Systematic Review and Meta-Analysis of Randomised Controlled Trials Testing the Effects of Vitamin C Supplementation on Blood Lipids." *Clinical Nutrition* 35 (3): 626–37. https://doi.org/10.1016/j.clnu.2015.05.021.

Babish, John G., Clinton J. Dahlberg, Joseph J. Ou, William J. Keller, Wei Gao, Mohan R. Kaadige, Holly Brabazon, et al. 2016. "Synergistic in Vitro Antioxidant Activity and Observational Clinical Trial of F105, a Phytochemical Formulation Including Citrus Bergamia, in Subjects with Moderate Cardiometabolic Risk Factors." *Canadian Journal of Physiology and Pharmacology* 94 (12): 1257–66. https://doi.org/10.1139/cjpp-2016-0062.

Baigent, C., Keech A, Kearney PM, Blackwell L, Buck G, Pollicino C, Kirby A, et al. 2005. "Efficacy and Safety of Cholesterol-Lowering Treatment: Prospective Meta-Analysis of Data from 90 056 Participants in 14 Randomised Trials of Statins." *Lancet* 366 (9493): 1267–78. https://doi.org/10.1016/S0140-6736(05)67394-1.

Banach, MacIej, Piotr Jankowski, Jacek Joíwiak, Barbara Cybulska, Adam Windak, Tomasz Guzik, Artur Mamcarz, et al. 2017. "PoLA/ CFPiP/PCS Guidelines for the Management of Dyslipidaemias for Family Physicians 2016." *Archives of Medical Science*. Termedia Publishing House Ltd. https://doi.org/10.5114/aoms.2017.64712.

Booth, John N., Lisandro D. Colantonio, George Howard, Monika M. Safford, MacIej Banach, Kristi Reynolds, Mary Cushman, and Paul Muntner. 2016. "Healthy Lifestyle Factors and Incident Heart Disease and Mortality in Candidates for Primary Prevention with Statin Therapy." *International Journal of Cardiology* 207 (March): 196–202. https://doi.org/10.1016/j.ijcard.2016.01.001.

Brown, Lisa, Bernard Rosner, Walter W Willett, and Frank M Sacks. 1999. "Cholesterol-Lowering Effects of Dietary Fiber: A Meta-Analysis." *The American Journal of Clinical Nutrition* 69 (1): 30–42. https://doi.org/10.1093/ajcn/69.1.30.

Bundy, Rafe, Ann F. Walker, Richard W. Middleton, Carol Wallis, and Hugh C.R. Simpson. 2008. "Artichoke Leaf Extract (Cynara Scolymus) Reduces Plasma Cholesterol in Otherwise Healthy Hypercholesterolemic Adults: A Randomized, Double Blind Placebo Controlled Trial." *Phytomedicine* 15 (9): 668–75. https://doi.org/ 10.1016/j.phymed.2008.03.001.

Bursill, Christina A., Mavis Abbey, and Paul D. Roach. 2007. "A Green Tea Extract Lowers Plasma Cholesterol by Inhibiting Cholesterol Synthesis and Upregulating the LDL Receptor in the Cholesterol-Fed Rabbit." *Atherosclerosis* 193 (1): 86–93. https://doi.org/10.1016/j. atherosclerosis.2006.08.033.

Cabral, Carlos Eduardo, and Márcia Regina Simas Torres Klein. 2017. "Phytosterols in the Treatment of Hypercholesterolemia and Prevention of Cardiovascular Diseases." *Arquivos Brasileiros de Cardiologia*. https://doi.org/10.5935/abc.20170158.

Chen, CH, JC Yang, YS Uang, and CJ Lin. 2013. "Improved Dissolution Rate and Oral Bioavailability of Lovastatin in Red Yeast Rice

Products." *International Journal of Pharmaceutics* 444 (1–2): 18–24. https://doi.org/10.1016/j.ijpharm.2013.01.028.

Chen, M, Li Yanping, Qi Sun, An Pan, Joann E. Manson, Kathryn M. Rexrode, Walter C. Willett, Eric B. Rimm, and Frank B. Hu. 2016. "Dairy Fat and Risk of Cardiovascular Disease in 3 Cohorts of US Adults1-3." *American Journal of Clinical Nutrition* 104 (5): 1209–17. https://doi.org/10.3945/ajcn.116.134460.

Cicero, Arrigo F.G., Alessandro Colletti, Gani Bajraktari, Olivier Descamps, Dragan M. Djuric, Marat Ezhov, Zlatko Fras, et al. 2017. "Lipid Lowering Nutraceuticals in Clinical Practice: Position Paper from an International Lipid Expert Panel." *Archives of Medical Science* 5: 965–1005. https://doi.org/10.5114/aoms.2017.69326.

Cicero, Arrigo F.G., Federica Fogacci, and Alessandro Colletti. 2017. "Potential Role of Bioactive Peptides in Prevention and Treatment of Chronic Diseases: A Narrative Review." *British Journal of Pharmacology*. John Wiley and Sons Inc. https://doi.org/10.1111/bph.13608.

Covas, María Isabel, Kristiina Nyyssönen, Henrik E. Poulsen, Jari Kaikkonen, Hans Joachim F. Zunft, Holger Kiesewetter, Antonio Gaddi, et al. 2006. "The Effect of Polyphenols in Olive Oil on Heart Disease Risk Factors: A Randomized Trial." *Annals of Internal Medicine* 145 (5): 333–41. https://doi.org/10.7326/0003-4819-145-5-200609050-00006.

Czernichow, Sébastien, Daniel Thomas, and Eric Bruckert. 2010. "N-6 Fatty Acids and Cardiovascular Health: A Review of the Evidence for Dietary Intake Recommendations." *British Journal of Nutrition*. https://doi.org/10.1017/S0007114510002096.

Dong, Bin, Hai Li, Amar Bahadur Singh, Aiqin Cao, and Jingwen Liu. 2015. "Inhibition of PCSK9 Transcription by Berberine Involves Down-Regulation of Hepatic HNF1α Protein Expression through the Ubiquitin-Proteasome Degradation Pathway." *Journal of Biological Chemistry* 290 (7): 4047–58. https://doi.org/10.1074/jbc.M114.597229.

Duffy, S J, J F Jr Keaney, M Holbrook, N Gokce, P L Swerdloff, B Frei, and J A Vita. 2001. "Short- and Long-Term Black Tea Consumption Reverses Endothelial Dysfunction in Patients with Coronary Artery Disease." *Circulation* 104 (2): 151–56. https://doi.org/10.1161/01.cir.104.2.151.

Englisch, Werner, Christoph Beckers, Markus Unkauf, Michel Ruepp, and Vsevolod Zinserling. 2000. "Efficacy of Artichoke Dry Extract in Patients with Hyperlipoproteinemia." *Arzneimittel-Forschung/Drug Research* 50 (3): 260–65. https://doi.org/10.1055/s-0031-1300196.

Eslick, Guy D., Peter R.C. Howe, Caroline Smith, Ros Priest, and Alan Bensoussan. 2009. "Benefits of Fish Oil Supplementation in Hyperlipidemia: A Systematic Review and Meta-Analysis." *International Journal of Cardiology* 136 (1): 4–16. https://doi.org/10.1016/j.ijcard.2008.03.092.

Fechner, Anita, Michael Kiehntopf, and Gerhard Jahreis. 2014. "The Formation of Short-Chain Fatty Acids Is Positively Associated with the Blood Lipid–Lowering Effect of Lupin Kernel Fiber in Moderately Hypercholesterolemic Adults." *The Journal of Nutrition* 144 (5): 599–607. https://doi.org/10.3945/jn.113.186858.

Ferguson, Jessica J.A., Elizabeth Stojanovski, Lesley MacDonald-Wicks, and Manohar L. Garg. 2016. "Fat Type in Phytosterol Products Influence Their Cholesterol-Lowering Potential: A Systematic Review and Meta-Analysis of RCTs." *Progress in Lipid Research* 64 (October): 16–29. https://doi.org/10.1016/j.plipres.2016.08.002.

Ferrucci, Luigi, Antonio Cherubini, Stefania Bandinelli, Benedetta Bartali, Annamaria Corsi, Fulvio Lauretani, Antonio Martin, Cristina Andres-Lacueva, Umberto Senin, and Jack M. Guralnik. 2006. "Relationship of Plasma Polyunsaturated Fatty Acids to Circulating Inflammatory Markers." *Journal of Clinical Endocrinology and Metabolism* 91 (2): 439–46. https://doi.org/10.1210/jc.2005-1303.

Fitó, M., M. Cladellas, R. De La Torre, J. Martí, M. Alcántara, M. Pujadas-Bastardes, J. Marrugat, et al. 2005. "Antioxidant Effect of Virgin Olive Oil in Patients with Stable Coronary Heart Disease: A Randomized,

Crossover, Controlled, Clinical Trial." *Atherosclerosis* 181 (1): 149–58. https://doi.org/10.1016/j.atherosclerosis.2004.12.036.

Flammer, Andreas J, Frank Hermann, Isabella Sudano, Lukas Spieker, Matthias Hermann, Karen A Cooper, Mauro Serafini, et al. 2007. "Dark Chocolate Improves Coronary Vasomotion and Reduces Platelet Reactivity." *Circulation* 116 (21): 2376–82. https://doi.org/10.1161/CIRCULATIONAHA.107.713867.

Fleenor, Bradley S., Amy L. Sindler, Natasha K. Marvi, Kate L. Howell, Melanie L. Zigler, Mutsuko Yoshizawa, and Douglas R. Seals. 2013. "Curcumin Ameliorates Arterial Dysfunction and Oxidative Stress with Aging." *Experimental Gerontology* 48 (2): 269–76. https://doi.org/10.1016/j.exger.2012.10.008.

Gao, Linggen, Qunxia Mao, Jian Cao, Yutang Wang, Xianliang Zhou, and Li Fan. 2012. "Effects of Coenzyme Q10 on Vascular Endothelial Function in Humans: A Meta-Analysis of Randomized Controlled Trials." *Atherosclerosis*. https://doi.org/10.1016/j.atherosclerosis.2011.10.027.

Gerards, Maaike C., Ruben J. Terlou, Huixin Yu, C. H.W. Koks, and V. E.A. Gerdes. 2015. "Traditional Chinese Lipid-Lowering Agent Red Yeast Rice Results in Significant LDL Reduction but Safety Is Uncertain - A Systematic Review and Meta-Analysis." *Atherosclerosis* 240 (2): 415–23. https://doi.org/10.1016/j.atherosclerosis.2015.04.004.

Giugliano, Dario, Antonio Ceriello, and Katherine Esposito. 2006. "The Effects of Diet on Inflammation. Emphasis on the Metabolic Syndrome." *Journal of the American College of Cardiology*. Elsevier. https://doi.org/10.1016/j.jacc.2006.03.052.

Goodfellow, J, M F Bellamy, M W Ramsey, C J Jones, and M J Lewis. 2000. "Dietary Supplementation with Marine Omega-3 Fatty Acids Improve Systemic Large Artery Endothelial Function in Subjects with Hypercholesterolemia." *Journal of the American College of Cardiology* 35 (2): 265–70. https://doi.org/10.1016/s0735-1097(99)00548-3.

Harris, William S, and Deepti Bulchandani. 2006. "Why Do Omega-3 Fatty Acids Lower Serum Triglycerides?" *Current Opinion in Lipidology* 17 (4): 387–93. https://doi.org/10.1097/01.mol.0000236363.63840.16.

Heiss, Christian, Petra Kleinbongard, Andre Dejam, Sandra Perre, Hagen Schroeter, Helmut Sies, and Malte Kelm. 2005. "Acute Consumption of Flavanol-Rich Cocoa and the Reversal of Endothelial Dysfunction in Smokers." *Journal of the American College of Cardiology* 46 (7): 1276–83. https://doi.org/10.1016/j.jacc.2005.06.055.

Heitzer, T, H Just, and T Munzel. 1996. "Antioxidant Vitamin C Improves Endothelial Dysfunction in Chronic Smokers." *Circulation* 94 (1): 6–9. https://doi.org/10.1161/01.cir.94.1.6.

Heitzer, T, S Yla Herttuala, E Wild, J Luoma, and H Drexler. 1999. "Effect of Vitamin E on Endothelial Vasodilator Function in Patients with Hypercholesterolemia, Chronic Smoking or Both." *Journal of the American College of Cardiology* 33 (2): 499–505. https://doi.org/10.1016/s0735-1097(98)00584-1.

Hobbs, FD Richard, Maciej Banach, Dimitri P. Mikhailidis, Aseem Malhotra, and Simon Capewell. 2016. "Is Statin-Modified Reduction in Lipids the Most Important Preventive Therapy for Cardiovascular Disease? A pro/Con Debate." *BMC Medicine* 14 (1). https://doi.org/10.1186/s12916-016-0550-5.

Houston, Mark. 2012. "The Role of Nutraceutical Supplements in the Treatment of Dyslipidemia." *Journal of Clinical Hypertension.* https://doi.org/10.1111/j.1751-7176.2011.00576.x.

Hsu, Shih-ping, Ming-shiou Wu, Chih-ching Yang, Kuo-chin Huang, Shaw-yih Liou, and Su-ming Hsu. 2007. "Chronic Green Tea Extract Supplementation Reduces Hemodialysis- Enhanced Production of Hydrogen Peroxide and Hypochlorous Acid," *American Journal of Clinical Nutrition* 86 (5): 1539–47. https://doi.org/86/5/1539 [pii].

Juraschek, Stephen P., Eliseo Guallar, Lawrence J. Appel, and Edgar R. Miller. 2012. "Effects of Vitamin c Supplementation on Blood Pressure: A Meta-Analysis of Randomized Controlled Trials." *American Journal of Clinical Nutrition* 95 (5): 1079–88. https://doi.org/10.3945/ajcn.111.027995.

Karimian, Maryam S., Matteo Pirro, Thomas P. Johnston, Muhammed Majeed, and Amirhossein Sahebkar. 2017. "Curcumin and Endothelial Function: Evidence and Mechanisms of Protective Effects." *Current*

Pharmaceutical Design 23 (17): 2462–73. https://doi.org/10.2174/1381612823666170222122822.

Khaw, Kay-tee, Sheila Bingham, Ailsa Welch, and Robert Luben. 2001. *Relation between Plasma Ascorbic Acid and Mortality in Men and Women in EPIC- ...* 357: 657–63.

Koppen, Laura M., Andrea Whitaker, Audrey Rosene, and Robert D. Beckett. 2017. "Efficacy of Berberine Alone and in Combination for the Treatment of Hyperlipidemia: A Systematic Review." *Journal of Evidence-Based Complementary and Alternative Medicine*. SAGE Publications Ltd. https://doi.org/10.1177/2156587216687695.

Kumar, Pradeep, Pooja Malhotra, Ke Ma, Amika Singla, Omar Hedroug, Seema Saksena, Pradeep K. Dudeja, Ravinder K. Gill, and Waddah A. Alrefai. 2011. "SREBP2 Mediates the Modulation of Intestinal NPC1L1 Expression by Curcumin." *American Journal of Physiology-Gastrointestinal and Liver Physiology* 301 (1): G148–55. https://doi.org/10.1152/ajpgi.00119.2011.

Kuriyama, Shinichi, Taichi Shimazu, Kaori Ohmori, Nobutaka Kikuchi, Naoki Nakaya, Yoshikazu Nishino, Yoshitaka Tsubono, and Ichiro Tsuji. 2006. "Green Tea Consumption and Mortality Due to Cardiovascular Disease, Cancer, and All Causes in Japan: The Ohsaki Study." *Journal of the American Medical Association* 296 (10): 1255–65. https://doi.org/10.1001/jama.296.10.1255.

Lammi, Carmen, Chiara Zanoni, Graziana M. Scigliuolo, Alfonsina D'Amato, and Anna Arnoldi. 2014. "Lupin Peptides Lower Low-Density Lipoprotein (LDL) Cholesterol through an Up-Regulation of the LDL Receptor/Sterol Regulatory Element Binding Protein 2 (SREBP2) Pathway at HepG2 Cell Line." *Journal of Agricultural and Food Chemistry* 62 (29): 7151–59. https://doi.org/10.1021/jf500795b.

Lerman, Amir, and Andreas M. Zeiher. 2005. "Endothelial Function: Cardiac Events." *Circulation* 111 (3): 363–68. https://doi.org/10.1161/01.CIR.0000153339.27064.14.

Li, Hai, Bin Dong, Sahng Wook Park, Hyun Sook Lee, Wei Chen, and Jingwen Liu. 2009. "Hepatocyte Nuclear Factor 1α Plays a Critical Role in PCSK9 Gene Transcription and Regulation by the Natural

Hypocholesterolemic Compound Berberine." *Journal of Biological Chemistry* 284 (42): 28885–95. https://doi.org/10.1074/jbc.M109.052407.

Li, Y C, J Kong, M Wei, Zhou-Feng Chen, Shu Q. Liu, and Li-Ping Cao. 2002. "1,25-Dihydroxyvitamin D3 Is a Negative Endocrine Regulator of the Renin-Angiotensin System." *Journal of Clinical Investigation* 110 (2): 229–38. https://doi.org/10.1172/jci15219.

Li, Yinhua, Long Jiang, Zhangrong Jia, Wei Xin, Shiwei Yang, Qiu Yang, and Luya Wang. 2014. "A Meta-Analysis of Red Yeast Rice: An Effective and Relatively Safe Alternative Approach for Dyslipidemia." *PLoS ONE* 9 (6): e98611. https://doi.org/10.1371/journal.pone.0098611.

London, Barry, Christine Albert, Mark E. Anderson, Wayne R. Giles, David R. Van Wagoner, Ethan Balk, George E. Billman, et al. 2007. "Omega-3 Fatty Acids and Cardiac Arrhythmias: Prior Studies and Recommendations for Future Research - A Report from the National Heart, Lung, and Blood Institute and Office of Dietary Supplements Omega-3 Fatty Acids and Their Role in Cardiac Arrhythmogenesis Workshop." *Circulation*. https://doi.org/10.1161/CIRCULATION AHA.107.712984.

Mach, François, Colin Baigent, Alberico L Catapano, Konstantinos C Koskinas, Manuela Casula, Lina Badimon, M John Chapman, et al. 2019. "2019 ESC/EAS Guidelines for the Management of Dyslipidaemias: Lipid Modification to Reduce Cardiovascular Risk: The Task Force for the Management of Dyslipidaemias of the European Society of Cardiology (ESC) and European Atherosclerosis Society (EAS)." *European Heart Journal* 41 (1): 111–88. https://doi.org/10.1093/eurheartj/ehz455.

Maithilikarpagaselvi, Nachimuthu, Magadi Gopalakrishna Sridhar, Rathinam Palamalai Swaminathan, Ramalingam Sripradha, and Bhawana Badhe. 2016. "Curcumin Inhibits Hyperlipidemia and Hepatic Fat Accumulation in High-Fructose-Fed Male Wistar Rats." *Pharmaceutical Biology* 54 (12): 2857–63. https://doi.org/10.1080/13880209.2016.1187179.

McCully, K S. 1969. "Vascular Pathology of Homocysteinemia: Implications for the Pathogenesis of Arteriosclerosis." *The American Journal of Pathology* 56 (1): 111–28.

Merchant, Anwar T., Frank B. Hu, Donna Spiegelman, Walter C. Willett, Eric B. Rimm, and Alberto Ascherio. 2003. "Dietary Fiber Reduces Peripheral Arterial Disease Risk in Men." *The Journal of Nutrition* 133 (11): 3658–63. https://doi.org/10.1093/jn/133.11.3658.

Miceli, Natalizia, Maria R. Mondello, Maria T. Monforte, Vasileios Sdrafkakis, Paola Dugo, Maria L. Crupi, Maria F. Taviano, Rita De Pasquale, and Ada Trovato. 2007. "Hypolipidemic Effects of Citrus Bergamia Risso et Poiteau Juice in Rats Fed a Hypercholesterolemic Diet." *Journal of Agricultural and Food Chemistry* 55 (26): 10671–77. https://doi.org/10.1021/jf071772i.

Mollace, Vincenzo, Miriam Scicchitano, Sara Paone, Francesca Casale, Carla Calandruccio, Micaela Gliozzi, Vincenzo Musolino, et al. 2018. "Hypoglycemic and Hypolipemic Effects of a New Lecithin Formulation of Bergamot Polyphenolic Fraction: A Double Blind, Randomized, Placebo- Controlled Study." *Endocrine, Metabolic & Immune Disorders - Drug Targets* 19 (2): 136–43. https://doi.org/10.2174/1871530319666181203151513.

Mozaffarian, Dariush. 2016. "Natural Trans Fat, Dairy Fat, Partially Hydrogenated Oils, and Cardiometabolic Health: The Ludwigshafen Risk and Cardiovascular Health Study." *European Heart Journal* 37 (13): 1079–81. https://doi.org/10.1093/eurheartj/ehv595.

Mukamal, Kenneth J., Malcolm Maclure, James E. Muller, Jane B. Sherwood, and Murray A. Mittleman. 2002. "Tea Consumption and Mortality after Acute Myocardial Infarction." *Circulation* 105 (21): 2476–81. https://doi.org/10.1161/01.CIR.0000017201.88994.F7.

Nagaya, N, H Yamamoto, M Uematsu, T Itoh, K Nakagawa, T Miyazawa, K Kangawa, and K Miyatake. 2004. "Green Tea Reverses Endothelial Dysfunction in Healthy Smokers." *Heart (British Cardiac Society)* 90 (12): 1485–86. https://doi.org/10.1136/hrt.2003.026740.

Nauman, C Mirielle, and J Jeremy Johnson. 2019. "Clinical Application of Bergamot (*Citrus Bergamia*) for Reducing High Cholesterol and

Cardiovascular Disease Markers." *Integrative Food, Nutrition and Metabolism* 6 (2). https://doi.org/10.15761/ifnm.1000249.

O'Keeffe, Celine, Zubair Kabir, Martin O'flaherty, Janette Walton, Simon Capewell, and Ivan J. Perry. 2013. "Modelling the Impact of Specific Food Policy Options on Coronary Heart Disease and Stroke Deaths in Ireland." *BMJ Open* 3 (7). https://doi.org/10.1136/bmjopen-2013-002837.

Patti, Angelo M., Peter P. Toth, Rosaria V. Giglio, Maciej Banach, Marcello Noto, Dragana Nikolic, Giuseppe Montalto, and Manfredi Rizzo. 2017. "Nutraceuticals as an Important Part of Combination Therapy in Dyslipidaemia." *Current Pharmaceutical Design* 23 (17). https://doi.org/10.2174/1381612823666170317145851.

Pereira, Mark A., Eilis O'Reilly, Katarina Augustsson, Gary E. Fraser, Uri Goldbourt, Berit L. Heitmann, Goran Hallmans, et al. 2004. "Dietary Fiber and Risk of Coronary Heart Disease: A Pooled Analysis of Cohort Studies." *Archives of Internal Medicine* 164 (4): 370–76. https://doi.org/10.1001/archinte.164.4.370.

Perez-Vizcaino, Francisco, and Juan Duarte. 2010. "Flavonols and Cardiovascular Disease." *Molecular Aspects of Medicine* 31: 478–94. https://doi.org/10.1016/j.mam.2010.09.002.

Pfister, Roman, Stephen J. Sharp, Robert Luben, Nick J. Wareham, and Kay Tee Khaw. 2011. "Plasma Vitamin C Predicts Incident Heart Failure in Men and Women in European Prospective Investigation into Cancer and Nutrition-Norfolk Prospective Study." *American Heart Journal* 162 (2): 246–53. https://doi.org/10.1016/j.ahj.2011.05.007.

Potter, Kathleen, Graeme J Hankey, Daniel J Green, John Eikelboom, Konrad Jamrozik, and Leonard F Arnolda. 2008. "The Effect of Long-Term Homocysteine-Lowering on Carotid Intima-Media Thickness and Flow-Mediated Vasodilation in Stroke Patients: A Randomized Controlled Trial and Meta-Analysis." *BMC Cardiovascular Disorders* 8 (September): 24. https://doi.org/10.1186/1471-2261-8-24.

Pulido-Moran, Mario, Jorge Moreno-Fernandez, Cesar Ramirez-Tortosa, and M. Carmen Ramirez-Tortosa. 2016. "Curcumin and Health." *Molecules*. MDPI AG. https://doi.org/10.3390/molecules21030264.

Qin, Si, Lifan Huang, Jiaojiao Gong, Shasha Shen, Juan Huang, Hong Ren, and Huaidong Hu. 2017. "Efficacy and Safety of Turmeric and Curcumin in Lowering Blood Lipid Levels in Patients with Cardiovascular Risk Factors: A Meta-Analysis of Randomized Controlled Trials." *Nutrition Journal* 16 (1): 68. https://doi.org/10.1186/s12937-017-0293-y.

Rangboo, Vajiheh, Mostafa Noroozi, Roza Zavoshy, Seyed Amirmansoor Rezadoost, and Asghar Mohammadpoorasl. 2016. "The Effect of Artichoke Leaf Extract on Alanine Aminotransferase and Aspartate Aminotransferase in the Patients with Nonalcoholic Steatohepatitis." *International Journal of Hepatology* 2016. https://doi.org/10.1155/2016/4030476.

Rapola, Janne M., Jarmo Virtamo, Samuli Ripatti, Jussi K. Huttunen, Demetrius Albanes, Philip R. Taylor, and Olli P. Heinonen. 1997. "Randomised Trial of α-Tocopherol and β-Carotene Supplements on Incidence of Major Coronary Events in Men with Previous Myocardial Infarction." *Lancet* 349 (9067): 1715–20. https://doi.org/10.1016/S0140-6736(97)01234-8.

Ras, Rouyanne T., Harry Hiemstra, Yuguang Lin, Mario A. Vermeer, Guus S.M.J.E. Duchateau, and Elke A. Trautwein. 2013. "Consumption of Plant Sterol-Enriched Foods and Effects on Plasma Plant Sterol Concentrations – A Meta-Analysis of Randomized Controlled Studies." *Atherosclerosis* 230 (2): 336–46. https://doi.org/10.1016/j.atherosclerosis.2013.08.012.

Ried, Karin, Catherine Toben, and Peter Fakler. 2013. "Effect of Garlic on Serum Lipids: An Updated Meta-Analysis." *Nutrition Reviews* 71 (5): 282–99. https://doi.org/10.1111/nure.12012.

Rigacci, Stefania, and Massimo Stefani. 2016. "Nutraceutical Properties of Olive Oil Polyphenols. An Itinerary from Cultured Cells through Animal Models to Humans." *International Journal of Molecular Sciences*. MDPI AG. https://doi.org/10.3390/ijms17060843.

Rizos, Evangelos C., Evangelia E. Ntzani, Eftychia Bika, Michael S. Kostapanos, and Moses S. Elisaf. 2012. "Association between Omega-3 Fatty Acid Supplementation and Risk of Major Cardiovascular

Disease Events: A Systematic Review and Meta-Analysis." *JAMA - Journal of the American Medical Association*. https://doi.org/10.1001/2012.jama.11374.

Sahebkar, Amirhossein. 2014. "A Systematic Review and Meta-Analysis of Randomized Controlled Trials Investigating the Effects of Curcumin on Blood Lipid Levels." *Clinical Nutrition* 33 (3): 406–14. https://doi.org/10.1016/j.clnu.2013.09.012.

Slavin, Joanne L., and Beate Lloyd. 2012. "Health Benefits of Fruits and Vegetables." *Advances in Nutrition* 3 (4): 506–16. https://doi.org/10.3945/an.112.002154.

Sood, Nitesh, William L. Baker, and Craig I. Coleman. 2008. "Effect of Glucomannan on Plasma Lipid and Glucose Concentrations, Body Weight, and Blood Pressure: Systematic Review and Meta-Analysis." *American Journal of Clinical Nutrition* 88 (4): 1167–75. https://doi.org/10.1093/ajcn/88.4.1167.

Sorrentino, Sajoscha A, Christian Besler, Lucia Rohrer, Martin Meyer, Kathrin Heinrich, Ferdinand H Bahlmann, Maja Mueller, et al. 2010. "Endothelial-Vasoprotective Effects of High-Density Lipoprotein Are Impaired in Patients with Type 2 Diabetes Mellitus but Are Improved after Extended-Release Niacin Therapy." *Circulation* 121 (1): 110–22. https://doi.org/10.1161/CIRCULATIONAHA.108.836346.

Stangl, Verena, Henryk Dreger, Karl Stangl, and Mario Lorenz. 2007. "Molecular Targets of Tea Polyphenols in the Cardiovascular System." *Cardiovascular Research*. https://doi.org/10.1016/j.cardiores.2006.08.022.

Stasi, Delia Di, Roberto Bernasconi, Roberto Marchioli, Rosa Maria Marfisi, Giovanna Rossi, Gianni Tognoni, and Maria Teresa Tacconi. 2004. "Early Modifications of Fatty Acid Composition in Plasma Phospholipids, Platelets and Mononucleates of Healthy Volunteers after Low Doses of n-3 Polyunsaturated Fatty Acids." *European Journal of Clinical Pharmacology* 60 (3): 183–90. https://doi.org/10.1007/s00228-004-0758-8.

Stefanutti, Claudia, Fabio Mazza, Dario Mesce, Claudia Morozzi, Serafina Di Giacomo, Marco Vitale, and Mario Pergolini. 2017. "*Monascus*

Purpureus for Statin and Ezetimibe Intolerant Heterozygous Familial Hypercholesterolaemia Patients: A Clinical Study." *Atherosclerosis Supplements* 30 (November): 86–91. https://doi.org/10.1016/j.atherosclerosissup.2017.05.021.

Stephens, N G, A Parsons, M J Brown, P M Schofield, F Kelly, K Cheeseman, and Mj Mitchinson. 1996. "Randomised Controlled Trial of Vitamin E in Patients with Coronary Disease: Cambridge Heart Antioxidant Study (CHAOS)." *The Lancet* 347 (9004): 781–86. https://doi.org/10.5555/uri:pii:S0140673696908661.

Sugden, J A, J I Davies, M D Witham, A D Morris, and A D Struthers. 2008. "Vitamin D Improves Endothelial Function in Patients with Type 2 Diabetes Mellitus and Low Vitamin D Levels." *Diabetic Medicine : A Journal of the British Diabetic Association* 25 (3): 320–25. https://doi.org/10.1111/j.1464-5491.2007.02360.x.

Tagawa, H, H Shimokawa, T Tagawa, M Kuroiwa-Matsumoto, Y Hirooka, and A Takeshita. 1999. "Long-Term Treatment with Eicosapentaenoic Acid Augments Both Nitric Oxide-Mediated and Non-Nitric Oxide-Mediated Endothelium-Dependent Forearm Vasodilatation in Patients with Coronary Artery Disease." *Journal of Cardiovascular Pharmacology* 33 (4): 633–40. https://doi.org/10.1097/00005344-199904000-00017.

Tai, Mi-Hsueh, Po-Kong Chen, Pei-Yi Chen, Ming-Jiuan Wu, Chi-Tang Ho, and Jui-Hung Yen. 2014. "Curcumin Enhances Cell-Surface LDLR Level and Promotes LDL Uptake through Downregulation of PCSK9 Gene Expression in HepG2 Cells." *Molecular Nutrition & Food Research* 58 (11): 2133–45. https://doi.org/10.1002/mnfr.201400366.

Tarcin, Ozlem, Dilek Gogas Yavuz, Beste Ozben, Ahu Telli, Ayliz Velioglu Ogunc, Meral Yuksel, Ahmet Toprak, et al. 2009. "Effect of Vitamin D Deficiency and Replacement on Endothelial Function in Asymptomatic Subjects." *The Journal of Clinical Endocrinology and Metabolism* 94 (10): 4023–30. https://doi.org/10.1210/jc.2008-1212.

Tokede, Oluwabunmi A., Temilola A. Onabanjo, Alfa Yansane, J. Michael Gaziano, and Luc Djoussé. 2015. "Soya Products and Serum Lipids: A

Meta-Analysis of Randomised Controlled Trials." *British Journal of Nutrition* 114 (6): 831–43. https://doi.org/10.1017/S0007114515002603.

Toth, Peter P., Angelo M. Patti, Dragana Nikolic, Rosaria V. Giglio, Giuseppa Castellino, Teresa Biancucci, Fabiana Geraci, et al. 2016. "Bergamot Reduces Plasma Lipids, Atherogenic Small Dense LDL, and Subclinical Atherosclerosis in Subjects with Moderate Hypercholesterolemia: A 6 Months Prospective Study." *Frontiers in Pharmacology* 6 (JAN): 299. https://doi.org/10.3389/fphar.2015.00299.

Vanuzzo, Diego, Lorenza Pilotto, Rossana Lombardi, Guido Lazzerini, Marisa Carluccio, Silvia Diviacco, Franco Quadrifoglio, et al. 2007. "Both Vitamin B6 and Total Homocysteine Plasma Levels Predict Long-Term Atherothrombotic Events in Healthy Subjects." *European Heart Journal* 28 (4): 484–91. https://doi.org/10.1093/eurheartj/ehl470.

Wall, Rebecca, R. Paul Ross, Gerald F. Fitzgerald, and Catherine Stanton. 2010. "Fatty Acids from Fish: The Anti-Inflammatory Potential of Long-Chain Omega-3 Fatty Acids." *Nutrition Reviews*. https://doi.org/10.1111/j.1753-4887.2010.00287.x.

Wei, Z. H., H. Wang, X. Y. Chen, B. S. Wang, Z. X. Rong, B. S. Wang, B. H. Su, and H. Z. Chen. 2009. "Time- and Dose-Dependent Effect of Psyllium on Serum Lipids in Mild-to-Moderate Hypercholesterolemia: A Meta-Analysis of Controlled Clinical Trials." *European Journal of Clinical Nutrition* 63 (7): 821–27. https://doi.org/10.1038/ejcn.2008.49.

Ye, Ping, Zong Liang Lu, Bao Min Du, Zuo Chen, Yang Feng Wu, Xue Hai Yu, and Yu Cheng Zhao. 2007. "Effect of Xuezhikang on Cardiovascular Events and Mortality in Elderly Patients with a History of Myocardial Infarction: A Subgroup Analysis of Elderly Subjects from the China Coronary Secondary Prevention Study." *Journal of the American Geriatrics Society* 55 (7): 1015–22. https://doi.org/10.1111/j.1532-5415.2007.01230.x.

Younes, Maged, Peter Aggett, Fernando Aguilar, Riccardo Crebelli, Birgit Dusemund, Metka Filipič, Maria Jose Frutos, et al. 2018. "Scientific Opinion on the Safety of Monacolins in Red Yeast Rice." *EFSA Journal* 16 (8). https://doi.org/10.2903/j.efsa.2018.5368.

Zeb, Irfan, Naser Ahmadi, Khurram Nasir, Jigar Kadakia, Vahid Nabavi Larijani, Ferdinand Flores, Dong Li, and Matthew J. Budoff. 2012. "Aged Garlic Extract and Coenzyme Q10 Have Favorable Effect on Inflammatory Markers and Coronary Atherosclerosis Progression: A Randomized Clinical Trial." *Journal of Cardiovascular Disease Research* 3 (3): 185–90. https://doi.org/10.4103/0975-3583.98883.

Zheng, Xin-Xin, Yan-Lu Xu, Shao-Hua Li, Xu-Xia Liu, Rutai Hui, and Xiao-Hong Huang. 2011. "Green Tea Intake Lowers Fasting Serum Total and LDL Cholesterol in Adults: A Meta-Analysis of 14 Randomized Controlled Trials." *The American Journal of Clinical Nutrition* 94 (2): 601–10. https://doi.org/10.3945/ajcn.110.010926.

Zhu, X., X. Sun, M. Wang, C. Zhang, Y. Cao, G. Mo, J. Liang, and S. Zhu. 2015. "Quantitative Assessment of the Effects of Beta-Glucan Consumption on Serum Lipid Profile and Glucose Level in Hypercholesterolemic Subjects." *Nutrition, Metabolism and Cardiovascular Diseases* 25 (8): 714–23. https://doi.org/10.1016/j.numecd.2015.04.008.

Zuchi, Cinzia, Giuseppe Ambrosio, Thomas F. Lüscher, and Ulf Landmesser. 2010. "Nutraceuticals in Cardiovascular Prevention: Lessons from Studies on Endothelial Function." *Cardiovascular Therapeutics* 28 (4): 187–201. https://doi.org/10.1111/j.1755-5922.2010.00165.x.

In: Nutraceuticals: Evidence Based Use
Editor: Raluca Maria Pop
ISBN: 978-1-53618-052-7

Chapter 4

The Role of Probiotics in Cardiovascular Diseases

Maria Adriana Neag[1,*] and Adrian Catinean[2]
[1]Department of Pharmacology, Toxicology and Clinical pharmacology, "Iuliu Hatieganu" University of Medicine and Pharmacy, Cluj-Napoca, Romania
[2]Department of Internal Medicine, "Iuliu Hatieganu" University of Medicine and Pharmacy, Cluj-Napoca, Romania

Abstract

Nowadays, one of the major health problems worldwide, with a negative impact on the morbidity and mortality or the cost of the patient's health, is cardiovascular disease. Often, this problem is associated with metabolic problems (obesity, dyslipidemia, etc). Lately, great importance has been given to the role of the gut microbiota in many diseases, from gastrointestinal problems to allergies or even cancer. In this regard, the relationship between microbiota and cardiovascular diseases has been

[*] Corresponding Author's E-mail: maria.neag@umfcluj.ro.

studied. Therefore, in this chapter, we highlight the role of the microbiota in cardiovascular diseases and also the role of probiotics in this type of disease.

Keywords: cardiovascular diseases, gut microbiota, probiotics

INTRODUCTION

Hypertension (HTN), atherosclerosis (ATS), ischemic heart diseases (IHD) and heart failure (HF) are the most common cardiovascular diseases (CVDs) (Nakata and Inoue 2014). These represent the leading causes of death in the world. For this reason, interest in the prevention and treatment of CVDs has increased in the last years (Roth et al. 2017). Reducing the risk of CVDs includes the use of preventive drugs (eg. aspirin for patients with ischemic heart disease), control of risk factors (eg dyslipidemia, obesity, diabetes) and lifestyle changes (eg. smoking, diet, exercise) (Chow et al. 2015).

Lately, in addition to these CVDs prevention trends, the use of many nutraceuticals has increased. The nutraceutical class includes either functional foods or dietary supplements with physiological benefits or prophylaxis against chronic diseases (Momtazi et al. 2017; Nasri et al. 2014). In the last 10 years, the interest towards a subclass of nutraceuticals - probiotics has also increased. Probiotics are strains of bacteria or yeast that administered can improve or protect against many diseases (Rahmani, Moradzadeh, and Farahmand 2019). In this chapter, we want to identify the link between probiotics and CVDs.

PROBIOTICS AND HYPERTENSION

Management of HTN prevention and treatment is essential, as lowering BP reduces morbidity and mortality. Nowadays, the control of BP values is poor worldwide (Mancia et al. 2007). For this reason, researchers are looking for new targets for the prevention and treatment of this disease.

It is known that gut microbiota plays a crucial role in maintaining the balance of the body's homeostasis. The first studies highlighting the link between gut microbiota and high blood pressure (BP) were conducted in the 1960s and 1970s (Marques, Mackay, and Kaye 2018). Subsequent studies have identified differences between the abundance and diversity of the gut microbiota in hypertensive patients *vs* healthy individuals. An increase of *Prevotella* and a decrease of *Faecalibacterium*, *Roseburia,* and *Bifidobacterium* were observed in hypertensive patients, whereas *Bacteroidetes* were predominant in healthy individuals (Li et al. 2017).

The pathogenesis of HTN is complex and many triggers have been studied, from genetic predisposition to environmental factors. Although it is known that BP is caused by three main factors (smooth vascular muscle cell tone, cardiac function, renal sodium excretion), the microbiota has also been studied as a possible factor involved in the pathology of high BP. Some bacteria in the gut microbiota are capable of degrading various food proteins resulting in peptides capable of inhibiting the renin-angiotensin system (RAS), blocking vasoconstriction, and lowering BP (Kang and Cai 2018). Also, some metabolites of the gut microbiota, such as short-chain fatty acids (SCFA), trimethylamine N-oxide, p-cresol sulfate are capable of affecting the cardiovascular system (Afsar et al. 2016). Furthermore, the gut microbiota by certain species of bacteria (eg, *Bacillus*, *Bifidobacterium*, *Lactobacillus*, *Morganella*, *Escherichia coli*) influences the production of neurotransmitters such as dopamine, norepinephrine, serotonin, gamma-aminobutyric acid (GABA), which may influence central nervous system activity but also BP values (Strandwitz 2018). Another agent involved in lowering BP is nitric oxide (NO), a well-known vasodilation agent (Figure 1). Initially, the researchers considered 3 possible ways to produce NO through the gut microbiota: L-arginine as a substrate for NO synthase, denitrification of nitrogen gas and reduction of dissimilatory nitrate to ammonium. However, only the last one was demonstrated by Vermeiren and colleagues (Vermeiren et al. 2009). Also, it is known that the gut microbiota modulates T-helper 17 cells that are associated with high blood pressure and vascular dysfunction (Robles-Vera et al. 2017).

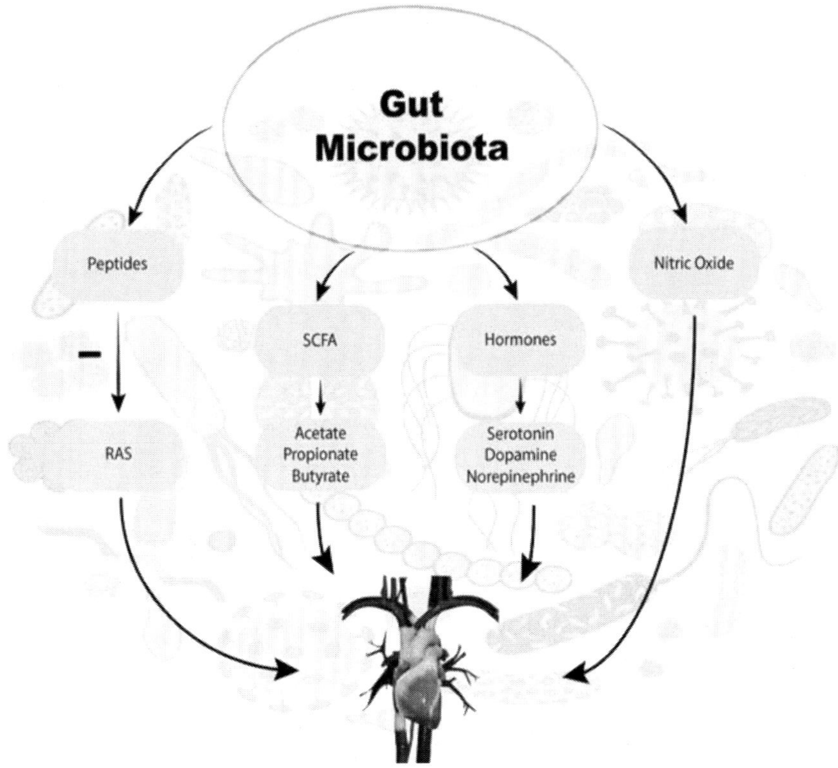

Figure 1. Microbiota mechanisms involved in regulating blood pressure. Abbreviations: RAS, renin-angiotensin system; SCFA, short-chain fatty acids.

The main SCFA produced by the gut microbiota are acetate, propionate, and butyrate. *Streptococcus spp., Prevotella spp., Bifidobacterium spp., Clostridium spp., A. muciniphila* are the main acetate-producing bacteria, *Bacteroides spp., Salmonella spp., Veillonella spp.* are the main propionate-producing bacteria and *Lachnospiraceae, Ruminococcaceae,* and *Acidaminococcaceae* families produce especially butyrate (Ma and Li 2018).

These SCFAs act on SCFA receptors: GPR41 (Free Fatty Acid Receptor 3), GPR43 (Free Fatty Acid Receptor 2) or Olfr78 (olfactory receptor 78). GPR41 is coupled to Gi, while GPR43 is coupled to Gi or Gq. These receptors are expressed in the gastrointestinal tract, immune cells, etc. The third type (Olfr78) belongs to the receptors of the same class (receptors coupled with G protein), is expressed in the kidneys, especially in renal afferent arteriole and the juxtaglomerular cells (Miyamoto et al. 2016). However, Olfr78 and GPR41 are the major SCFA-activated receptors involved in BP regulation. But, activating of these receptors has different effects: activation of Olfr78 leads to an increase in BP, and that of GPR41 leads to a decrease in BP. The reduction of BP after Olfr78 activation is partially a consequence of renin stimulation from the juxtaglomerular cells (Pluznick et al. 2013).

Marques et al., (2017) in an experimental mice study, showed that high fiber diet or supplementation with SCFA acetate decreases the development of hypertension induced by mineralocorticoid (deoxycorticosterone acetate) excess.

Table 1. Effect of probiotic strains of arterial blood pressure

Tested strains	Type of study	Effects/Mechanisms	References
Lactobacillus fermentum CECT5716 (LC40)	Experimental (male C57Bl/6J mice)	Decreases hypertension induced by tacrolimus: -favoring butyrate-producing bacteria -ameliorates vascular oxidative stress -independently of RAS inhibition	(Toral et al. 2019)
Lactobacillus fermentum CECT5716 (LC40) *L. coryniformis* (K8) *L. gasseri* (LC9)	Experimental (spontaneously hypertensive rats and Wistar Kyoto)	Reduce high blood pressure in genetic hypertension rats: -decrease *Bacteroides* and *Clostridium sp* -increase *Lactobacillus sp* -improve endothelial dysfunction -ameliorate vascular inflammation -reduce vascular oxidative stress	(Gómez-Guzmán et al. 2015)
Lactobacillus helveticus LBK-16H	Clinical	Decreases systolic blood pressure (-7.6 mmHg) and diastolic blood pressure (-4.1 mmHg) in 10 weeks	(Jauhiainen et al. 2005)
Lactobacillus plantarum	Clinical (pilot study)	Decreases systolic blood pressure and BMI	(Sharafedtinov et al. 2013)
Lactobacillus casei strain C1	Experimental (spontaneously hypertensive rats and Wistar)	Decreases systolic and diastolic blood pressure -increase NO level -decrease oxidative stress	(Yap et al. 2016)

On the other hand, the microbiota can also contribute to the bioavailability of antihypertensive drugs. For example, Yoo et al., (2016) studied the effect of antibiotic treatment on the bioavailability of amlodipine, one of the most used calcium channel blockers for the treatment of hypertension. Ampicillin has been used as antibiotics because it is a broad-spectrum antibiotic with an impact on gut microbiota. The results of the study revealed that ampicillin increases the bioavailability of amlodipine. The AUC of the calcium channel blocker increased by 42 and 133% for 10 and 20 mg of amlodipine (Yoo et al. 2016).

The effect of a few probiotics strains on arterial blood pressure are shown in Table 1.

PROBIOTICS AND ATHEROSCLEROSIS

ATS is one of the most important risk factors for CVDs. Cholesterol accumulation and abnormal inflammatory and immune responses characterize this silent progressive disease (Abdolmaleki et al. 2019).

There is the idea that future therapy in CVD, including ATS, can be an anti-inflammatory therapy focused on the gut microbiota. It is an important immunologically active organ in the body considered an essential part of the body with a role in regulating inflammatory reactions (Yamashita et al. 2015).

We mentioned above that SCFAs are directly related to HTN, but they are also closely related to ATS. Propionate could reduce the expression of vascular cell adhesion molecule 1 (VCAM-1) and intercellular adhesion molecule 1 (ICAM-1) by inhibiting nuclear factor kappa B (NF-kB). These two endothelial adhesion molecules interfere with leukocyte adhesion and play an important role in ATS (Zapolska-Downar and Naruszewicz 2009). In an experimental study Fotis and collab. demonstrated that ICAM-1 is one of the first molecules involved in the formation of ATS lesion and the high cholesterol diet increases ICAM-1 expression, but VCAM-1 expression is not influenced by this type of diet, in the early stage of ATS (Bauke et al. 2018; Fotis et al. 2012).

It is known that the production of SCFA in the intestine depends on the composition of the microbiota and one of the bacteria that play an important role in the production of SCFA, especially acetate, is *Akkermansia muciniphila*, an important mucin-degrading bacterium. Acetate also plays an important role in inflammation and restoration of the metabolic balance (Dao et al. 2016). Due to its potential benefit, Li et al., (2016) studied the role of *Akkermansia muciniphila* in the pathogenesis of ATS in E apolipoprotein (Apoe $^{-/-}$) deficient mice. This bacterium inhibits the expression of ICAM-1, MCP-1 and TNF-alpha, chemokines involved in macrophage adhesion to the endothelium. It decreases ATS lesions in Apoe $^{-/-}$ mice but does not influence Western diet-induced hypercholesterolemia. Thus, the antiatherosclerotic effect of *Akkermansia muciniphila* is most likely due to its anti-inflammatory activity (Li et al. 2016).

Akkermansia muciniphila is also involved in the prevention of metabolic endotoxemia. Lipopolysaccharide (LPS), known as lipoglycan and endotoxin, is found in the outer membrane of Gram-negative bacteria and is an important trigger for the inflammatory response. Under normal conditions, LPS is detected in the intestinal lumen, but under harsh conditions (eg low pH, inflammation), LPS passes into the circulatory system. LPS elevated plasmatic concentration is known as metabolic endotoxemia and is correlated with abnormal immune and inflammatory response (Jeong et al. 2015). *Akkermansia muciniphila* prevents metabolic endotoxemia, having the ability to restore the thickness of the mucus layer, regulate the production of antimicrobial peptides, and increase the production of endogenous molecules with an anti-inflammatory effect, regulate the production of intestinal peptides that play a role in maintaining glucose homeostasis and the intestinal barrier (Cani et al. 2016). In this regard, Reunanen et al., (2015) demonstrated that *Akkermansia muciniphila* can adhere to intestinal cells and strengthen the epithelial cell layer.

Choline, betaine, and trimethylamine N-oxide (TMAO), metabolites of dietary lipid phosphatidylcholine, are associated with atherosclerotic cardiovascular disease (Tang and Hazen 2017). TMAO is an important agent involved in ATS, inflammation and platelet hyperactivity. It is formed in the liver by trimethylamine (TMA) oxidation by flavin-containing monooxygenases (FMOs), in particular FMO3. TMA results in the intestine from precursors such as choline, phosphatidylcholine, and L-carnitine by the action of the gut microbiota (Roth et al. 2017) (Figure 2).

Thus, the level of TMAO is regulated by FMOs activity, but other experimental studies have shown that FMO3 can directly influence blood glucose and lipid levels, independent of TMA/TMAO. FMO3 also influences the two major regulators involved in metabolism and inflammation, such as peroxisome proliferator-activated receptors (PPAR) and Kruppel-like factor 15 (KLF15) (Shih et al. 2015).

One of the causes of the inflammatory process is the change in intestinal microbiota diversity and composition (Tang and Hazenb 2017).

The inflammatory process involved in ATS could be influenced by conjugated linoleic acid (CLA). CLA is a polyunsaturated fatty acid with a role in reducing the inflammatory response. *In vivo* and *in vitro* studies have shown that CLA inhibits cyclooxygenase-2 (COX-2) and thus influences PG synthesis and exerts an anti-inflammatory effect. On the other hand, CLA activates PPAR-ɣ and reduces the formation of nuclear factor kappa-B (NF-kB) and pro-inflammatory cytokines (TNF-α, Il-1β, IL-8), demonstrating its anti-inflammatory effect (Yang et al. 2015). But CLA production depends on the activity of the gut microbiota. The diversity and high composition of the gut microbiota increase the amount of CLA (Druart et al. 2014).

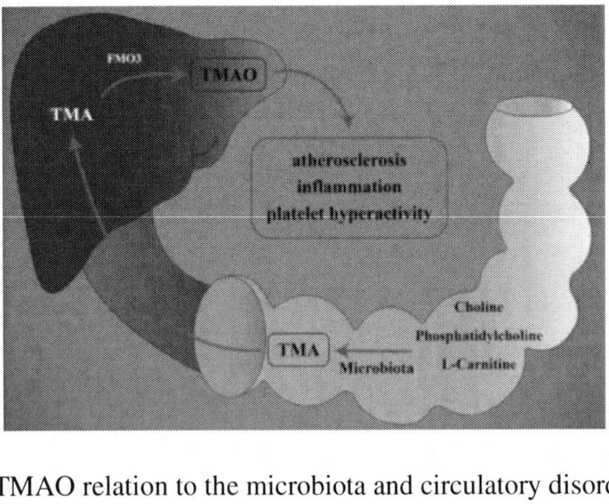

Figure 2. TMA/TMAO relation to the microbiota and circulatory disorders.
Abbreviations: TMA, trimethylamine; TMAO, trimethylamine N-oxide; FMO3, flavin-containing monooxygenases.

The effect of a few probiotic strains on ATS are shown in Table 2.

Table 2. Effect of probiotic strains in atherosclerosis

Tested strains	Type of study	Effects/Mechanisms	References
Enterococcus faecium CRL183	Experimental (New Zealand white male rabbits)	No effects on atherosclerosis development Improve the lipid profile	(Cavallini et al. 2009)
Lactobacillus reuteri ATCC 4659, L. reuteri DSM 1798 or L. reuteri L6798	Experimental ($Apoe^{-/-}$ mice)	Protected against diet-induced obesity Lowered blood insulin level No effects on atherosclerosis development	(Fåk and Bäckhed 2012)
Lactobacillus acidophilus ATCC 435	Experimental ($Apoe^{-/-}$ mice)	Decreased lesion size of atherosclerosis (most likely by inhibiting the activation of NF-κB and also by an antioxidant effect).	(Lihua Chen et al. 2013)
Lactobacillus casei	Experimental ($Apoe^{-/-}$ mice)	Reduced atherosclerotic plaque (improve oxidative stress, inflammatory response and lipid profile	(Z. Tang et al. 2014)
Lactobacillus rhamnosus GR-1	Experimental (Apoe-/- mice)	Attenuated the formation of atherosclerotic lesions (reduces oxidative stress – decreases SOD, MDA, inhibits inflammation – modulates the level of TNF-α, IL-6, MCP-1	(Fang et al. 2019)
Bifidobacterium animalis subsp. *lactis LKM512*	Clinical study (healthy adults at risk of atherosclerosis)	Decreased the risk of atherosclerosis development (decreased TMA concentration after consumption of LKM512 vs placebo)	(Matsumoto et al. 2017)

PROBIOTICS AND HEART FAILURE (HF)

The relationship between HF and the intestinal microbiota is caused by intestinal changes (e.g., hypoperfusion and congestion) in patients with HF, which may influence the permeability/absorption in the intestine and may also affect the function of the epithelial barrier (Nagatomo and Tang 2015). This "leaky gut" is associated with increased endotoxemia and increased inflammatory response (Sandek et al. 2007; Kamo et al. 2017).

Table 3. Effect of probiotic strains on chronic heart failure

Tested strains	Type of study	Effects/Mechanisms	References
Lactobacillus rhamnosus GR-1	Experimental (male Sprague–Dawley rats)	Attenuates postinfarction remodeling and heart failure in rats Leptin: adiponectin ratio was normalized by probiotic administration	(Gan et al. 2014)
Saccharomyces boulardii	Clinical study (chronic systolic HF outpatients)	Reduction of biochemical and inflammatory biomarkers (creatinine, uric acid, hsCRP), and also an improvement in cardiovascular functioning (left atrial diameter, LVEF)	(Costanza et al. 2015)
Lactobacillus plantarum 299v	Clinical study (men with stable coronary artery disease)	Improves endothelium-dependent vasodilation and reduced systemic inflammation (increased NO bioavailability and reduced IL-8, IL-12, leptin level)	(Malik et al. 2018)
Lactobacillus reuteri GMNL-263	Experimental (male Sprague–Dawley rats)	Decreased the levels of TG and T-CHO and reversed the cardiac systolic function. Reversed high-calorie diet associated cardiac inflammation, cardiac eccentric hypertrophy and fibrosis.	(Liao et al. 2016)

Patients with HF have changes in the abundance and composition of the intestinal microbiota. The gut is a huge microorganism reservoir, with good and bad bacteria, which in normal conditions, in healthy people, are in balance. Intestinal dysbiosis occurs in patients with HF: potential pathogenic bacteria such as *Campylobacter*, *Shigella*, *Salmonella*, *Yersinia enterocolitica,* and *Candida species* increase and the bacterial genera *Blautia* and *Collinsella*, and especially *Faecalibacterium prausnitzii* decrease (Ahmad, Ward, and Dwivedi 2019).

It is known that chronic heart failure (CHF) is associated with a systemic inflammatory state. The data on inflammation in CVD has been aforementioned.

The effect of a few probiotics strains on CHF is shown in Table 3.

Microbiota and Cardiovascular Risk Factors

Two of the major CV risk factors are obesity and diabetes mellitus. The prevalence of HTN is higher in obese patients compared to lean subjects. Obesity seems to influence the incidence, prognosis, and mortality of CVD (Ortega et al. 2016). It has been shown that the intestinal microbiota is involved in the extraction of nutrients from the diet and in the intestinal permeability. It thus regulates the level of LPS with an important role in chronic inflammation. Any level of chronic inflammation is associated with leptin and insulin resistance, obesity and diabetes (Saad, Santos, and Prada 2016). Patients with obesity are characterized by the accumulation of lipids in adipocytes and by a degree of inflammation. Lipid accumulation activates two major inflammatory pathways NF-κB (nuclear factor-B) and c-Jun N-terminal kinase (JNK) and may affect insulin resistance, including hypothalamic insulin resistance (Li Chen et al. 2015; Zand, Morshedzadeh, and Naghashian 2017).

It is known that in patients with obesity and/or diabetes the microbiota composition is altered, especially the *Bacteroidetes/Firmicutes* ratio (Table 4).

Table 4. Studies on the increase in Bacteroidetes/ Firmicutes ratio and its effects

Gut microbiota changes	Type of study	Effects/Mechanisms	References
Increase Firmicutes/ Bacteroidetes ratio	Clinical study (obese vs normal-weight children, 6 to 16 years old)	Decreased SCFA absorption due to low-grade inflammation or more rapid gut transit	(Riva et al. 2017)
	Clinical study (79 subjects: 51 obese, 23 children and 28 adults; 28 subjects of normal weight, 17 children and 11 adults)	Increased of fermentation of indigestible carbohydrates Increased acetate production and subsequently increased energy absorption	(Ismail et al. 2011)
	Clinical study (175 subjects – 91 normal group, 62 overweight group, and 22 obesity group)	Gender differences in iron metabolism may affect the composition of the microbiota	(Xu et al. 2012)

Microbiota changes influence the differentiation of intestinal epithelial cells, the integrity of the intestinal barrier and the function of tight junctions. As a result of these changes, systemic endotoxemia increases, which results in the alteration of metabolic processes in the peripheral organs (liver, adipose tissue): lipogenesis, inflammation, oxidative stress, insulin resistance (Figure 3) (Patterson et al. 2016).

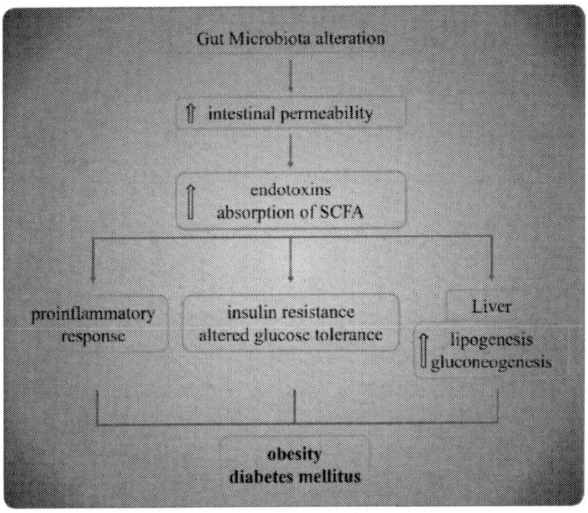

Figure 3. Gut microbiota alterations concerning obesity and diabetes. Abbreviations: SCFA, short-chain fatty acids.

Also, obese and diabetic patients have a reduced amount of *Akkermansia muciniphila*. The role of this bacterium in intestinal integrity has been mentioned previously. Researchers have shown that metformin, one of the most commonly used oral antidiabetic agents, increase *Akkermansia muciniphila* levels and thus improve metabolic function and systemic inflammation (Chobot et al. 2018; Miele et al. 2015).

Moreover, there is a relationship between the intestine, obesity and the brain through the gut-brain axis. Appetite and satiety through hypothalamic neuroendocrine pathways are influenced by some bacterial strains. They can modulate the secretion of intestinal hormones, such as GLP-1, leptin, YY peptide, etc. Also, intestinal bacteria influence the production of serotonin and aminobutyric acid, substances that play a role in central appetite regulation (Torres-Fuentes et al. 2017).

The intestinal microbiota is not the only one involved in CVD. The presence of pathogens such as *Porphyromonas gingivalis, Prevotella intermediate* or *Tannerella forsyia* in patients with periodontal infection, increases endotoxemia by leaking endotoxins, such as LPS, in the serum and maintaining chronic systemic inflammation (Nagpal, Yamashiro, and Izumi 2015). A consequence of this chronic inflammatory condition is the increase of pro-inflammatory mediators such as TNF -α, IL-1β, IFN-γ, and others, thus contributing to insulin resistance and alteration of glucose metabolism. It may also contribute to the modification of lipid metabolism (Pizzo et al. 2010). Because the major role of inflammation and infection in CVD is well known, the possible role of periodontal infection and oral microbiota in these diseases may be underlined. Furthermore, the "bad bacteria" in the oral cavity, those that cause periodontitis, can affect other cells (eg, endothelial cells) and thus may be involved in platelet aggregation. To underline the role of the oral microbiota in BCV, some studies have shown that *P. gingivalis* and *A. actinomycetemcomitans* were present in the atheromatous plaque of CVD patients (Rivera et al. 2013; Pessi et al. 2013; Reyes et al. 2013). Therefore, chronic periodontitis and oral microbiota can have a significant impact on both CV and CVD risk factors and can be considered risk factors for them.

Conclusion

Studies show that the microbiota plays an important role in the host homeostasis. The risk of CVD is directly proportional to the connection between the intestinal microbiota and the physiological process in the body. The role of the oral microbiota in the development of CV risk factors and CVD has also been demonstrated. Thus, many researchers are exploring the efficacy of the pharmacological intervention by probiotics or other nutraceuticals, which target specific metabolic activities in the gut, and novel therapies including probiotics, which modulate the microbiota. These could be future options for treating or preventing CV disorders. Moreover, manipulation of the gut microbiota may lead to the development of novel CV therapies.

References

Abdolmaleki, Fereshte, Seyed Mohammad Gheibi Hayat, Vanessa Bianconi, Thomas P. Johnston, and Amirhossein Sahebkar. 2019. "Atherosclerosis and Immunity: A Perspective." *Trends in Cardiovascular Medicine* 29 (6): 363–71. https://doi.org/10.1016/j.tcm.2018.09.017.

Afsar, Baris, Nosratola D. Vaziri, Gamze Aslan, Kayhan Tarim, and Mehmet Kanbay. 2016. "Gut Hormones and Gut Microbiota: Implications for Kidney Function and Hypertension." *Journal of the American Society of Hypertension* 10 (12): 954–61. https://doi.org/10.1016/j.jash.2016.10.007.

Ahmad, Adilah F., Natalie C. Ward, and Girish Dwivedi. 2019. "The Gut Microbiome and Heart Failure." *Current Opinion in Cardiology* 34 (2): 225–32. https://doi.org/10.1097/HCO.0000000000000598.

Bauke, Stephan, Kai Golibrzuch, Hainer Wackerbarth, Peter Fendt, Lars Zigan, Stefan Seefeldt, Olaf Thiele, and Thomas Berg. 2018. "Optical Sensor System for Time-Resolved Quantification of Methane Concentrations: Validation Measurements in a Rapid Compression Machine." *Journal of Quantitative Spectroscopy and Radiative Transfer* 210: 101–10. https://doi.org/10.1016/j.jqsrt.2018.02.016.

Cani, Patrice D., Hubert Plovier, Matthias Van Hul, Lucie Geurts, Nathalie M. Delzenne, Céline Druart, and Amandine Everard. 2016. "Endocannabinoids-at the Crossroads between the Gut Microbiota and Host Metabolism." *Nature Reviews Endocrinology* 12 (3): 133–43. https://doi.org/10.1038/nrendo.2015.211.

Cavallini, Daniela Cu, Raquel Bedani, Laura Q. Bomdespacho, Regina C. Vendramini, and Elizeu A. Rossi. 2009. "Effects of Probiotic Bacteria, Isoflavones and Simvastatin on Lipid Profile and Atherosclerosis in

Cholesterol-Fed Rabbits: A Randomized Double-Blind Study." *Lipids in Health and Disease* 8: 1–8. https://doi.org/10.1186/1476-511X-8-1.

Chen, Li, Rui Chen, Hua Wang, and Fengxia Liang. 2015. "Mechanisms Linking Inflammation to Insulin Resistance." *International Journal of Endocrinology* 2015. https://doi.org/10.1155/2015/508409.

Chen, Lihua, Wenen Liu, Yanming Li, San Luo, Qingxia Liu, Yiming Zhong, Zijuan Jian, and Meihua Bao. 2013. "Lactobacillus Acidophilus ATCC 4356 Attenuates the Atherosclerotic Progression through Modulation of Oxidative Stress and Inflammatory Process." *International Immunopharmacology* 17 (1): 108–15. https://doi.org/10.1016/j.intimp.2013.05.018.

Chobot, Agata, Katarzyna Górowska-Kowolik, Magdalena Sokołowska, and Przemysława Jarosz-Chobot. 2018. "Obesity and Diabetes—Not Only a Simple Link between Two Epidemics." *Diabetes/Metabolism Research and Reviews* 34 (7): 1–9. https://doi.org/10.1002/dmrr.3042.

Chow, Clara K., Julie Redfern, Graham S. Hillis, Jay Thakkar, Karla Santo, Maree L. Hackett, Stephen Jan, et al. 2015. "Effect of Lifestyle-Focused Text Messaging on Risk Factor Modification in Patients with Coronary Heart Disease: A Randomized Clinical Trial." *JAMA - Journal of the American Medical Association* 314 (12): 1255–63. https://doi.org/10.1001/jama.2015.10945.

Costanza, Annelise C., Samuel D. Moscavitch, Hugo C C Faria Neto, and Evandro T. Mesquita. 2015. "Probiotic Therapy with Saccharomyces Boulardii for Heart Failure Patients: A Randomized, Double-Blind, Placebo-Controlled Pilot Trial." *International Journal of Cardiology* 179: 348–50. https://doi.org/10.1016/j.ijcard.2014.11.034.

Dao, Maria Carlota, Amandine Everard, Judith Aron-Wisnewsky, Nataliya Sokolovska, Edi Prifti, Eric O. Verger, Brandon D. Kayser, et al. 2016. "Akkermansia Muciniphila and Improved Metabolic Health during a Dietary Intervention in Obesity: Relationship with Gut Microbiome Richness and Ecology." *Gut* 65 (3): 426–36. https://doi.org/10.1136/gutjnl-2014-308778.

Druart, Céline, Audrey M. Neyrinck, Bruno Vlaeminck, Veerle Fievez, Patrice D. Cani, and Nathalie M. Delzenne. 2014. "Role of the Lower and Upper Intestine in the Production and Absorption of Gut Microbiota-Derived PUFA Metabolites." *PLoS ONE* 9 (1). https://doi.org/10.1371/journal.pone.0087560.

Fåk, Frida, and Fredrik Bäckhed. 2012. "Lactobacillus Reuteri Prevents Diet-Induced Obesity, but Not Atherosclerosis, in a Strain Dependent Fashion in Apoe-/- Mice." *PLoS ONE* 7 (10): 1–8. https://doi.org/10.1371/journal.pone.0046837.

Fang, Y., H. Q. Chen, X. Zhang, H. Zhang, J. Xia, K. Ding, and Z. Y. Fang. 2019. "Probiotic Administration of Lactobacillus Rhamnosus GR-1 Attenuates Atherosclerotic Plaque Formation in ApoE -/- Mice Fed with a High-Fat Diet." *European Review for Medical and Pharmacological Sciences* 23 (8): 3533–41. https://doi.org/10.26355/ eurrev_201904_17722.

Fotis, Lampros, Georgios Agrogiannis, Ioannis S. Vlachos, Alkistis Pantopoulou, Angeliki Margoni, Maria Kostaki, Christos Verikokos, Dimitrios Tzivras, Dimitri P. Mikhailidis, and Despina Perrea. 2012. "Intercellular Adhesion Molecule (ICAM)-1 and Vascular Cell Adhesion Molecule (VCAM)-1 at the Early Stages of Atherosclerosis in a Rat Model." *In Vivo* 26 (2): 243–50.

Gan, Xiaohong Tracey, Grace Ettinger, Cathy X. Huang, Jeremy P. Burton, James V. Haist, Venkatesh Rajapurohitam, James E. Sidaway, et al. 2014. "Probiotic Administration Attenuates Myocardial Hypertrophy and Heart Failure after Myocardial Infarction in the Rat." *Circulation: Heart Failure* 7 (3): 491–99. https://doi.org/10.1161/ CIRCHEART FAILURE.113.000978.

Gómez-Guzmán, Manuel, Marta Toral, Miguel Romero, Rosario Jiménez, Pilar Galindo, Manuel Sánchez, María José Zarzuelo, Mónica Olivares, Julio Gálvez, and Juan Duarte. 2015. "Antihypertensive Effects of Probiotics Lactobacillus Strains in Spontaneously Hypertensive Rats." *Molecular Nutrition and Food Research* 59 (11): 2326–36. https://doi.org/10.1002/mnfr.201500290.

Ismail, Nagwa Abdallah, Shadia H. Ragab, Abeer Abd ElBaky, Ashraf R S Shoeib, Yasser Alhosary, and Dina Fekry. 2011. "Frequency of Firmicutes and Bacteroidetes in Gut Microbiota in Obese and Normal

Weight Egyptian Children and Adults." *Archives of Medical Science* 7 (3): 501–7. https://doi.org/10.5114/aoms.2011.23418.

Jauhiainen, Tiina, Heikki Vapaatalo, Tuija Poussa, Sinikka Kyrönpalo, Martin Rasmussen, and Riitta Korpela. 2005. "Lactobacillus Helveticus Fermented Milk Lowers Blood Pressure in Hypertensive Subjects in 24-h Ambulatory Blood Pressure Measurement." *American Journal of Hypertension* 18 (12): 1600–1605. https://doi.org/10.1016/j.amjhyper.2005.06.006.

Jeong, Jin Ju, Kyung Ah Kim, Se Eun Jang, Jae Yeon Woo, Myung Joo Han, and Dong Hyun Kim. 2015. "Orally Administrated Lactobacillus Pentosus Var. Plantarum C29 Ameliorates Age-Dependent Colitis by Inhibiting the Nuclear Factor-Kappa B Signaling Pathway via the Regulation of Lipopolysaccharide Production by Gut Microbiota." *PLoS ONE* 10 (2). https://doi.org/10.1371/journal.pone.0116533.

Kamo, Takehiro, Hiroshi Akazawa, Wataru Suda, Akiko Saga-Kamo, Yu Shimizu, Hiroki Yagi, Qing Liu, et al. 2017. "Dysbiosis and Compositional Alterations with Aging in the Gut Microbiota of Patients with Heart Failure." *PLoS ONE* 12 (3): 1–14. https://doi.org/10.1371/journal.pone.0174099.

Kang, Yongbo, and Yue Cai. 2018. "Gut Microbiota and Hypertension: From Pathogenesis to New Therapeutic Strategies." *Clinics and Research in Hepatology and Gastroenterology* 42 (2): 110–17. https://doi.org/10.1016/j.clinre.2017.09.006.

Li, Jin, Shaoqiang Lin, Paul M. Vanhoutte, Connie W. Woo, and Aimin Xu. 2016. "Akkermansia Muciniphila Protects against Atherosclerosis by Preventing Metabolic Endotoxemia-Induced Inflammation in Apoe-/- Mice." *Circulation* 133 (24): 2434–46. https://doi.org/10.1161/CIRCULATIONAHA.115.019645.

Li, Jing, Fangqing Zhao, Yidan Wang, Junru Chen, Jie Tao, Gang Tian, Shouling Wu, et al. 2017. "Gut Microbiota Dysbiosis Contributes to the Development of Hypertension." *Microbiome* 5 (1): 1–19. https://doi.org/10.1186/s40168-016-0222-x.

Liao, Po Hsiang, Wei Wen Kuo, Chia Hua Kuo, Yu Lan Yeh, Chia Yao Shen, Ya Hui Chen, Ray Jade Chen, V. Vijaya Padma, Yi Hsing Chen, and Chih Yang Huang. 2016. "Lactobacillus Reuteri GMNL-263 Reduces Hyperlipidaemia and the Heart Failure Process in High-Calorie Diet-Fed Induced Heart Dysfunction in Rats." *Journal of Functional Foods* 20 (91): 226–35. https://doi.org/10.1016/j.jff. 2015.11.009.

Ma, Junli, and Houkai Li. 2018. "The Role of Gut Microbiota in Atherosclerosis and Hypertension." *Frontiers in Pharmacology* 9 (SEP): 1–14. https://doi.org/10.3389/fphar.2018.01082.

Malik, Mobin, Tisha M. Suboc, Sudhi Tyagi, Nita Salzman, Jingli Wang, Rong Ying, Michael J. Tanner, Mamatha Kakarla, John E. Baker, and Michael E. Widlansky. 2018. "Lactobacillus Plantarum 299v Supplementation Improves Vascular Endothelial Function and Reduces Inflammatory Biomarkers in Men with Stable Coronary Artery Disease." *Circulation Research* 123 (9): 1091–1102. https://doi. org/10.1161/CIRCRESAHA.118.313565.

Mancia, Giuseppe, Guy De Backer, Anna Dominiczak, Renata Cifkova, Robert Fagard, Giuseppe Germano, Guido Grassi, et al. 2007. *2007 Guidelines for the Management of Arterial Hypertension: The Task Force for the Management of Arterial Hypertension of the European Society of Hypertension (ESH) and of the European Society of Cardiology (ESC). Journal of Hypertension.* Vol. 25. https://doi. org/10.1097/HJH.0b013e3281fc975a.

Marques, Francine Z., Charles R. Mackay, and David M. Kaye. 2018. "Beyond Gut Feelings: How the Gut Microbiota Regulates Blood Pressure." *Nature Reviews Cardiology* 15 (1): 20–32. https://doi.org/ 10.1038/nrcardio.2017.120.

Matsumoto, Mitsuharu, Yusuke Kitada, Yumi Shimomura, and Yuji Naito. 2017. "Bifidobacterium Animalis Subsp. Lactis LKM512 Reduces Levels of Intestinal Trimethylamine Produced by Intestinal Microbiota in Healthy Volunteers: A Double-Blind, Placebo-Controlled Study." *Journal of Functional Foods* 36: 94–101. https://doi.org/10.1016/j.jff. 2017.06.032.

Miele, Luca, Valentina Giorgio, Maria Adele Alberelli, Erica De Candia, Antonio Gasbarrini, and Antonio Grieco. 2015. "Impact of Gut Microbiota on Obesity, Diabetes, and Cardiovascular Disease Risk."

Current Cardiology Reports 17 (12). https://doi.org/10.1007/s11886-015-0671-z.

Momtazi, Amir Abbas, Maciej Banach, Matteo Pirro, Niki Katsiki, and Amirhossein Sahebkar. 2017. "Regulation of PCSK9 by Nutraceuticals." *Pharmacological Research* 120: 157–69. https://doi.org/10.1016/j.phrs.2017.03.023.

Nagatomo, Yuji, and W. H.Wilson Tang. 2015. "Intersections between Microbiome and Heart Failure: Revisiting the Gut Hypothesis." *Journal of Cardiac Failure* 21 (12): 973–80. https://doi.org/10.1016/j.cardfail.2015.09.017.

Nagpal, Ravinder, Yuichiro Yamashiro, and Yuichi Izumi. 2015. "The Two-Way Association of Periodontal Infection with Systemic Disorders: An Overview." *Mediators of Inflammation* 2015. https://doi.org/10.1155/2015/793898.

Nakata, Rieko, and Hiroyasu Inoue. 2014. "Resveratrol and Cardiovascular Disease." *Current Nutrition Reports* 3 (3): 163–69. https://doi.org/10.1007/s13668-014-0084-4.

Nasri, Hamid, Azar Baradaran, Hedayatollah Shirzad, and Mahmoud Rafieian Kopaei. 2014. "New Concepts in Nutraceuticals as Alternative for Pharmaceuticals." *International Journal of Preventive Medicine* 5 (12): 1487–99.

Ortega et al. 2016. "Obesity, Diabetes, and Cardiovascular Diseases Compendium." *Obesity and Cardiovascular Disease*, 1752–70. https://doi.org/10.1161/CIRCRESAHA.115.306883.

Patterson, Elain E., Paul M. Ryan, John F. Cryan, Timothy G. Dinan, R. Paul Ross, Gerald F. Fitzgerald, and Cath Erine Stanton. 2016. "Gut Microbiota, Obesity and Diabetes." *Postgraduate Medical Journal* 92 (1087): 286–300. https://doi.org/10.1136/postgradmedj-2015-133285.

Pessi, Tanja, Vesa Karhunen, Pasi P. Karjalainen, Antti Ylitalo, Juhani K. Airaksinen, Matti Niemi, Mikko Pietila, et al. 2013. "Bacterial Signatures in Thrombus Aspirates of Patients with Myocardial Infarction." *Circulation* 127 (11): 1219–28. https://doi.org/10.1161/CIRCULATIONAHA.112.001254.

Pizzo, Giuseppe, Rosario Guiglia, Lucio Lo Russo, and Giuseppina Campisi. 2010. "Dentistry and Internal Medicine: From the Focal Infection Theory to the Periodontal Medicine Concept." *European Journal of Internal Medicine* 21 (6): 496–502. https://doi.org/10.1016/j.ejim.2010.07.011.

Rahmani, Parisa, Alireza Moradzadeh, and Fatemeh Farahmand. 2019. "Of." *Biochemical Pharmacology*, 100164. https://doi.org/10.1016/j.phanu.2019.100164.

Reunanen, Justus, Veera Kainulainen, Laura Huuskonen, Noora Ottman, Clara Belzer, Heikki Huhtinen, Willem M. de Vos, and Reetta Satokaria. 2015. "Akkermansia Muciniphila Adheres to Enterocytes and Strengthens the Integrity of the Epithelial Cell Layer." *Applied and Environmental Microbiology* 81 (11): 3655–62. https://doi.org/10.1128/AEM.04050-14.

Reyes, Leticia, David Herrera, Emil Kozarov, Silvia Roldán, and Ann Progulske-Fox. 2013. "Periodontal Bacterial Invasion and Infection: Contribution to Atherosclerotic Pathology." *Journal of Clinical Periodontology* 40 (SUPPL. 14). https://doi.org/10.1111/jcpe.12079.

Riva, Alessandra, Francesca Borgo, Carlotta Lassandro, Elvira Verduci, Giulia Morace, Elisa Borghi, and David Berry. 2017. "Pediatric Obesity Is Associated with an Altered Gut Microbiota and Discordant Shifts in Firmicutes Populations." *Environmental Microbiology* 19 (1): 95–105. https://doi.org/10.1111/1462-2920.13463.

Rivera, Mercedes F., Ju Youn Lee, Monika Aneja, Vishalkant Goswami, Liying Liu, Irina M. Velsko, Sasanka S. Chukkapalli, et al. 2013. "Polymicrobial Infection with Major Periodontal Pathogens Induced Periodontal Disease and Aortic Atherosclerosis in Hyperlipidemic ApoEnull Mice." *PLoS ONE* 8 (2). https://doi.org/10.1371/journal.pone.0057178.

Robles-Vera, Iñaki, Marta Toral, Miguel Romero, Rosario Jiménez, Manuel Sánchez, Francisco Pérez-Vizcaíno, and Juan Duarte. 2017. "Antihypertensive Effects of Probiotics." *Current Hypertension Reports* 19 (4). https://doi.org/10.1007/s11906-017-0723-4.

Roth, Gregory A., Catherine Johnson, Amanuel Abajobir, Foad Abd-Allah, Semaw Ferede Abera, Gebre Abyu, Muktar Ahmed, et al. 2017. "Global, Regional, and National Burden of Cardiovascular Diseases for 10 Causes, 1990 to 2015." *Journal of the American College of Cardiology* 70 (1): 1–25. https://doi.org/10.1016/j.jacc.2017.04.052.

Saad, M. J. A., A. Santos, and P. O. Prada. 2016. "Linking Gut Microbiota and Inflammation to Obesity and Insulin Resistance." *Physiology* 31 (4): 283–93. https://doi.org/10.1152/physiol.00041.2015.

Sandek, Anja, Juergen Bauditz, Alexander Swidsinski, Sabine Buhner, Jutta Weber-Eibel, Stephan von Haehling, Wieland Schroedl, et al. 2007. "Altered Intestinal Function in Patients With Chronic Heart Failure." *Journal of the American College of Cardiology* 50 (16): 1561–69. https://doi.org/10.1016/j.jacc.2007.07.016.

Sharafedtinov, Khaider K., Oksana A. Plotnikova, Ravilay I. Alexeeva, Tatjana B. Sentsova, Epp Songisepp, Jelena Stsepetova, Imbi Smidt, and Marika Mikelsaar. 2013. "Hypocaloric Diet Supplemented with Probiotic Cheese Improves Body Mass Index and Blood Pressure Indices of Obese Hypertensive Patients - A Randomized Double-Blind Placebo-Controlled Pilot Study." *Nutrition Journal* 12 (1): 1–11. https://doi.org/10.1186/1475-2891-12-138.

Shih, Diana M., Zeneng Wang, Richard Lee, Yonghong Meng, Nam Che, Sarada Charugundla, Hannah Qi, et al. 2015. "Flavin Containing Monooxygenase 3 Exerts Broad Effects on Glucose and Lipid Metabolism and Atherosclerosis." *Journal of Lipid Research* 56 (1): 22–37. https://doi.org/10.1194/jlr.M051680.

Strandwitz, Philip. 2018. "Neurotransmitter Modulation by the Gut Microbiota." *Brain Research* 1693: 128–33. https://doi.org/10.1016/j.brainres.2018.03.015.

Tang, W. H.Wilson, and Stanley L. Hazen. 2017. "Microbiome, Trimethylamine N-Oxide, and Cardiometabolic Disease." *Translational Research* 179: 108–15. https://doi.org/10.1016/j.trsl. 2016.07.007.

Tang, Zhihong, Jing Ma, Zhen Zeng, Siping Wu, and Mengjun Hou. 2014. "Inhibition of Atherosclerotic Plaque Formation in ApoE-Deficient Mice by Dietary Supplementation with *Lactobacillus Casei*."

Functional Foods in Health and Disease 4 (4): 147. https://doi.org/10.31989/ffhd.v4i4.19.

Toral, Marta, Inãki Robles-Vera, Néstor De La Visitación, Miguel Romero, Tao Yang, Manuel Sánchez, Manuel Gómez-Guzmán, Rosario Jiménez, Mohan K. Raizada, and Juan Duarte. 2019. "Critical Role of the Interaction Gut Microbiota-Sympathetic Nervous System in the Regulation of Blood Pressure." *Frontiers in Physiology* 10 (MAR): 1–14. https://doi.org/10.3389/fphys.2019.00231.

Torres-Fuentes, Cristina, Harriët Schellekens, Timothy G. Dinan, and John F. Cryan. 2017. "The Microbiota–Gut–Brain Axis in Obesity." *The Lancet Gastroenterology and Hepatology* 2 (10): 747–56. https://doi.org/10.1016/S2468-1253(17)30147-4.

Vermeiren, Joan, Tom Van De Wiele, Willy Verstraete, Pascal Boeckx, and Nico Boon. 2009. "Nitric Oxide Production by the Human Intestinal Microbiota by Dissimilatory Nitrate Reduction to Ammonium." *Journal of Biomedicine and Biotechnology* 2009. https://doi.org/10.1155/2009/284718.

Xu, Peiru, Min Li, Jihong Zhang, and Tao Zhang. 2012. "Correlation of Intestinal Microbiota with Overweight and Obesity in Kazakh School Children." *BMC Microbiology* 12. https://doi.org/10.1186/1471-2180-12-283.

Yamashita, Tomoya, Kazuyuki Kasahara, Takuo Emoto, Takuya Matsumoto, Taiji Mizoguchi, Naoki Kitano, Naoto Sasaki, and Kenichi Hirata. 2015. "Intestinal Immunity and Gut Microbiota as Therapeutic Targets for Preventing Atherosclerotic Cardiovascular Diseases." *Circulation Journal* 79 (9): 1882–90. https://doi.org/10. 1253/circj.CJ-15-0526.

Yang, Bo, Haiqin Chen, Catherine Stanton, R. Paul Ross, Hao Zhang, Yong Q. Chen, and Wei Chen. 2015. "Review of the Roles of Conjugated Linoleic Acid in Health and Disease." *Journal of Functional Foods* 15: 314–25. https://doi.org/10.1016/j.jff.2015. 03.050.

Yap, Wei Boon, Faisal Malau Ahmad, Yi Cheng Lim, and Satirah Zainalabidin. 2016. "*Lactobacillus Casei* Strain C1 Attenuates Vascular Changes in Spontaneously Hypertensive Rats." *Korean Journal of*

Physiology and Pharmacology 20 (6): 621–28. https://doi.org/10.4196/kjpp.2016.20.6.621.

Yoo, Hye Hyun, In Sook Kim, Dae Hyeong Yoo, and Dong Hyun Kim. 2016. "Effects of Orally Administered Antibiotics on the Bioavailability of Amlodipine: Gut Microbiota-Mediated Drug Interaction." *Journal of Hypertension* 34 (1): 156–62. https://doi.org/ 10.1097/HJH.0000000000000773.

Zand, Hamid, Nava Morshedzadeh, and Farnush Naghashian. 2017. "Signaling Pathways Linking Inflammation to Insulin Resistance." *Diabetes and Metabolic Syndrome: Clinical Research and Reviews* 11: S307–9. https://doi.org/10.1016/j.dsx.2017.03.006.

Zapolska-Downar, Danuta, and M. Naruszewicz. 2009. "Propionate Reduces the Cytokine-Induced VCAM-1 and ICAM-1 Expression by Inhibiting Nuclear Factor-?? B (NF-??B) Activation." *Journal of Physiology and Pharmacology* 60 (2): 123–31.

In: Nutraceuticals: Evidence Based Use ISBN: 978-1-53618-052-7
Editor: Raluca Maria Pop

Chapter 5

THE USE OF NUTRACEUTICALS IN DIABETES

***Anca-Elena Crăciun*[1,*] *and Cristian-Ioan Crăciun*[2]**

[1]Department of Diabetes and Nutrition Diseases, "Iuliu Hatieganu," University of Medicine and Pharmacy, Cluj-Napoca, Romania
[2]Department of Pharmacology, Toxicology and Clinical Pharmacology, "Iuliu Hatieganu" University of Medicine and Pharmacy, Cluj-Napoca, Romania

ABSTRACT

One in eleven people around the globe has diabetes mellitus (DM). In type 1 DM (T1DM), due to the autoimmune destruction of the insulin-producing beta-cells of the pancreas and the severe insulin deficiency, the only form of treatment is exogen insulin-therapy for survival and no nutraceutical substance can replace exogenous administration of insulin. On the other hand, type 2 DM (T2DM) is a heterogeneous disease, ranging from severe insulin-resistance and increased levels of insulinemia to less severe insulin-resistance, but relative insulin-deficiency. In T2DM

* Corresponding Author's E-mail: anca.craciun@umfcluj.ro.

nutraceuticals can decrease insulin resistance and ameliorate hyperglycemic status. In both T1DM and T2DM nutraceuticals can reduce oxidative stress, playing a role in the occurrence of chronic complications. Recent studies bring new data about the role of nutraceuticals in the prevention of DM. In the following chapter, we want to update the information about the potential role of nutraceuticals in the prevention and adjuvant use in the treatment of hyperglycemia and chronic complications of DM. Due to the lake of evidence regarding efficiency and safety, the use of nutraceuticals in gestational diabetes is very limited.

Keywords: diabetes mellitus, nutraceutical, prevention, vitamin, herbals

INTRODUCTION

Diabetes mellitus (DM) is a growing global health problem, with 463 million people living with DM in 2019. It is estimated to increase with 51% by 2045 (International Diabetes Federation 2019). The majority of the cases are represented by T2DM (90-95%), but an increased number of cases is observed in both types.

Diabetes mellitus is a complex metabolic disease associated with hyperglycemia due to insulin resistance and/or insulin deficiency. First was considered to be a disorder of glucose metabolism, but actually, it is associated with impaired metabolism of lipids, subclinical inflammation and increased oxidative stress (Nimesh and Ashwlayan 2018).

In T1DM the primary defect is severe insulin deficiency due to autoimmune destruction of the beta-cells from pancreatic islets. The patients are usually young people with no other cardiovascular risk factors but can have another associated autoimmune disease (e.g., autoimmune thyroiditis). Exogenous insulin-therapy is the only evidence-based therapy accepted for treatment (for survival and prevention of chronic complications). The Diabetes Control and Complications Trial (DCCT) and The Epidemiology of Diabetes Interventions and Complications (EDIC) – the observational follow-up of DCCT - proved that intensive glycemic control can prevent the development of chronic complications and improve the perspectives of a healthy life span in T1DM (Nathan et al. 2014).

The pathophysiology of hyperglycemia in T2DM is more complex and heterogeneous, represented by a combination of insulin resistance and a relative deficit of insulin secretion, often accompanied by visceral obesity, impaired lipid metabolism, increased oxidative stress, subclinical inflammation, and high or very high cardiovascular risk. The UK Prospective Diabetes Study (UKPDS) proved that intensive glycemic control from the diagnosis of DM decreases the risk of chronic microvascular complications, but not the macrovascular ones in the first 10 years of follow-up (UKPDS Group 1998), probably due to multifactorial pathophysiology of the macrovascular disease. The impact of early glycemic control (in the first year after diagnosis) was analyzed in The Diabetes & Aging Study (Laiteerapong et al. 2019). The study demonstrated that achieving an HbA1c less than 6.5% in the first year after the diagnosis of T2DM is associated with significantly reduced risk of microvascular complications and mortality in the first 10 years of follow-up (Laiteerapong et al. 2019).

Globally, the use of nutraceuticals is increasing, becoming part of the daily diet as preventive healthcare measures. In this context is important to know which are the evidence of benefits and possible harms of nutraceuticals used in the context of a very serious metabolic disease such as DM.

Classes of Nutraceuticals Used in Diabetes

Currently, there are more definitions of nutraceuticals, the one provided by European Nutraceutical Association (ENA) is: "nutraceuticals are nutritional products that provide health and medical benefits, including the prevention and treatment of disease" (Santini et al. 2018). The literature search on nutraceuticals and diabetes revealed that the most common nutraceuticals used in this disease are chemical constituents (nutrients, herbals, and phytochemicals), probiotic microorganisms (prebiotics and probiotics) and dietary fibers.

Within this chapter, the most relevant and recent articles with important results regarding their evidence and role in T2D will be further discussed.

CHEMICAL CONSTITUENTS – THE NUTRIENTS

Vitamin C

Vitamin C (ascorbate) is required for several important biological functions, acting as an enzymatic cofactor and reducing agent. Chronic low-grade inflammation and increase oxidative stress has a role in the development of insulin resistance present in T2DM and vitamin C, a known antioxidant agent, can play a benefic role in these cases. In the EPIC-Norfolk cohort of more than 20,000 participants in 12 years of follow-up, there were 735 newly diagnosed T2DM cases. It was found a strong inverse association between plasma vitamin C concentration and the risk for T2DM, with an odds ratio for diabetes of 0.38 in the highest quartile versus the lowest one (95% confidence interval, 0.28-0.52) (Harding et al. 2008). Further research is required to establish if eating more vitamin C-rich foods, like fruits and vegetables, or taking vitamin C supplements, may reduce the risk for prediabetes, diabetes, and its complications.

Vitamin D

In vitro studies proved that pancreatic beta cells of rats express 1α-hydroxylase mRNA, the enzyme which is also present in renal cells, with important role in complete activation of vitamin D. These findings suggest that the local production of the enzyme explains the autocrine link between vitamin D status and pancreatic function (Bland et al. 2004). In observational and prospective studies, in patients at risk for T2DM, a low level of 25-hydroxyvitamin D was independently associated with impaired pancreatic beta-cell function and insulin resistance (Kayaniyil et al. 2010) and higher level of 25-hydroxyvitamin D predicted a better beta-cell function in 3 years of follow-up (Kayaniyil et al. 2011). In a recent published randomized, placebo-controlled trial with 2423 participants at high risk for T2DM, not selected for vitamin D insufficiency, a daily supplement of 4000 IU vitamin D3 did not significantly reduced the occurrence of T2DM (hazard ratio of 0.88; 95% confidence interval, 0.75 to 1.04; $p = 0.12$) (Pittas et al. 2019).

In T1DM, a systematic review of 7 randomized controlled trials, involving 287 newly diagnosed T1DM patients (a period of 4 weeks to 1 year since diagnosis), showed that supplementation of vitamin D in the form of alfacalcidol or cholecalciferol appears to attenuate the natural history of the disease (Gregoriou et al. 2017).

The role of vitamin D supplementation for gestational diabetes prevention failed to prove a major benefit in recently published data of DALI study, a European, multicenter, randomized, placebo-controlled trial, in which were enrolled 154 overweight and obese pregnant women without gestational diabetes at inclusion in the study (Corcoy et al. 2020).

Chromium

In a euglycemic- hyperinsulinemic human study was proved that plasma chromium level decreases after the first bolus of glucose and insulin and increases after insulin action stops (Morris et al. 1993). Further studies showed that the chromium-binding oligopeptide chromodulin plays an important role in the auto-amplification of insulin signaling (Vincent 2000). Up to date chromium emerge as a nutritional supplement, not as a therapeutic agent and only in patients suffering from chromium deficiency (such as prolonged parenteral nutrition cases) can expect to have beneficial effects. A biomimetic of chromodulin may have therapeutic effects and potential therapeutics are underway.

Magnesium

Low plasma levels of magnesium have been associated with several chronic diseases, such as insulin resistance and T2DM, hypertension, or cardiovascular disease (Gröber et al. 2015). Magnesium deficiency appears to be associated with increased risk of diabetic retinopathy and poor glycemic control in T2DM patients (Kumar et al. 2019). In a systematic review and meta-analysis of double-blind randomized controlled trials investigating the effect of magnesium supplementation on glucose metabolism in people with or at risk of diabetes (12 studies with T2DM patients and 6 studies in people with high risk for T2DM) found that magnesium supplementation appears to have a beneficial role. This effect was obtained by the improvement of glucose parameters in T2DM patients and by correcting insulin-sensitivity parameters in those at high risk for T2DM (Veronese et al. 2016). Another systematic review and meta-analysis analyzing the effect of magnesium supplementation on T2DM associated cardiovascular risk factors showed that magnesium supplementation has a beneficial effect on glycemic and lipid profiles and also on the systolic blood pressure values. The authors concluded that magnesium can be very useful in decreasing the risk of cardiovascular disease in T2D, although more evidence supported by large RCTs are still needed (Verma and Garg 2017).

Chemical Constituents – The Herbals

The use of botanicals for treatment of diabetes is described from ancient times in classical Indian, Chinese, Egyptian, Greek, Roman and Syrian texts. Before modern pharmacotherapy, man has used plants to treat DM, more than 700 recipes, containing more than 400 plants, are known for their antidiabetic activity (Saravanamuttu and Sudarsanam 2012). Due to the large number of plant species used along history to treat diabetes, it is impossible to cover extensively the topic in a book chapter. Next, we will present current evidence for some of these herbals and spices, in alphabetical order.

Acacia Arabica (Lam.) Muhl. (family Fabaceae), with the common name babul is known as an efficient agent to treat malaria and has a known hypoglycemic effect. The hypoglycemic effect of the bark decoct has a similar mechanism of action as standard drug tolbutamide (Rajvaidhya et al. 2012). The bark extract increased the levels of insulin in streptozotocin-induced diabetic rats, had beneficial effects on lipid profile (decreasing total cholesterol, LDL-cholesterol and triglycerides and increasing HDL-cholesterol) and increased the blood level of coenzyme Q10. The effect was dose-dependent (Hegazy et al. 2013).

Aegle marmelos (L.) Correa (family Rutaceae) (AM), with the common names Indian Quince, holy fruit, golden apple or bael, is used in Indian medicine to treat diverse symptoms. The bark extract contains umbelliferone β-D-galactopyranoside, a compound with antihyperglycemic, antihyperlipidemic and antioxidant activity demonstrated in streptozotocin-induced diabetic rats (Kumar et al. 2013). Ethanol extract of *AM* leaves has a beneficial effect on metabolic and renal parameters in alloxan-induced early-stage diabetic nephropathy in rats (Bhatti et al. 2013).

Allium cepa L. (family Amaryllidaceae) (AC), with the common name onion, is known to contain vitamins, minerals, and bioactive compounds like sulfur amino acids (such as S-methyl cysteine), flavonoids (mainly quercetin), anthocyanin, phytosterols, and saponins. These compounds proved to have beneficial effects in preventing obesity and related co-morbidities. The anti-diabetic potential was observed in rats and mice when bulbs juice, 80% ethanol extract of peeled onions and bulbs infusion were used (Marrelli et al. 2018). The antihyperglycemic effect of onion extracts was explained by the regulation of the liver hexokinase and glucose 6-phosphatase activity while the hypolipidemic effect by the normalization of HMG coenzyme-A reductase activity (Akash et al 2014). In a study conducted to investigate the clinical effects of *AC* on glycemic status in type 1 and type 2 diabetic patients, the ingestion of 100g crude red onion produced a significant reduction in fasting and postprandial glycemia in both types of DM (Taj Eldin et al. 2010).

Allium sativum L. (family Amaryllidaceae), with the common name garlic, is one of the best researched herbal products. The active compounds contained in garlic include enzymes (e.g., alliinase), sulfur-containing compounds (e.g., alliin) and compounds produced enzymatically from alliin (e.g., allicin) (Majewski 2014). In T2DM patients the adjunct use of garlic extract to metformin improved glycemic control in addition to antihyperlipidemic effect (Ashraf et al. 2011). In addition to promoting antihyperglycemic and hypotriglyceridemic effects, the potential beneficial effects of the garlic can be supported by total antioxidant levels, catalase activity, antiglycation and anti-lipid-peroxidation actions (Jung et al. 2011; Ota and Ulrih 2017).

Aloe vera (family Aloaceae), with the common name aloe, is used as an antidiabetic remedy in the Indian and Arabic peninsula (Cefalu et al. 2008). The gel obtained from the inner portion of the leaves contain a water-soluble fiber named glucomannan. As other viscous fibers, glucomannan improves glycemic control and lipid profile, suggesting a therapeutic potential in the treatment of the diseases accompanied by insulin resistance, such as T2DM (Vuksan et al. 2000). The leaf latex extract of *Aloe megalacantha Baker* (family Aloaceae) showed also significant antihyperglycemic and antihyperlipidemic potential in streptozotocin-induced diabetic model (Hammeso et al. 2019).

Cinnamomum cassia (family Lauraceae), with the common name Chinese cassia or Chinese cinnamon, is part of a genus of 250 species. The spice used in human nutrition typically contains a combination of different forms of cinnamon in the bark or powdered form, with C. *cassia*, usually predominating (Costello et al. 2016). In streptozotocin-induced diabetic rats, a combination of *Nigella sativa* and *Cinnamomum cassia* extracts significantly reduced blood glucose and lipid parameters, compared to the diabetic control group (Kaur et al. 2018). The evidence provided to date by RCTs in humans does not suggest that the addition of a cinnamon supplement can produce a clinically significant drop in fasting plasma glucose or HbA1c in patients with T2DM (Costello et al. 2016).

Coccinia indica Wight & Arn. (family Cucurbitaceae) (CI), with the common name ivy gourd, has been used in Ayurvedic medicine to treat glycosuria. The most used extract for antidiabetic use is the leaves extract, which may mimic the action of insulin and may suppress the activity of some enzymes involved in glucose production (Cefalu et al. 2008). In experimentally induced diabetic rats the diet supplementation with *CI* fruits and leaves had a renoprotective effect (Gurukar et al. 2013). In a placebo-controlled trial that involved 60 newly-diagnosed T2DM patients, the use of 1 g alcoholic extract of the *CI* had a significant decrease in the fasting, postprandial blood glucose and HbA1c levels (Kuriyan et al. 2008).

Curcuma longa L. (family Zingiberaceae), with the common name turmeric, is usually used as a spice in food preparation. Curcumin, the most representative active compound of this plant is found in the plant rhizome. Because curcumin has a low bioavailability, various formulations with enhanced bioavailability are currently being developed (Tsuda 2018). In a recently published review, Pivari and his collab. stated that:

> "curcumin has antioxidant, cardio-protective, anti-inflammatory, anti-microbial, nephro-protective, anti-neoplastic, hepato-protective, immunomodulatory, hypoglycaemic and anti-rheumatic effects. In animal models, curcumin extract delays diabetes development, improves β-cell functions, prevents β-cell death, and decreases insulin resistance." (Pivari et al. 2019)

Curcumin might also have beneficial effects on the eye diseases of DM patients, such as diabetic keratopathy (by promoting corneal wound healing), diabetic retinopathy or corneal neovascularization (as it has antiangiogenic action on the choroidal vasculature) (Dulbecco and Savarino 2013).

Gymnema sylvestre (Retz.) R. Br. (family Apocynaceae), with the common name gurmar (means "sugar destroyer"), is a leaf extract. The leaves belong to a native African and Indian tree. These leaves are rich in active compounds such as glycosides, gurmarin peptides, resins, gymnemic acids, saponins, stigmasterol, and quercitol (Cefalu et al. 2008; Marles and Farnsworth 1995; Saravanamuttu and Sudarsanam 2012). Gymnemagenin was found to dock well with crystallographic structures with 7 of the 15 proteins having a key role in carbohydrate metabolism (Rathore et al. 2016).

Momordica charantia L. (family Cucurbitaceae) (MC), with the common name bitter gourd or bitter melon, is one of the most used nutraceuticals in diabetes, especially in Asia, Africa, South American and Caribbean regions. *Momordica charantia* contains more than 200 bioactive compounds and some of them have a hypoglycemic effect. One of them is polypeptide p, a protein with a similar structure to animal insulin. Recently was proved that a protein from *Momordica Charantia* seeds, named inhibitor against trypsin, binds the insulin receptors and triggers the insulin signaling pathway (Lo et al. 2013). The *in vitro* and *in vivo* studies in animals and humans proved that MC has been shown to have an antihyperglycemic effect, but the results from clinical trials in patients with DM or prediabetes are inconsistent. Not all participants responded to the intervention, but in responders, a better antihyperglycemic effect was observed in those with higher initial blood glucose levels (Krawinkel et al. 2018). Also, in a pilot study in T2DM patients was observed that the use of dried-fruit pulp of MC for 16 weeks significantly decreased the levels of circulating total advanced glycation end products (AGEs) (Trakoon-osot et al. 2013). However, human studies included small numbers of participants, with various MC extraction techniques and with no extract standardization (Cefalu et al. 2008). Also, a series of severe adverse reaction were reported. Administration of MC extracts in various animal studies identified reduced fertility in mice and alkaline phosphatase and gamma-glutamyl-transferase increased levels. Regarding human studies, convulsions and hypoglycemic coma were reported in children (Basch et al. 2003). The use of MC extract should be done with caution in patients with DM taking pharmacologic treatment with the risk of hypoglycemia (such as insulin or sulfonylurea) to prevent severe hypoglycemic events.

Trigonella foenum graecum (family Fabaceae), with the common name fenugreek, is used in traditional medicine from ancient times in China, India, and Nepal. The active ingredients of the plant are represented by saponins (diosgenin), alkaloids, coumarins, and 4-hydroxyisoleucine and the soluble dietary fibers (Ota and Ulrih 2017). Diosgenin has potential beneficial effects on diabetes by reducing fasting plasma glucose in streptozotocin-induced diabetic rats by increasing insulin secretion and beta-cell regeneration. It also has positive effects on lipid panel (Kalailingam et al. 2014). The soluble dietary fiber fraction from the seeds had an antihyperglycemic effect in T1DM and T2DM patients by delaying carbohydrate digestion and absorption and by enhancing the peripheral insulin action (Hannan et al. 2007). In another study, the adjunct use of fenugreek seeds in T2DM patients improved glycemic control, decreased insulin resistance and hypertriglyceridemia (Gupta et al. 2001).

Wolffia globosa (Mankai strain) (family Araceae) (WG), with the common name Asian watermeal, duckweed, is an edible green aquatic plant, with important protein content, up to 45% of the dry matter and includes all 9 essential and 6 conditional amino acids. In a recently published randomized cross-over controlled trial, the dinner was composed of a shake of WG duckweed or yogurt, and the results showed that the plant may serve as an emerging alternative high-quality protein with potential beneficial postprandial glycemic effects (Zelicha et al. 2019).

Other nutraceuticals with *in vitro* and *in vivo* evidence of antidiabetic action are: *Abelmoschus esculentus* (family Malvaceae), common name okra; *Annona squamosa L.* (family Annonaceae), common name sugar apple; *Azadirachta indica A. Juss.* (family Meliaceae), common name neem; *Andrographis paniculata Nees* (family Acanthaceae), common name king of bitter; *Artemisia pallens Wall. Ex DC.* (family Compositae), common name davana; *Asparagus racemosus* (family Lileaceae); *Biophytum sensitivum (L.) DC.* (family Oxalidaceae), common name life plant; *Beta vulgaris L.* (family Chenopodiaceae), common name garden beet; *Brassica juncea (L.) Czern.* (family Brassicaceae) common name brown mustard; *Cassia auriculata L.* (family Leguminosae), common name Tanner's Cassia; *Boerhavia diffusa L.* (family Nyctaginaceae), common name Tar vine; *Carica papaya* (family Caricaceae), common name papaya; *Citrullus colocynthis (L.)* (family Cucurbitaceae), common name bitter apple; *Cajanus cajan (L.) Millsp.* (family Fabaceae), common name pigeon pea; *Casearia esculenta Roxb.* (family Flacourtiaceae), common name carilla; *Catharanthus roseus (L.) G. Don* (family Apocynaceae), common name Madagascar periwinkle; *Camellia sinensis Kuntze* (family Theaceae), common name green tea; *Enicostemma littorale Blume* (family Gentiaceae), common name nahi; *Eugenia jambolana Lam* (family Myrtaceae), common name indian black berry; *Helicteres isora L.* (family Sterculiaceae), common name screw tree; *Ipomoea batatas* (L.) Lam. (family Convolvulaceae), common name sweet potato; *Morus alba L.* (family Moraceae), common name white mulberry; *Murraya koenigii (L.) Spreng.* (family Rutaceae), common name curry-leaf tree; *Nigella sativa L.* (family Ranunculaceae), common name black cumin; *Ocimum sanctum L.* (family Lamiaceae), common name holy basil; *Plantago ovata* (family Plantaginaceae), common name psyllium; *Punica granatum L. (family Punicaceae),* common name pomegranate; *Scoparia dulcis L.* (family Scrophulariaceae), common name sweet broomweed (Cefalu et al. 2008; Saravanamuttu and Sudarsanam 2012; Ota and Ulrih 2017).

Chemical Constituents – The Phytochemicals

Resveratrol

Resveratrol is a non-flavonoid polyphenol found in various foods, the richest source being the shell and stem of *Vitis vinifera L.* (Vitaceae), and consecutively red wine. In T2DM patients, resveratrol might have beneficial exercise-mimetic effects, decreases insulin resistance, decreases oxidative stress, but has no effect on parameters that are related to β-cell function (Öztürk et al. 2017). Red wine, a rich source of resveratrol in combination with other flavonoids and alcohol, might have a beneficial effect in DM by multiple effects: inhibition of α-glucosidase, inhibition of glucose absorption in the intestine by sodium-dependent glucose transporter-1, improve insulin sensitivity, lipid profiles and endothelial function, but further intervention studies are needed to clarify the conflicting findings (Fernandes et al. 2017). However, moderate alcohol intake should not be promoted as a means to protect against cardiovascular disease in diabetes or prediabetes (Cosentino et al. 2020).

Probiotic Microorganisms

The most used probiotics in studies regarding their effect on T2DM belonged to the genera *Lactobacillus* and *Bifidobacterium*. A critical review analyzing the clinical application of probiotics in T2DM, involving seven randomized placebo-controlled clinical trials and thirteen experimental studies showed that the use of probiotics in glycemic control presented conflicting results and further research is needed to assess the effects of probiotics in T2DM (Bordalo Tonucci et al. 2017).

Dietary Fibers (soluble and insoluble)

In the last years, gut microbiota emerges as an important contributor to the pathogenesis of T2DM. Gut microbiota in human adults with T2DM differs from non-diabetic adults, by a reduction of phylum *Firmicutes* and class *Clostridia* and an increased number of *Betaproteobacteria* (Larsen et al. 2010). In a case-control study gut microbiota differed in composition and functionality between children with type 1 diabetes and maturity-onset diabetes of the young 2 and healthy control subjects. They had a significantly higher relative abundance of *Bacteroides* and a lower relative abundance of *Bifidobacterium*, and a significantly higher *Prevotella* abundance and a lower *Ruminococcus and Bacteroides* abundance, respectively (Leiva-Gea et al. 2018). Dietary fibers (DF), as indigestible part of the diet, have positive effect on gut microbiome, promote the production of short-chain fatty acids (such as acetate, propionate, and butyrate) as potentially active metabolites, with improved hemoglobin A1c levels by increasing the secretion of glucagon-like peptide -1 (Gowd et al. 2019).

European Food Safety Authority defined DF ”as non-digestible carbohydrates plus lignin, including non-starch polysaccharides (NSP) – cellulose, hemicelluloses, pectins, hydrocolloids (i.e., gums, mucilages, -glucans), resistant oligosaccharides – fructooligosaccharides (FOS), galactooligosaccharides (GOS), other resistant oligosaccharides, resistant starch – consisting of physically enclosed starch, some types of raw starch granules, retrograded amylose, chemically and/or physically modified starches, and lignin associated with the dietary fiber polysaccharides" (EFSA Panel 2010).

Diet, especially its content of soluble DF (sDF), has a strong impact on gut microbiota. sDF are with high viscosity and fermentable in the colon, having a good influence on the glycemic index, improving insulin sensitivity and decreasing inflammation and the development of non-alcoholic fatty liver disease (Gowd et al. 2019). Insoluble DF (iDF) are less fermentable in the colon and are very good in constipation treatment. Studies have found that iDF improve insulin sensitivity and have strong long-term antidiabetic activity, but less effect on postprandial hyperglycemia (Gowd et al. 2019).

Conclusion

Nutraceuticals have been used as medicinal agents since ancient times and many botanical extracts are now available as commercial supplements. Although there is a lot of historical evidence of using popular botanicals in treating hyperglycemia, there is a paucity of strong clinical data. There are insufficient evidence-based results to actively recommend the use of nutraceuticals to treat either high blood glucose or other related risk factors. Thus, well designed randomized clinical trials are warranted in this area.

References

Akash, M. S., Rehman, K. and Chen, S. (2014). "Spice Plant Allium Cepa: Dietary Supplement for Treatment of Type 2 Diabetes Mellitus." *Nutrition,* 30 (10): 1128–37.

Ashraf, R., Khan, R. A. and Ashraf, I. (2011). "Garlic (Allium Sativum) Supplementation with Standard Antidiabetic Agent Provides Better Diabetic Control in Type 2 Diabetes Patients." *Pakistan Journal of Pharmaceutical Sciences,* 24 (4): 565–70.

Basch, E., Gabardi, S. and Ulbricht, C. (2003). "Bitter Melon (Momordica Charantia): A Review of Efficacy and Safety." *American Journal of Health-System Pharmacy,* 60 (4): 356–9.

Bhatti, R., Sharma, S., Singh, J., Singh, A. and Ishar. M. P. (2013). "Effect of Aegle Marmelos (L.) Correa on Alloxan Induced Early Stage Diabetic Nephropathy in Rats." *Indian Journal of Experimental Biology,* 51 (6): 464–9.

Bland, R., Markovic, D., Hills, C. E., Hughes, S. V., Chan, S. L., Squires, P. E. and Hewison, M. (2004). "Expression of 25-hydroxyvitamin D3-1alpha-hydroxylase in Pancreatic Islets." *Journal of Steroid Biochemistry and Molecular Biology,* 89–90 (1-5): 121–5.

Bordalo Tonucci, L., Dos Santos, K. M., De Luces Fortes Ferreira, C. L., Ribeiro, S. M., De Oliveira, L. L. and Martino, H. S. (2017). "Gut Microbiota and Probiotics: Focus on Diabetes Mellitus." *Critical Reviews in Food Science and Nutrition,* 57 (11): 2296–2309.

Cefalu, W. T., Ye, J. and Wang, Z. Q. (2008). "Efficacy of Dietary Supplementation with Botanicals on Carbohydrate Metabolism in Humans." *Endocrine, Metabolic & Immune Disorders-Drug Targets,* 8(2): 78–81.

Corcoy, R., Mendoza, L. C., Simmons, D., Desoye, G., Adelantado, J. M., Chico, A., et al. (2020). "The DALI Vitamin D Randomized Controlled Trial for Gestational Diabetes Mellitus Prevention: No Major Benefit Shown besides Vitamin D Sufficiency." *Clinical Nutrition,* 39 (3): 976-84.

Cosentino, F., Grant, P. J., Aboyans, V., Bailey, C. J., Ceriello, A., Delgado, V., et al. (2020). "2019 ESC Guidelines on Diabetes, Pre-Diabetes, and Cardiovascular Diseases Developed in Collaboration with the EASD." *European Heart Journal*, 41 (2): 255-323.

Costello, R. B., Dwyer, J. T., Saldanha, L., Bailey, R. L., Merkel, J., Wambogo, E. (2016). "Do Cinnamon Supplements Have a Role in Glycemic Control in Type 2 Diabetes? A Narrative Review." *Journal of the Academy Nutrition and Dietetics,* 116 (11): 1794–1802.

Dulbecco, P. and Savarino, V. (2013). "Therapeutic Potential of Curcumin in Digestive Diseases." *World Journal of Gastroenterology,* 19 (48): 9256–70.

EFSA Panel on Dietetic Products, Nutrition, and Allergies (NDA). (2010). "Scientific Opinion on Dietary Reference Values for Carbohydrates and Dietary Fibre." *EFSA Journal,* 8 (3): 1462.

Fernandes, I., Pérez-Gregorio, R., Soares, S., Mateus, N., De Freitas, V. (2017). "Wine Flavonoids in Health and Disease Prevention." *Molecules,* 22(2). pii: E292.

Gowd, V., Xie, L., Zheng, X. and Chen, W. (2019). "Dietary Fibers as Emerging Nutritional Factors against Diabetes: Focus on the Involvement of Gut Microbiota." *Critical Reviews in Biotechnology*, 39(4): 524–40.

Gregoriou, E., Mamais, I., Tzanetakou, I., Lavranos, G. and Chrysostomou, S. (2017). "The Effects of Vitamin D Supplementation in Newly Diagnosed Type 1 Diabetes Patients: Systematic Review of Randomized Controlled Trials." *Review of Diabetic Studies,* 14 (2–3): 260–68.

Gröber, U., Schmidt, J. and Kisters, K. (2015). "Magnesium in Prevention and Therapy." *Nutrients,* 7 (9): 8199–226.

Gupta, A., Gupta, R. and Lal, B. (2001). "Effect of Trigonella Foenum-Graecum (Fenugreek) Seeds on Glycaemic Control and Insulin Resistance in Type 2 Diabetes Mellitus : A Double Blind Placebo Controlled Study." *Journal of Association of Physicians of India* 49: 1057–61.

Gurukar, M. S., Mahadevamma, S. and Chilkunda, N. D. (2013). "Renoprotective Effect of Coccinia Indica Fruits and Leaves in Experimentally Induced Diabetic Rats." *Journal of Medicinal Food,* 16(9): 839–46.

Hammeso, W. W., Emiru, Y. K., Ayalew Getahun K. and Kahaliw, W. (2019). "Antidiabetic and Antihyperlipidemic Activities of the Leaf Latex Extract of Aloe Megalacantha Baker (Aloaceae) in Streptozotocin-Induced Diabetic Model." *Evidence-Based Complementary and Alternative Medicine*, 2019:8263786.

Hannan, J. M., Ali, L., Rokeya, B., Khaleque, J., Akhter, M., Flatt, P. R. and Abdel-Wahab, Y. H. (2007). "Soluble Dietary Fibre Fraction of Trigonella Foenum-Graecum (Fenugreek) Seed Improves Glucose Homeostasis in Animal Models of Type 1 and Type 2 Diabetes by Delaying Carbohydrate Digestion and Absorption, and Enhancing Insulin Action." *British Journal of Nutrition* 97 (3): 514–21.

Harding, A. H., Wareham, N. J., Bingham, S.A., Khaw, K., Luben, R., Welch, A., Forouhi, N. G. (2008). "Plasma Vitamin C Level, Fruit and Vegetable Consumption, and the Risk of New-Onset Type 2 Diabetes MellitusThe European Prospective Investigation of Cancer – Norfolk Prospective Study." *Archives of Internal Medicine,* 168 (14): 1493–9.

Hegazy, G. A., Alnoury, A. M. and Gad, H. G. (2013). "The Role of Acacia Arabica Extract as an Antidiabetic, Antihyperlipidemic, and Antioxidant in Streptozotocin- Induced Diabetic Rats." *Saudi Medical Journal.* 34 (7): 727–33.

International Diabetes Federation (IDF). (2019). *Ninth Edition 2019. IDF Diabetes Atlas, 9th Edition*. https://diabetesatlas.org/en/resources/

Jung, Y. M., Lee, S. H., Lee, D. S., You, M. J., Chung, I. K. and Cheon, W. H. (2011). "Fermented Garlic Protects Diabetic, Obese Mice When Fed a High-Fat Diet by Antioxidant Effects." *Nutrition Research,* 31 (5): 387–96.

Kalailingam, P., Kannaian, B., Tamilmani, E. and Kaliaperumal, R. (2014). "Efficacy of Natural Diosgenin on Cardiovascular Risk, Insulin Secretion, and Beta Cells in Streptozotocin (STZ)-Induced Diabetic Rats." *Phytomedicine* 21 (10): 1154–61.

Kaur, G., Invally, M., Khan, M. K. and Jadhav, P. (2018). "A Nutraceutical Combination of Cinnamomum Cassia & Nigella Sativa for Type 1 Diabetes Mellitus." *Journal of Ayurveda and Integrative Medicine* 9 (1): 27–37.

Kayaniyil, S., Retnakaran, R., Harris, S. B., Vieth, R., Knight, J. A., Gerstein, H. C., et al. (2011). "Prospective Associations of Vitamin D with β-Cell Function and Glycemia: The PROspective Metabolism and ISlet Cell Evaluation (PROMISE) Cohort Study." *Diabetes,* 60 (11): 2947–53.

Kayaniyil, S., Vieth, R., Retnakaran, R., Knight, J. A., Qi, Y., Gerstein, H. C., et al. (2010). "Association of Vitamin D with Insulin Resistance and β-Cell Dysfunction in Subjects at Risk for Type 2 Diabetes." *Diabetes Care,* 33 (6): 1379–81.

Krawinkel, M. B., Ludwig, C., Swai, M. E., Yang, R. Y., Chun, K. P. and Habicht, S. D. (2018). "Bitter Gourd Reduces Elevated Fasting Plasma Glucose Levels in an Intervention Study among Prediabetics in Tanzania." *Journal of Ethnopharmacology,* 216: 1–7.

Kumar, P., Bhargava, S., Agarwal, P. K., Garg, A., Khosla, A. (2019). "Association of Serum Magnesium with Type 2 Diabetes Mellitus and Diabetic Retinopathy." *Journal of Family Medicine and Primary Care* 8(5):1671-7.

Kumar, V., Ahmed, D., Verma, A., Anwar, F., Ali, M., Mujeeb, M. (2013). "Umbelliferone β-D-Galactopyranoside from Aegle Marmelos (L.) Corr. An Ethnomedicinal Plant with Antidiabetic, Antihyperlipidemic and Antioxidative Activity." *BMC Complementary and Alternative Medicine,* 13:273.

Kuriyan, R., Rajendran, R., Bantwal, G., Kurpad, A. V. (2008). "Effect of Supplementation of Coccinia Cordifolia Extract on Newly Detected Diabetic Patients." *Diabetes Care,* 31 (2): 216–20.

Laiteerapong, N., Ham, S. A., Gao, Y., Moffet, H. H., Liu, J. Y., Huang, E.S., Karter, A. J. (2019). "The Legacy Effect in Type 2 Diabetes:Impactofearlyglycemic Control on Future Complications (the Diabetes & Aging Study)." *Diabetes Care* 42 (3): 416–26.

Larsen, N., Vogensen, F. K., van den Berg, F. W., Nielsen, D.S., Andreasen, A. S., Pedersen, B. K., et al. 2010. "Gut Microbiota in Human Adults with Type 2 Diabetes Differs from Non-Diabetic Adults." *PLoS One,* 5(2):e9085.

Leiva-Gea, I., Sánchez-Alcoholado, L., Martín-Tejedor, B., Castellano-Castillo, D., Moreno-Indias, I., Urda-Cardona, A., et al. (2018). "Gut Microbiota Differs in Composition and Functionality between Children with Type 1 Diabetes and MODY2 and Healthy Control Subjects: A Case-Control Study." *Diabetes Care,* 41 (11): 2385–95.

Lo, H. Y., Ho, T. Y., Lin, C., Li, C. C., Hsiang, C. Y. (2013). "Momordica Charantia and Its Novel Polypeptide Regulate Glucose Homeostasis in Mice via Binding to Insulin Receptor." *Journal of Agricultural and Food Chemistry,* 61 (10): 2461–8.

Majewski, M. (2014). "Allium Sativum: Facts And Myths Regarding Human Health". *Roczniki Państwowego Zakładu Higieny*, 65 (1): 1–8.

Marles, R. J., and Farnsworth, N. R. (1995). "Antidiabetic Plants and Their Active Constituents." *Phytomedicine* 2 (2): 137–89.

Marrelli, M., Amodeo, V., Statti, G., Conforti, F. (2018). "Biological Properties and Bioactive Components of Allium Cepa L.: Focus on Potential Benefits in the Treatment of Obesity and Related Comorbidities." *Molecules,* 24(1). pii: E119.

Morris, B. W., MacNeil, S., Stanley, K., Gray, T. A., Fraser, R. (1993). "The Inter-Relationship between Insulin and Chromium in Hyperinsulinaemic Euglycaemic Clamps in Healthy Volunteers." *Journal of Endocrinology,* 139 (2): 339–45.

Nathan, D. M., DCCT/EDIC Research Group. (2014). "The Diabetes Control and Complications Trial/Epidemiology of Diabetes Interventions and Complications Study at 30 Years: Overview." *Diabetes Care,* 37 (1): 9–16.

Nimesh, S. and Ashwlayan, V. D. (2018). "Nutraceuticals in the Management of Diabetes Mellitus." *Pharmacy & Pharmacology International Journal*, 6 (2): 114–20.

Ota, A. and Ulrih, N. P. (2017). "An Overview of Herbal Products and Secondary Metabolites Used for Management of Type Two Diabetes." *Frontiers in Pharmacology,* 8:436.

Öztürk, E., Arslan, A. K. K., Yerer, M. B., Bishayee, A. (2017). "Resveratrol and Diabetes: A Critical Review of Clinical Studies." *Biomedicine and Pharmacotherapy*, 95:230-4.

Pittas, A. G., Dawson-Hughes, B., Sheehan, P., Ware, J. H., Knowler, W. C., Aroda, V. R., et al. (2019). "Vitamin D Supplementation and Prevention of Type 2 Diabetes." *New England Journal of Medicine*, 381(6): 520-30.

Pivari, F., Mingione, A., Brasacchio, C., Soldati, L. (2019). "Curcumin and Type 2 Diabetes Mellitus: Prevention and Treatment." *Nutrients*. 11 (8): pii: E1837.

Rajvaidhya, S., Nagori, B. P., Singh, G. K., Dubey, B. K., Desai, P., Alok, S., et al. (2012). "A Review On Acacia Arabica - An Indian Medicinal Plant." *International Journal of Pharmaceutical Sciences and Research*, 3 (7): 1995–2005.

Rathore, P. K., Arathy, V., Attimarad, V. S., Kumar, P., Roy, S. (2016). "In-Silico Analysis of Gymnemagenin from Gymnema Sylvestre (Retz.) R.Br. with Targets Related to Diabetes." *Journal of Theoretical Biology* 391: 95–101.

Santini, A., Cammarata, S. M., Capone, G., Ianaro, A., Tenore, G. C., Pani, L., Novellino, E. (2018). "Nutraceuticals: Opening the Debate for a Regulatory Framework." *British Journal of Clinical Pharmacology,* 84 (4): 659–72.

Saravanamuttu, S. and Sudarsanam, D. (2012). "Antidiabetic Plants and Their Active Ingredients: a Review." *International Journal of Pharmaceutical Sciences and Research,* 3 (10): 3639–50.

Taj Eldin, I. M., Ahmed, E. M. and Elwahab H. M. A. (2010). "Preliminary Study of the Clinical Hypoglycemic Effects of Allium Cepa (Red Onion) in Type 1 and Type 2 Diabetic Patients." *Environmental Health Insights,* 4: 71-7.

Trakoon-osot, W., Sotanaphun, U., Phanachet, P., Porasuphatana, S., Udomsubpayakul, U. and Komindr, S. (2013). "Pilot Study: Hypoglycemic and Antiglycation Activities of Bitter Melon (Momordica Charantia L.) in Type 2 Diabetic Patients." *Journal of Pharmacy Research,* 6 (8): 859–64.

Tsuda, T. (2018). "Curcumin as a Functional Food-Derived Factor: Degradation Products, Metabolites, Bioactivity, and Future Perspectives." *Food and Function,* 9 (2): 705–14.

UK Prospective Diabetes Study (UKPDS) Group. (1998). "Intensive Blood-Glucose Control with Sulphonylureas or Insulin Compared with Conventional Treatment and Risk of Complications in Patients with Type 2 Diabetes (UKPDS 33)." *Lancet,* 352 (9131): 837–53.

Verma, H. and Garg, R. (2017). "Effect of Magnesium Supplementation on Type 2 Diabetes Associated Cardiovascular Risk Factors: A Systematic Review and Meta-Analysis." *Journal of Human Nutrition and Dietetics* 30 (5): 621–33.

Veronese, N., Watutantrige-Fernando, S., Luchini, C., Solmi, M., Sartore, G., Sergi, G., et al. (2016). "Effect of Magnesium Supplementation on Glucose Metabolism in People with or at Risk of Diabetes: A Systematic Review and Meta-Analysis of Double-Blind Randomized Controlled Trials." *European Journal of Clinical Nutrition,* 70(12):1463.

Vincent, J. B. (2000). "Elucidating a Biological Role for Chromium at a Molecular Level." *Accounts of Chemical Research* 33 (7): 503–10.

Vuksan, V., Sievenpiper, J. L,, Owen, R., Swilley, J. A., Spadafora, P., Jenkins, D. J., et al. (2000). "Beneficial Effects of Viscous Dietary Fiber From Konjac-Mannan in Subjects Results of a Controlled Metabolic Trial." *Diabetes Care,* 23 (1): 9–14.

Zelicha, H., Kaplan, A., Yaskolka Meir, A., Tsaban, G., Rinott, E., Shelef, I., et al. (2019). "The Effect of Wolffia Globosa Mankai, a Green Aquatic Plant, on Postprandial Glycemic Response: A Randomized Crossover Controlled Trial." *Diabetes Care,* 42 (7): 1162–69.

In: Nutraceuticals: Evidence Based Use ISBN: 978-1-53618-052-7
Editor: Raluca Maria Pop

Chapter 6

ANTIOXIDANTS IN ASTHMA: FRIENDS OR FOES

Ioana Corina Bocșan[1,*], Gavriela Feketea[2], Adriana Muntean[3], Claudia Militaru[1] and Raluca Maria Pop[1]

[1]Department of Pharmacology, Toxicology and Clinical Pharmacology, "Iuliu Hațieganu" University of Medicine and Pharmacy, Cluj Napoca, Romania

[2]Pediatric Allergy Outpatient Clinic, Department of Pediatrics, Karamandaneio Children's Hospital, Patras, Greece

[3]Department of Allergology and Clinical Immunology, "Iuliu Hațieganu" Romania University of Medicine and Pharmacy, Cluj Napoca, Romania

---

[*] Corresponding Author's E-mail:corinabocsan@yahoo.com.

ABSTRACT

Asthma and allergic rhinitis are chronic respiratory disorders associated with airway inflammation. Environmental factors, including diet in association with genetic susceptibility, play a critical role in their pathophysiology. Systemic inflammation can be a feature of asthma and allergic rhinitis and is associated with worse disease outcomes.

Several components found in foods may have antioxidant, anti-allergic and anti-inflammatory properties, which can improve the asthma control or could prevent its occurrence. It has been proposed that oxidative stress which is present in association with chronic inflammation in asthma can be reverted by supplemental or dietary intake of antioxidant vitamins (A, C, D, and E) or selenium. A few small clinical trials suggest that antioxidants might improve asthma control or lung function in asthmatic children or adults.

Vitamin C acts on airways through a prostaglandinic effect and may have a protective effect on lung function. Regarding vitamin E data are contradictory. While some studies have proven that α-tocopherol has beneficial effects on lung function, wheezing, and adult-onset asthma in some populations, in other studies dietary supplementation with α-tocopherol did not affect lung function and asthma control. Beta-carotene may improve lung function and lower asthma prevalence.

In this chapter, we focused on a few dietary supplements with antioxidant effects that may play a significant role in respiratory allergic diseases prevention and control.

Keywords: allergic inflammation, vitamins, antioxidants

INTRODUCTION

Allergy is a complex systemic immune-mediated disease. Under the umbrella of allergy, different clinical manifestations are included: allergic rhinitis, asthma, urticaria, atopic dermatitis. The common characteristic is actually the pathophysiology of allergy, an IgE mediated process that can affect several organs and tissues.

Respiratory allergies, like asthma and allergic rhinitis, are commonly observed now in the general population. Allergic rhinitis is the most common IgE mediated disease (Khan, 2014), and represents a risk factor for asthma development. The evolution from allergic rhinitis to asthma was reported in several studies, and it is a part of "atopic march" (Bousquet et al., 2003; Leynaert et al., 2004; Spergel, 2010).

Asthma is one of the top chronic diseases, affecting 334 million people globally. Asthma affects people of all ages and has a substantial impact on the community (Pathinayake et al., 2018).

In asthma development, both genetic and environmental factors are involved. The prevalence of allergic diseases is increasing continuously, especially in western countries and in younger persons (children and adolescents) (Moreno-Macias and Romieu, 2014; Patelarou et al., 2011). Even if human genomic sequencing permitted to understand better the genetic risk factors for allergic asthma occurrence, there are a lot of unknown environmental factors that may contribute to the appearance and progression of respiratory allergies. The rapid increase of asthma prevalence in the last decades could not be attributable only to genetic factors, but could be a consequence of external, environmental factors, associated maybe to epigenetic changes (Burbank et al., 2017).

Several environmental factors could interfere with immune system maturation, generating different dysfunctions in immune processes that may finally lead to allergic diseases (Holt and Sly, 2007). If modifiable risk factors could be identified, the asthma paradigm could be changed, from the chronic standard treatment to a preventable attitude, to reduce the occurrence of allergic diseases in childhood. In this way, we can switch from pharmacological therapy to primary prevention (Polk and Bacharier, 2019).

Diet is considered an important external factor that may influence the occurrence and progression of chronic diseases, including allergic bronchial asthma. Published data suggested that a Mediterranean diet, enriched in fruits and vegetables and poor in saturated fat protects against asthma and recurrent wheezing occurrence, while a western diet increases their risk (Han et al., 2013, 2015). Western diet augments systemic inflammation and reduces the level of antioxidants respectively. These changes are associated with asthma occurrence, poor asthma control, reduction of lung function and increases of the plasmatic level of proinflammatory cytokines, like IL-6 in adults (Thorburn et al., 2014; Wood et al., 2015). It seems that low consumption of antioxidants may predispose to asthma or other allergic features.

The epidemiological studies are extremely variable and report conflicting results. While some studies revealed that a low intake of antioxidants is associated with asthma diagnoses (recurrent wheezing, bronchial hyperreactivity, reduced pulmonary function) (Thorburn et al., 2014), others reported no association (Fogarty et al., 2003), or on the contrary, some studies reported adverse effects of antioxidants supplementation (Neurauter G et, 2004).

Diet in the first years of life is important in preventing asthma occurrence, but the diet during the antenatal period is also important. Data from literature revealed the role of mothers' diet in the modulation of child susceptibility to develop allergic diseases (Han et al., 2013; Thorburn et al., 2014). Therefore supplementation of diet and changes in lifestyle may represent a cheap, efficient and accessible method for primary prevention of several inflammatory chronic conditions (Devereux, 2006). Regarding allergic diseases, the most frequent and consistent published data are related to vitamin D and fish oil supplementations to mothers during pregnancy and reduction of risk for asthma in offspring. But also other vitamins, like, vitamin C, E, and A that are contained in different amounts in fruits and vegetables may play a role in preventing and control of chronic allergic conditions.

INFLAMMATION AND OXIDATIVE STRESS IN ALLERGY

Allergic reaction represents a cascade of inflammatory processes generated by exposure to allergens, which involve many cells and mediators. In allergic rhinitis the inflammatory process is located in the nasal mucosa and produces acute or chronic symptoms like sneezing, rhinorrhea, nasal congestion, and nasal itching and sometimes ocular symptoms, depending on the amount of exposure and the type of allergen involved in. Asthma is a chronic inflammatory airways disease, in which clinical manifestations result from airway obstruction, airway hyperresponsiveness (AHR) and airway inflammation. Those two diseases often coexist and represent „a single airways allergic disease" (Bousquet et al., 2003; Khan, 2014).

The relationship between upper and lower airways inflammation is not completely understood yet, but allergic inflammation is the key to understanding these diseases and the evolution of allergic rhinitis to asthma (Canonica and Compalati, 2009; Khan, 2014; Kowalski and Alam, 2001).

The allergic reaction is a cascade of inflammatory processes generated by exposure to allergens, which involve many cells and mediators. First exposure to a specific allergen induces sensitization and activation of Th2 lymphocytes, which secret a specific pattern of cytokine, stimulating IgE synthesis. Subsequent exposure to allergens determines its interaction with specific IgE that already exists on the mast cell surface, inducing their degranulation and release of several preformed or newly synthesized mediators, like histamine, cytokines, prostaglandins, leukotrienes, kinins. All these mediators activate adhesion cell molecules, which attract other inflammatory cells (eosinophils, neutrophils) and the process progresses from an acute to a chronic one (Adkinson et al., 2013; Gore and Custovic, 2004).

Oxidative stress is defined as an imbalance between reactive oxygen species (ROS) production and their inactivation by the biological system. The biological systems can detoxify the reactive intermediates and to repair the damage caused by ROS (Betteridge, 2000). Reactive oxygen species like hypochlorous acid (HOCl-), hydrogen peroxide (H2O2), superoxide radical (O2•-), nitric oxide (NO), peroxynitrite (ONOO) that are produced by neutrophils and macrophages will be detected during the normal inflammatory metabolic processes (Halliwell and Gutteridge, 1985). But, in chronic inflammation these oxidants (ROS) are excessively produced which can lead to oxidative stress (Sahiner et al., 2018; Yildiz et al., 2015).

The imbalance between ROS and antioxidants, known as oxidative stress, may cause dysfunctions of both intracellular signalling systems and arachidonic acid metabolism, causing a chronic inflammatory process located in the airways, which could become a systemic one (Figure 1). Even if oxidative stress may enhance the inflammatory process, increasing Th1 and Th2 cytokines synthesis, in some conditions it may orient the immune response towards a Th2 one (King et al., 2006; Moreno-Macias and Romieu, 2014; Sahiner et al., 2018).

The oxidative stress is, therefore, a continuous process, needed to destroy pathogenic microorganisms or allergens, thus preventing overactivation of T lymphocytes. The most important endogenous sources of ROS are mitochondrial processes, peroxisomes, inflammatory cells, and some enzymes (cytochrome P450, COX) (Halliwell, 2012). But also exogenous ROS add to the endogenous ones.

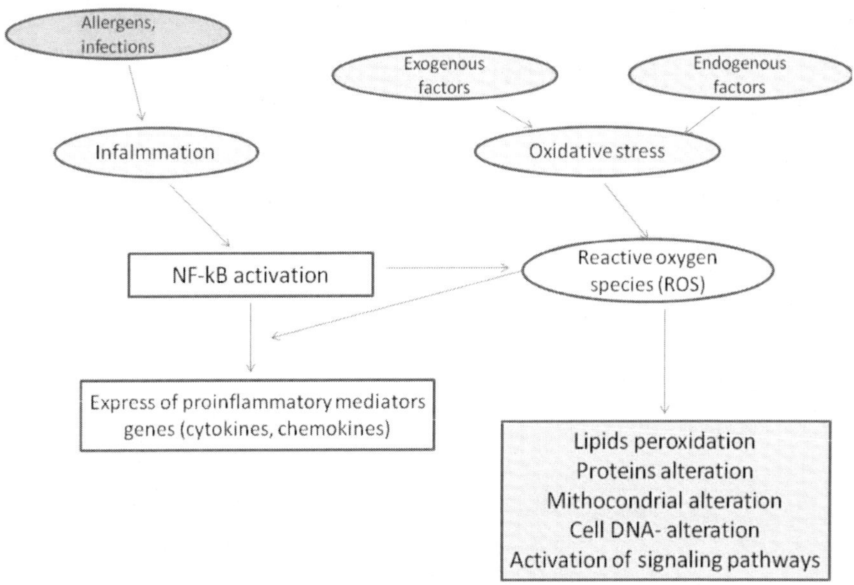

Figure 1. The link between chronic inflammation and oxidative stress in allergic disease (adapted after (Halliwell, 2012; King et al., 2006; Sahiner et al., 2018)).

The oxidative stress stimulates immune response and determines the occurrence of allergic diseases, like bronchial asthma, allergic rhinitis, atopic dermatitis or food allergy, because the protective antioxidant system is overwhelmed in these patients compared to healthy persons (Sackesen et al., 2008). Thus, inflammation and oxidative stress, at some point during their progression will induce each other, being tightly interconnected (Halliwell, 2006; Wood et al., 2003).

Since both processes are strongly linked and simultaneously described to initiate and maintain many pathological conditions, their prevention and control could provide a safe alternative in chronic disease management.

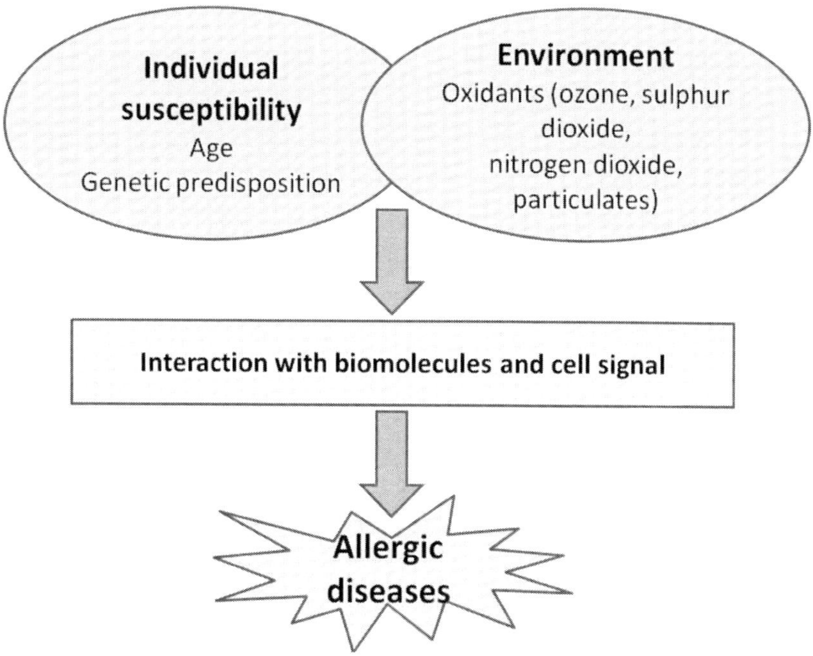

Figure 2. The link between genetic factors and oxidants in allergic diseases. (adapted over (Han et al., 2013; King et al., 2006)).

Antioxidants supplementation may compensate for the inflammatory and oxidative stress processes, which are enhanced in patients with asthma. Contrary, Murr et al., (2005) revealed that excessive supplementation with antioxidants could increase the susceptibility to develop allergic disorders, including bronchial asthma, through a reduced Th1 response and an increased Th2 one, with immunoglobulin production (Murr et al., 2005). Because of the existing contradictory data regarding antioxidants supplementation more research is needed.

ANTIOXIDANTS

Antioxidants are molecules stable enough to neutralize ROS and to prevent tissue damages. They represent the first line of defense against resulted products, acting at every level of oxidative stress pathway. Antioxidants could be endogenous (synthesized into the body) or exogenous (could come from the diet) as observed in table 1.

Antioxidant molecules are observed in variable concentrations in both extra- and intracellular spaces, with an unequal distribution among tissues. Pulmonary tissue and airways contain a large spectrum of antioxidant molecules (Ciencewicki et al., 2008). In normal conditions, when oxidative stress processes are reduced, the lung tissue can neutralize all prooxidant species, preventing their accumulation (Kelly et al., 1995).

Table 1. Endogenous and exogenous antioxidant molecules

Endogenous antioxidants		**Exogenous antioxidants**
Enzymatic	Non-enzymatic	
Catalase Superoxide dismutase Glutathione peroxidase Peroxiredoxin Thioredoxin	Glutathione Coenzyme Q10 Bilirubin Uric acid	Vitamin A Vitamin E Vitamin C Polyphenols Selenium Phytochemicals

Adapted after (Han et al., 2013; Moreno-Macias and Romieu, 2014).

Vitamins A, C, and E are antioxidants derived from the diet. They prevent lipid peroxidations. Among them, vitamin C is the best represented in the lungs (Sackesen et al., 2008). It is a hydrosoluble vitamin, with a proven antioxidant effect, being involved in scavenging free ROS and in suppressing the production of superoxide anions in macrophages (Romieu and Trenga, 2001). Vitamin C cannot be synthesized in tissues, so it should be obtained from external sources. The richest sources in vitamin C are cherries, citrus, kiwi, broccoli and cabbage. Vitamin C is also involved in restoring the antioxidant capacity of oxidized α-tocopherol from cell membranes, another vitamin with antioxidant effect (Cook-Mills, 2013).

Vitamin A is derived from the conversion of beta-carotene from plants and fruits. Beta-carotene accumulates in cell membranes and it is involved in the scavenging process of superoxide anions and the direct neutralization process of peroxyl free radicals (Kelly et al., 2003). It is a liposoluble antioxidant, found in high amounts in carrots, cabbage, mango and peaches (Han et al., 2015).

Vitamin E is found in 8 different forms of α, β, γ, and δ tocopherols and tocotrienols, having both prooxidant and antioxidants effects. From the entire group, α-tocopherol and γ-tocopherol represent the most important natural forms of vitamin E. These two forms have opposite regulatory effects in inflammation and oxidative stress. α-tocopherol has anti-inflammatory properties, while γ-tocopherol has proinflammatory effects (Cook-Mills, 2013; Cook-Mills et al., 2013). α-tocopherol is the best-studied form of vitamin E and it is considered the most important liposoluble antioxidant (Moreno-Macias and Romieu, 2014). It inhibits lipid peroxidation derived from cell membrane destruction. It could be found in green vegetables, olive or sunflower seed oil. γ-tocopherol has proinflammatory effects (Cook-Mills, 2013; Cook-Mills et al., 2013) and it is found in soy and rapeseed oils.

ANTIOXIDANTS AND ASTHMA CONTROL

Asthmatic patients have an exaggerated inflammatory response in airways, induced by exposure to both allergens and different viral infections. In this inflammatory process, several cells and mediators are involved, which also release free ROS (Wood et al., 2003). In this way, both inflammatory and oxidative stress pathways are activated, being closely connected in asthma pathogenesis. Oxidative stress, due to overproduction of ROS, has multiple negative effects on lung function, producing bronchospasm, mucus hypersecretion, increases of vascular exudation with inflammatory edema. On a second pathway, through activation of NF-kB and hypersecretion of proinflammatory cytokines, it also enhances inflammation (Biagioli et al., 1999) (Figure 1). Endogenous antioxidants cannot suppress oxidative stress already initiated, so supplementation with exogenous antioxidants seems like a promising approach.

The experimental studies revealed that antioxidants vitamins may reduce both inflammation and oxidative stress in airways, so they can improve symptoms of asthma or allergic rhinitis. Vitamin C reduces NF-kB expression in the lungs, with a secondary anti-inflammatory effect (Santanam et al., 2013), while vitamin D reduces neutrophils and eosinophils infiltration, reduces peribronchial inflammatory cells and NF-kB expression in pulmonary tissue (Kianian et al., 2019). Bansal et al., (2014) reported that an association of antiinflammatory agent (dexamethasone) and antioxidants (vitamin C and selenium) in nasal administration reduced bronchial hyperreactivity, IL-4, IL-5, IgE and NF-kB levels and restored glutathione, IL-10 and peroxidase concentrations in the lungs (Bansal et al., 2014). The authors concluded that anti-inflammatory and antioxidant effects of the aforementioned combination are a consequence of high synthesis of IL-10 in FOXP3(+) cells, being more intense than those of dexamethasone and α-lipoic acid. This combination could become a reasonable alternative in allergic diseases (Bansal et al., 2014).

But not always the results reported by experimental studies were also observed in clinical trials. Even if some research reported benefic effects of antioxidants supplementation on symptoms severity, on functional and biological parameters, more studies showed that a combination of them not a single one may have a better role, due to an additive or synergic effect.

On the other hand, it is difficult to measure the exact correlation between antioxidants levels, asthma control, and lung function, because it is difficult to estimate their intake from diet. Some studies revealed that patients with asthma had a lower intake of vitamins A, E and C (Anetor et al., 2003; Arora et al., 2002; Baker et al., 1999; De Luis et al., 2005; Misso et al., 2005; Patel et al., 2006) and a lower level of these antioxidants in serum (Baker et al., 1999; Kalayci et al.; Patel et al., 2006) or lungs (Kelly et al., 1999; Kongerud et al., 2003). Opposite, another study showed that intake or levels of antioxidants are not associated with asthma severity (Picado et al., 2001). The aforementioned observations were also documented by a systematic review and meta-analysis that included 40 studies relating to vitamins A, C, and E to asthma (Allen et al., 2009). The authors concluded that a low dietary intake of vitamin A and C, together with lower blood vitamin C levels were associated with asthma presence in both adults and children. The dietary intake of vitamins is negatively correlated with asthma severity (symptoms score or lung function). Vitamin E intake was significantly lower in people with severe asthma but was unrelated to asthma status (Allen et al., 2009).

A general overview concerning antioxidant intake effects on asthma control is presented in Table 2.

Table 2. The effects of vitamins on clinical symptoms and lung function

	Type of study	**Method**	**Medical evidence**	**References**
Vitamin C	Observational study	71 pts with respiratory allergy 7.5 g i.v., 2-3 times/week, 3-11 wks	Reduction of specific symptoms with 4.5 points Reduction of nonspecific symptoms with 4 points Reduction of antiallergic medication usage	(Vollbracht et al., 2018)
	Randomized cross over trial	8 pts with EIA 1500 mg/day orally *vs* placebo, 2 wks	Reduction of the maximum fall in post-exercise FEV1 Improvement of asthma symptoms score Reduction of FENO, LTC4-E4, and 9alpha, 11beta-PGF2 concentrations in postexercise	(Tecklenburg et al., 2007)
	Randomized cross over trial	20 pts with asthma 2 g orally 1 hour before exercise *vs* placebo	No influence on pulmonary function after exercise	(Cohen et al., 1997)
	Randomized controlled trial	300 pts with asthma 1g/day *vs* Mg *vs* placebo, 16 wks	Increase of serum level of vitamin C No changes in asthma symptoms and bronchial hyperreactivity No changes in FEV1 and FCV	(Fogarty et al., 2003)
	Randomized controlled trial	92 pts with asthma 1g/day *vs* Mg *vs* placebo, 16 wks	Corticosteroid sparing effects (equivalent to a decrease in 38 μg beclomethasone)	(Fogarty et al., 2006)

Table 2. (Continued)

	Type of study	Method	Medical evidence	References
Vitamin C	Randomized self-control crossing over trial	60 children with asthma 200 mg/day, 6 wks in combined phase (1000 mg triglyceridic oil, 15 mg Zn and 200 mg vitamin C), 6 wks	Improvement of the ACT score Improvement of pulmonary functions (FEV_1 and FEV_1/FVC ratio) Reduction of sputum total WBC count, eosinophil ratio Reduction of sputum ECP, LTC4-LTE4, IL-1β and PGD2	(Biltagi et al., 2009)*
	Metaanalysis	3 placebo-controlled studies 40 pts with EIA	Reduction of postexercise FEV1 decline	(Hemilä, 2013b)
	Metaanalysis	3 studies – 2 RCT + 1 cross over study 79 pts with cold-induced asthma	Improvement of common cold symptoms and alleviate cold-induced asthma exacerbations	(Hemilä, 2013a)
	Metaanalysis	8 randomized controlled study	No significant effect on asthma symptoms or lung function	(Ram et al., 2004)
	Metanalysis	11 randomized controlled trials – 4 crossing over + 7 vs placebo, 419 pts adults and children	No significant effect on asthma symptoms or lung function	(Milan et al., 2013)
	Metaanalysis	21 randomized studies, dietary intake and vitamins intake	Vitamin C is associated with less decline in lung function	(Forte et al., 2018)
Vitamin E	Randomized controlled trial	72 pts with asthma, 500 mg vitamin E *vs* placebo, 6 wks	No effect in bronchial reactivity to methacholine No changes in bronchodilator use, night time or day time symptom scores No changes in FEV1, FVC, morning peak flow No changes in serum IgE	(Pearson et al., 2004)
	Open study	8 pts with asthma, 823 mg of tocopherol *vs* placebo, 8 days	No changes in the alpha-tocopherol level Increases of delta and gamma-tocopherol No changes in IL-1α, Il-1β, IL-12p40, GM-CSF, TNF, Eotaxin, Rantes, MIP1α and IP-10	(Wiser et al., 2008)

	Type of study	Method	Medical evidence	References
Vitamin E	Randomized control study	300 children with asthma vitamin E 50 mg/day *vs* placebo, 8 wks	Increases in serum level of vitamin E Improvement of FEV1 and FEV1/FVC ratio Improvement of clinical manifestations	(Ghaffari et al., 2014)
	Randomized control study	63 pts with allergic rhinitis 400 UI/day vs placebo, 4 wks	No changes in rhinitis symptoms score No changes in serum levels of specific IgE and lipid peroxides	(Montaño Velázquez et al., 2006)
Vitamin A	Randomized crossover study	15 children with asthma, vitamin A aerosols *vs* placebo, 2 wks	No differences in the occurrence of asthma at night, in wheezing No differences in spirometric indices	(Reifen et al., 2015)
Vitamin C and E	Double-blind crossover study	17 adults with asthma exposed to ozone, 400 IU vitamin E + 500 mg vitamin C	Protective effect against lung function deterioration after ozone exposure	(Trenga et al., 2001)
	Randomized controlled study	158 children with asthma exposed to ozone, vitamin C 250 mg/day + vitamin E 50 mg/day *vs* placebo, 12 wks	Ozone levels were negatively correlated with lung function parameters Vitamin supplementation prevents ozone-induced decrement in FEF_{25-75}, FEV_1, and PEF	(Romieu et al., 2002)
	Randomized controlled trial	117 children with asthma exposed to ozone, vitamin C 250 mg/day + vitamin E 50 mg/day *vs* placebo, 12 wks	Decreases nasal IL-6 after 12 weeks Increases glutathione	(Sienra-Monge et al., 2004)
	Randomized controlled trial	14 pts with asthma, 500 mg α-tocopherol + 2000 mg vitamin C, 12 wks	Increases in vitamin C plasmatic level, No effect on vitamin C levels in the respiratory-tract lining fluid Decreases serum γ-tocopherol and sputum α-tocopherol levels No influence on symptoms score or lung function (FEV_1)	(Hernandez et al., 2009)
	Metaanalysis	5 studies – 214 adults and 160 children with EIA, vitamin C and E in different doses	No conclusive evidence	(Wilkinson et al., 2014)

Table 2. (Continued)

	Type of study	Method	Medical evidence	References
Vitamin A and E	Single-blind randomized crossover study	8 patients with EIA vitamin A 500 mg + α-tocopherol 300 IU *vs* placebo, 3 wks	Improvement of lung function (FVC; FEV_1; PERF; $FEF_{25\text{-}75\%}$) in 5 and 15 minutes after exercise	(Kurti et al., 2016)
Vitamins C, E and A	Randomized controlled trial	54 adults with asthma beta-carotene (9 mg/day) + vitamin C (1500 mg/day) + vitamin E (130 mg/day) + zinc (45 mg/day) + selenium (76 microg/day) + garlic (150 mg/day) *vs* placebo, 4 wks	Increases in serum levels of vitamin C, vitamin E, beta-carotene and selenium No change in serum antioxidant capacity, NO, plasma F(2)-isoprostanes, eNO or immune responses	(Dunstan et al., 2007)

Abbreviations: ECP, eosinophilic cationic protein; EIA, exercise-induced asthma; FEV_1, Forced expiratory volume in 1 second; FVC, forced vital capacity; $FEF_{25\text{-}75\%}$, forced expiratory flow 25–75%, FENO, fractional exhaled nitric oxide, GM-CSF, Granulocyte macrophage colony-stimulating factor IgE, immunoglobulin E; IL-1α, interleukin 1α; IL-1β, interleukin 1β; IL-12p40, interleukin 12p40; IP-10, Interferon gamma-induced protein 10; LTC4, leukotriene C4; LTE4, leukotriene E4; MIP-1α, macrophage inflammatory protein 1α; NO, nitric oxide; PERF, peak expiratory flow rate; PGD2, prostaglandin D2; pts, patients; TNF, tumour necrosis factor; WBC, white blood; wks, weeks;

Based on these observations the hypothesis that vitamin supplementation might better control the symptoms and lung function in patients with asthma was raised.

Vitamin C

From the entire group of antioxidants, vitamin C is probably the best-studied one, in different forms of allergic diseases. The effects of vitamin C supplementation were evaluated in different studies, in both adults and children, but the methodologies were extremely variable. Most of the studies investigated the dietary intake based on questionnaires and few studies prospectively investigated the role of vitamin C supplementation in asthma.

Most of the studies included a small number of patients (Cohen et al., 1997; Fogarty et al., 2006; Tecklenburg et al., 2007), so their conclusion could not be extrapolated to the entire population with asthma. Not all the studies were randomized placebo-controlled studies; there were also self-control crossing over trials (Biltagi et al., 2009).

On the other hand, the recommended dose varies considerably, from 200 mg to 2 g/day, respectively (Cohen et al., 1997). The route of administration is also different. In most of the studies vitamin C was administered orally (Biltagi et al., 2009; Cohen et al., 1997; Tecklenburg et al., 2007), while in others was administered intermittently, intravenously (i.v.) 7.5 g, 2-3 times/week (Vollbracht et al., 2018). Duration of vitamin C supplementation is variable from 1 week, in exercise-induced asthma (Cohen et al., 1997) to 3 weeks in acute forms of asthma (Vollbracht et al., 2018) or 16 weeks (Fogarty et al., 2003, 2006).

The clinical results are also extremely variable. Some of the studies reported an improvement of specific or nonspecific symptoms (Biltagi et al., 2009; Tecklenburg et al., 2007; Vollbracht et al., 2018), while other studies did not (Fogarty et al., 2003). But the inclusion criteria were different. In some of them adult asthmatic patients were included (Fogarty et al., 2003), while others included only children with asthma (Biltagi et al., 2009) or patients with a specific phenotype of asthma (exercise-induced asthma) (Tecklenburg et al., 2007), which may explain variable results regarding improvement of symptoms. The endpoints of published studies are also extremely variable. The asthma control based on symptoms evaluation was assessed in most of them, only in two studies one of the endpoints was the usage of other asthma specific medication. In these studies the authors reported a reduction of inhaled corticosteroids (Fogarty et al., 2006) and antiallergic medication usage (Vollbracht et al., 2018) in patients who received vitamin C supplementation.

Regarding vitamin C effect on lung function, some of the authors reported an improvement in this sense (Biltagi et al., 2009; Tecklenburg et al., 2007), while others did not observe any improvement of FEV1 and FCV

(Cohen et al., 1997; Fogarty et al., 2003). The explanation could come from the fact that Cohen et al., (1997) studied the influence of vitamin C in exercise-induced asthma, while other authors in adult patients with asthma (Fogarty et al., 2003).

Based on the published data some metanalysis were published over the years. Some of them included only studies performed in specific phenotypes of asthma (exercise-induced asthma, cold-induced asthma (Hemilä, 2013c, 2013a, 2013b), while others included RCT (Forte et al., 2018; Ram et al., 2004). Overall, it seems that vitamin C supplementation may have a positive effect in improving clinical symptoms in cold-induced asthma, but not in all cases. Vitamin C is associated with less decline in lung function (Forte et al., 2018), but the authors included all studies with vitamin C supplementation, including dietary ones.

Only 2 studies investigated also the role of vitamin C on chronic inflammation and oxidative stress (Biltagi et al., 2009; Tecklenburg et al., 2007). In both studies, vitamin C supplementation reduced fractional exhaled nitric oxide (FeNO), leukotrienes and prostaglandin levels, in both adults and children.

It is not well established what level of vitamin C should be attended to control asthma symptoms and to prevent exacerbations. Volbracht et al., (2018) determined the basal plasmatic level of vitamin C. The patients with cutaneous manifestations had lower plasmatic values than patients with respiratory allergies (Vollbracht et al., 2018). The same observation was noticed in patients with chronic manifestations compared to those with acute ones. This study concluded that vitamin C deficiency is associated with chronic respiratory manifestations that could be improved by long term supplementation with vitamin C in high doses. The authors suggested maintaining an optimal plasmatic level of over 0.7 mg/dl in allergic patients to prevent chronic respiratory symptoms (Vollbracht et al., 2018).

In Moreno-Macías H et al., (2014) study it was evaluated the acute effect of ozone on lung function of 257 asthmatic children according to dietary intake of vitamin C. They observed that patients with persistent asthma and low daily vitamin C intake had a significant reduction in FEF_{25-75} associated with ozone exposure, while the decrement was less intense among children with vitamin C intake over 105 mg/day (Moreno-Macias and Romieu, 2014). They considered 45 mg/day as Recommended Daily Intake (RDI). Starting from this result the authors considered that children with persistent asthma exposed to ozone might benefit from vitamin C intake above the RDI. Within this study, it was reported that intake over 105 mg/day can be reached from diet alone. Thus, they suggested that a normal diet is enough to assure an optimal level of vitamin C, even if patients have chronic inflammatory conditions and it is not necessary to add another dose of vitamin C (Moreno-Macias and Romieu, 2014).

Vitamin E

In the last two decades, few studies were published that evaluated the role of vitamin E supplementation in patients with asthma. One study was an open study (Wiser et al., 2008), while most of them were randomized controlled ones (Ghaffari et al., 2014; Montaño Velázquez et al., 2006; Pearson et al., 2004). As in previously mentioned studies related to vitamin C supplementation, the included number of patients was extremely variable. It ranged from 8 patients (Wiser et al., 2008) to 300 children (Ghaffari et al., 2014). The studies were performed on both adults (Montaño Velázquez et al., 2006; Pearson et al., 2004; Wiser et al., 2008) and children (Ghaffari et al., 2014).

The administered dose of vitamin E varied from 50 mg/day in children (Ghaffari et al., 2014) to 500 mg/day in adults (Pearson et al., 2004). In Wiser et al., (2008) study the authors administered the equivalent of 823 mg of tocopherols (Wiser et al., 2008). The duration of therapy was almost similar 6-8 weeks in all aforementioned studies (Ghaffari et al., 2014; Pearson et al., 2004; Wiser et al., 2008). Montano et al., (2006) published the only study that investigated the role of vitamin E supplementation in allergic rhinitis, not in asthma (Montaño Velázquez et al., 2006).

In children, vitamin E supplementation improves serum levels of vitamin E, clinical manifestations and lung function (Ghaffari et al., 2014). But the studies performed in adults did not report similar results. Vitamin E supplementation did not increase the alpha-tocopherol level and did not improve symptoms score, lung function, bronchial hyperreactivity (Pearson et al., 2004) or allergic, inflammatory parameters (Pearson et al., 2004; Wiser et al., 2008). The same lack of response was noticed by Montano in patients with allergic rhinitis (Montaño Velázquez et al., 2006).

Vitamin A

Previous data reported an association between low levels of serum vitamin A and chronic airway diseases in adults (Riccioni et al., 2007). A low level of vitamin A was negatively correlated with asthma severity in children (Arora et al., 2002). Hypovitaminosis A may also predispose to respiratory infections (Chandra, 1989) which may induce an asthma exacerbation in children. Also, animal studies predict that vitamin A could become an attractive player in the control of allergy in humans. These findings suggest that vitamin A supplementation might help to control chronic respiratory disease, but few therapeutic trials were already done to confirm this hypothesis.

Only one randomized crossover study that investigated the role of vitamin A alone supplementation was published (Reifen et al., 2015). The authors included 15 asthmatic children who received vitamin A as aerosol or placebo for 2 weeks. They reported no differences between vitamin A and placebo regarding asthma symptoms during day or night or lung function (Reifen et al., 2015).

Combinations of Vitamins

Vitamins A, C, and E are antioxidant molecules derived from the diet. They could modulate innate and adaptive immune responses and may, therefore, influence the course of allergic diseases, each one in different degrees. Studies that investigated supplementation of only one vitamin revealed contradictory results, and most of them lack to prove their benefic effect in asthma control. For these reasons, it was raised the hypothesis that a combination of these may have a better effect than a single administration.

Most of the study reported the combination of vitamins C and E (Romieu et al., 2002; Sienra-Monge et al., 2004; Trenga et al., 2001). Both adults and children were included in these studies. Three of them were randomized controlled *vs* placebo, while one study was a crossover study. The number of patients was variable from 17 patients with asthma (Trenga et al., 2001) to 158 asthmatic children (Romieu et al., 2002). But in three of the above-mentioned studies, the asthmatic patients were exposed naturally or controlled to ozone (Romieu et al., 2002; Sienra-Monge et al., 2004; Trenga et al., 2001). The main outcome of the research was to investigate the effect of combined antioxidants in preventing the adverse effect of exposure to air pollutants on lung function. Only one study included asthmatic patients in real-life conditions without other types of exposures (Hernandez et al., 2009).

The used doses were quite similar in all studies, with 250-500 mg of vitamin C and 50 mg/400 UI vitamin E (Hernandez et al., 2009; Romieu et al., 2002; Sienra-Monge et al., 2004; Trenga et al., 2001). Therapy duration was in average 12 weeks, longer than other studies, where, both vitamins were used alone. The combination between vitamins C and E prevented ozone-induced decrement in lung function (Romieu et al., 2002; Sienra-Monge et al., 2004; Trenga et al., 2001), decreased inflammatory parameters, like IL-6 and restored the oxidative stress at a normal level (Sienra-Monge et al., 2004). In the Hernandez et al., (2009) study, the vitamin association did not prove the same benefic effect on lung function in asthmatic patients if they were not exposed to ozone and pollutants (Hernandez et al., 2009). Wilkinson et al., (2014) realized a metanalysis that included 5 studies with a cumulative number of patients. There were 214 adults and 160 children with asthma, which received vitamins C and E in different doses (Wilkinson et al., 2014). The authors reported no conclusive evidence in the relationship with asthma symptoms control and lung function, and they confirmed the heterogenicity of the studies (Wilkinson et al., 2014).

Based on these observations we may conclude that a supplementation of vitamins C and E could be beneficial in some conditions of exposure to pollutants of asthmatic patients, but not in all of them.

The supplementation of vitamins A and E in patients with exercise-induced asthma was also investigated (Kurti et al., 2016). The patients received vitamin A 500 mg and α-tocopherol 300 IU/day for 3 weeks. The combination improved lung function immediately after exercise (Kurti et al., 2016).

The combination of vitamins C, E and A plus selenium and zinc were investigated in another randomized controlled study (Dunstan et al., 2007). Within this study, 54 patients with asthma received this antioxidant association or placebo in different doses for 4 weeks (Dunstan et al., 2007). Even if vitamin supplementation increased the serum levels of vitamin C, vitamin E, and beta-carotene, the authors did not report any changes of oxidative stress or inflammatory markers after 4 weeks of therapy (Dunstan et al., 2007).

Antioxidants and Asthma Prevention

The imbalance between reactive oxygen species and antioxidants induces oxidative stress and augments systemic and pulmonary inflammatory processes, reducing Th1 immune response in favour of Th2 reaction, which could lead to asthma occurrence or exacerbations. The reported data from observational studies showed that a high vitamin intake from diet protects against asthma occurrence in childhood or adults age. A meta-analysis of 23 studies that included both adults and children reported that high consumption of fruits and vegetables enriched in antioxidants reduced the asthma risk with 16% in children and 12% in adults (Han et al., 2015; Seyedrezazadeh et al., 2014). A further hypothesis was raised based on these observations: if individual supplementation with antioxidant vitamins during pregnancy could significantly reduce the rate of asthma occurrence in offspring?

Vitamin C

The experimental studies on animal models showed that vitamin C reduced oxidative stress, airways inflammation and secondarily bronchial hyperreactivity. It also reduced IL-10 level, attenuating Th2 response (Bansal et al., 2014). The previously mentioned *in vivo* studies did not confirm the experimental results, so vitamin C could not be considered a therapeutic option in asthma management at this moment. Data related to vitamin C in asthma prevention are also contradictory.

Greenough et al., (2010) evaluated the effect of both vitamin C (100 mg/day) and vitamin E (400 IU) administered in asthma or other allergic conditions incidences in pregnant women (from week 16-22 till labor *vs* placebo) (Greenough et al., 2010). Only patients with a high risk of preeclampsia were included in the study. There were no differences between supplementation and placebo regarding the incidence of recurrent wheezing, asthma or atopic dermatitis within the first 2 years of life. Moreover, prenatal supplementation of high doses of vitamins C and E is associated with more frequent presentation of the child to a specialist or general practitioner, increasing overall medical costs (Greenough et al., 2010). Besides this result, the authors did not exclude that lower doses administered in the prenatal period to normal healthy women, without a high risk of preeclampsia, could have a beneficial effect and might prevent the occurrence of lung diseases in children.

Nurmatov et al., (2015) evaluated the effects of vitamins C, A and E supplementation on primary prevention of asthma. The authors included 21 cohort studies, 15 case-control studies, and 26 cross-sectional studies, but none of the randomized clinical studies (Nurmatov et al., 2011). Within these trials, 14 of them analyzed the association between vitamin C and asthma. Even if few studies reported that prenatal vitamin C supplementation could reduce wheezing incidence within the first 2 years of life and could improve lung function (Kalayci et al.; Martindale et al., 2005; Rubin et al., 2004), these affirmations were not sustained (Nurmatov et al., 2011). The main reasons were due to the high studies heterogeneity, their different methodology and inconsistent reported data.

Patelarou et al., (2011) published a systematic review following the correlation between the levels of antioxidants in pregnant women and the risk of allergic disease in offspring. The authors included 18 studies that evaluated dietary antioxidants intake, but not a specific quantitative supplementation, in which the levels of antioxidants were measured in both mothers and children (Patelarou et al., 2011). They concluded that antioxidants status during pregnancy prevents allergic diseases occurrence in children (Patelarou et al., 2011). However, the mentioned results are questionable. The evaluation of antioxidants exposure was based on questionnaires and diet history. This could not permit a real and quantitative evaluation of their daily intake because the variability of diets is well known. Such kind of evaluation might underestimate the diet-allergy relationship. The measurement of serum antioxidants levels reflects the recent intake and not the constant intake of vitamins, which is the desired outcome in chronic allergic diseases.

A recent study showed that vitamin C supplementation in pregnant women could reduce the incidence of wheezing in children in their first years of life (McEvoy et al., 2014). The authors performed a randomized controlled study that included smoker pregnant women (75 patients that received 500 mg/day vitamin C and 83 placebo) and 76 non-smoker pregnant women. The children were followed up for 1 year regarding the number of wheezing episodes and lung function at birth and after 1 year. The authors reported that vitamin C administration from week 22 of pregnancy until birth reduced wheezing episodes in children during the first year of life compared to placebo (21% vs 40%). The lung function within the first 72 hours of life is significantly improved in children born from mothers that received vitamin C compared to placebo. But 1-year measurements did not reflect significant differences in lung function of children born from mothers that received vitamin C compared to those that received placebo (McEvoy et al., 2014). Wheezing is a transient clinical manifestation, specifically in small children and it is more frequently induced by viral infections. These observations raised the hypothesis that the benefic effect of vitamin C supplementation on wheezing incidence could be attributable to a possible antiviral effect of vitamin C and less to a real protective effect in reducing asthma incidence.

Vitamin E

The published data referring to vitamin E – asthma relationships are contradictory and variable according to studied population and predominant intake of α- or γ- tocopherol. Allen et al., (2009) reported that a low serum vitamin E level was correlated with more severe forms of asthma, which raised the hypothesis that supplementation during pregnancy could reduce the incidence of disease (Allen et al., 2009).

One of the first published data from 2004 evaluated the level of vitamins in the umbilical cord at birth and further asthma development in children. They were followed for 6 years. In this study patients from the Avon Longitudinal Study of Parents and Children were included. The patients were first evaluated at the age of 3.5 years and secondly at the age of 6 (Shaheen et al., 2004). The authors did not report any correlation between vitamin E levels in the umbilical cord, as a marker of vitamin E intake by mother during pregnancy, and asthma or wheezing occurrence in first years of life in children.

In Martindale et al., (2005) prospective study, 2000 pregnant women were included. Vitamins intake was evaluated based on a questionnaire applied during the 34th week of pregnancy, while their offspring were followed until the age of 2 years old. The children were also evaluated based on questionnaires similar to those used in the ISAAC study (Martindale et al., 2005). The vitamin E level was determined in the umbilical cord at birth. The authors reported a negative correlation between vitamin E intake during pregnancy and wheezing occurrence in the second year of life in the absence of intercurrent illness, but no association of vitamin E and viral induced wheezing or other manifestations within first year of life (Martindale et al., 2005). These data suggested that vitamin E supplementation during pregnancy might influence fetal immune response and further development of asthma or other medical condition, rather than fetal lung development and immune response to infections of the newborns.

In another pilot study, the authors optimized vitamin E intake in pregnant women through dietician counselling (Clark et al., 2012). Forty-three pregnant women with a personal history of asthma were included in the study. They received recommendations on how to improve their diet to achieve a vitamin E intake of 15 mg/day. Diet optimization was realized between weeks 12-20 of pregnancy. Through dietician counselling, an increase of vitamin E intake from 7.13 mg/day to 17,4 mg/day was realized. But vitamin E supplementation was not correlated with better control of asthma in women or reduced incidence of disease in children (Clark et al., 2012). Probably this supplementation was too short and if prolonged until birth the authors might have obtained better results.

In one meta-analysis published by Nurmatov et al., (2011), 15 studies that evaluated the association between vitamin E and asthma were included. The analysis of 3 cohort studies showed that a high vitamin E intake during pregnancy is correlated with a significant reduction of wheezing incidence at the age of 2 years in children (Nurmatov et al., 2011). Similar results were also reported by Allan et al., (2015). They included 2000 pregnant women and their children that were monitored until 10 years old (Allan et al., 2015). The vitamin E intake during pregnancy was evaluated through an applied questionnaire in the 34th week of pregnancy which investigated the diet for a period of 3 months before evaluation. Children evaluation was done based on the ISAAC questionnaire at the age of 10 years old. A low intake of vitamin E is associated with a higher probability of wheezing, diagnose of asthma established by a specialist and mother reported asthma, in children in their first 10 years of life (Allan et al., 2015).

Data reported by Allan et al., (2011) were different compared to the study conducted by Nwaru et al., (2011). Within Nwaru study the data were obtained from a different cohort (Finnish Type 1 Diabetes Prediction and Prevention Nutrition Study), and children evaluation was performed at the age of 5 years. The authors evaluated mothers' diet from the 8th month of pregnancy and the individual supplementation of vitamin E over the entire pregnancy. The children's evaluation was performed at the age of 5 years based on the ISAAC questionnaire. The mother's vitamin intake does not influence the development of asthma, rhinitis or atopic eczema in children up to the age of 5 (Nwaru et al., 2011).

Recently another metaanalysis that included 10 studies which investigated maternal intake of vitamin E was published (Wu et al., 2018). An adequate vitamin E intake is associated with a lower risk of asthma or wheezing in childhood, but the low risk was not correlated with maternal serum vitamin E levels. The reported correlation between vitamin E intake and asthma was present despite sample size or duration of monitoring (Wu et al., 2018). Because the exact intake of vitamin E could not be established, the authors could not firmly stated what is the exact amount of vitamin E which can confer an optimal immune response in children.

Vitamin A

Experimental studies revealed that prenatal deficiency of vitamin A or retinoic acid (the active metabolite of vitamin A) determines an abnormal accumulation of smooth muscle fibres in airways and bronchial hyperreactivity (Han et al., 2015). Fenretinide, a semisynthetic analog of vitamin A, reduces oxidative stress, airways inflammation (decreases proinflammatory mediators) and bronchial hyperreactivity (Kanagaratham et al., 2014). Even if experimental studies reported promising results, the clinical trials could not sustain the same observations and could not clarify some inconclusive aspects.

Even if some clinical researches showed that vitamin A deficiency is correlated with severe asthma and lung dysfunctions (Allen et al., 2009; Arora et al., 2002), few studies analyzed the risk of asthma occurrence concerning prenatal vitamin A level or supplementation. Prenatal administration of vitamin A enhances Th1 immune response (Cox et al., 2006), raising the hypothesis that changing Th1-Th2 immune balance in uterine life may reduce the risk of asthma in childhood (Cox et al., 2006; Holgate, 2004).

One of the first studies from Nepal proved that supplementation of the diet with provitamin A or β-carotene during pregnancy is not associated with a lower incidence of asthma in children, even if an improvement of lung function was noticed (Checkley et al., 2011). The meta-analysis published by Nurmatov et al., (2011) also showed that a low level of vitamin A in children is associated with a higher risk of asthma, but vitamin A supplementation by mothers during pregnancy was not correlated with asthma or wheezing occurrence in childhood (Nurmatov et al., 2011). Some studies included in the aforementioned meta-analysis, however, reported benefic effects of vitamin A intake, so one cannot say with certainty whether the link between vitamin A and asthma may represent a coincidence or is a consequence of the oxidative stress that characterizes allergic diseases.

A cohort study analyzed the link between vitamins A, E and K intake by mothers starting from the second part of pregnancy and development of allergic diseases in children (Maslova et al., 2014). Data from 44594 mothers and children from the Danish National Birth Cohort were analyzed. Vitamins intake was established based on a questionnaire applied in the second trimester of pregnancy to mothers. Their offspring were monitored for 7 years, following the diagnoses of asthma or rhinitis at ages of 18 months and 7 years respectively. Vitamin A and E consumption during pregnancy are not correlated with asthma occurrence at the age of 18 months and 7 years. But a direct vitamin A supplementation not vitamin intake from diet was associated with the wheezing occurrence. β-carotene consumption reduced asthma risk, especially if it was administered in doses over 10 000 μg/day to mothers. Vitamin A consumption during pregnancy reduced the occurrence of rhinitis at age of 7. The authors concluded that not only quantity but also vitamin A quality may be an important factor to evaluate the risk of allergy occurrence (Maslova et al., 2014). But the authors did not correlate vitamin A intake with its serum level and the evaluation of vitamin consumption was based on a single questionnaire that analyzed the diet during 2nd trimester of pregnancy.

Conclusion

The lungs, which are exposed to air pollutants and oxygen, are highly susceptible to oxidative injury. In normal conditions, the lungs are well equipped_with strong antioxidant defense, but an imbalance between the oxidative factors and the antioxidant defense systems favouring the oxidative stress has been involved in many chronic diseases, including allergic asthma. Epidemiologic observational studies have suggested that vitamins with antioxidant effects may play a role in the pathogenesis of asthma and other allergic disorders.

Small clinical trials suggest that specific antioxidant vitamins from diet or vitamin supplements might improve asthma control or lung function in asthmatic children or adults. But most meta-analysis of interventional studies of variable methodological quality suggested a week or no association between specific vitamin supplementation and allergic disease control. Most of the favourable data refer to vitamin C supplementation in some asthma phenotypes, but not in all cases. Clinical trials of "whole diet" interventions or combinations of vitamins proved a better asthma control and are advisable on the basis of existing evidence. Therefore, a single antioxidant supplementation is unlikely to counteract the complex existing prooxidant/antioxidant imbalance in asthma.

Allergic diseases start in intrauterine life, so it is essential to identify and to modulate risk factors if it is possible, to limit the occurrence of allergies in childhood. Oxidative stress is a process that enhances chronic inflammation in asthma. A, C and E vitamins with antioxidant properties might limit the oxidative stress in airways if they are recommended as supplements even before the entire process will start. The meta-analysis of observational studies suggested an association between low intake of vitamins, especially vitamins C and E, and an increase of asthma and allergy prevalence in childhood.

The high discrepancy of reported data from different studies with variable methodology is also a consequence of different ways that were used to evaluate the intake of the vitamin (specific supplementation of variable amounts of vitamins or evaluation of diet based on questionnaire). Another conflicting issue that results from trials is related to the amount and duration of vitamin supplementation. It also raises the question of whether supplementation of a single vitamin or dietary enrichment which would bring all the vitamins described to play a role in modulating immune function in the newborn is necessary. The ingested amount of vitamins is variable from one study to another and from a population to others. These conflicting data raise the question of whether there would be a critical period during pregnancy, a "window of opportunity" when supplementation with vitamins could modulate the further development of asthma or other allergic diseases in children. Further longitudinal studies or randomized controlled trials that analyzed vitamin supplementations in mothers are needed to establish if prenatal supplementation with vitamins C and E would be needed, especially in the light of unknown adverse effects caused by their long-term administration.

REFERENCES

Adkinson, N. F., Bochner, B. S., Burks, A. W., Busse, William W. ., Holgate, S. T., Lemanske, R. F., et al. (2013). *Middleton's allergy : principles and practice.*

Allan, K. M., Prabhu, N., Craig, L. C. A., McNeill, G., Kirby, B., McLay, J., et al. (2015). Maternal vitamin D and E intakes during pregnancy are associated with asthma in children. *Eur. Respir. J.* 45, 1027–36. doi:10.1183/09031936.00102214.

Allen, S., Britton, J. R., and Leonardi-Bee, J. A. (2009). Association between antioxidant vitamins and asthma outcome measures: Systematic review and meta-analysis. *Thorax* 64, 610–619. doi:10.1136/thx.2008.101469.

Anetor, J. I., Ajose, O. A., Ige, O., Oyeleye, A. O., and Ojo, P. O. (2003). Antioxidant status of adult Nigerian asthmatics: Implications for prognosis. *Nutr. Health* 17, 221–229. doi:10.1177/02601060 0301700305.

Arora, P., Kumar, V., and Batra, S. (2002). Vitamin A status in children with asthma. *Pediatr. Allergy Immunol.* 13, 223–226. doi:10.1034/j.1399-3038.2002.00010.x.

Baker, J. C., Tunnicliffe, W. S., Duncanson, R. C., and Ayres, J. G. (1999). Dietary antioxidants and magnesium in type 1 brittle asthma: A case control study. *Thorax* 54, 115–118. doi:10.1136/thx.54.2.115.

Bansal, P., Saw, S., Govindaraj, D., and Arora, N. (2014). Intranasal administration of a combination of choline chloride, vitamin C, and selenium attenuates the allergic effect in a mouse model of airway disease. *Free Radic. Biol. Med.* 73, 358–365. doi:10.1016/j.freeradbiomed.2014.05.018.

Betteridge, D. J. (2000). What is oxidative stress? in *Metabolism: Clinical and Experimental* (W.B. Saunders), 3–8. doi:10.1016/S0026-0495(00)80077-3.

Biagioli, M. C., Kaul, P., Singh, I., and Turner, R. B. (1999). The role of oxidative stress in rhinovirus induced elaboration of IL-8 by respiratory epithelial cells. *Free Radic. Biol. Med.* 26, 454–462. doi:10.1016/S0891-5849(98)00233-0.

Biltagi, M. Al, Baset, A. A., Bassiouny, M., Kasrawi, M. Al, and Attia, M. (2009). Omega-3 fatty acids, vitamin C and Zn supplementation in asthmatic children: a randomized self-controlled study. *Acta Paediatr.* 98, 737–42. doi:10.1111/j.1651-2227.2008.01213.x.

Bousquet, J., Vignola, A. M., and Demoly, P. (2003). Links between rhinitis and asthma. *Allergy Eur. J. Allergy Clin. Immunol.* 58, 691–706. doi:10.1034/j.1398-9995.2003.00105.x.

Burbank, A. J., Sood, A. K., Kesic, M. J., Peden, D. B., and Hernandez, M. L. (2017). Environmental determinants of allergy and asthma in early life. *J. Allergy Clin. Immunol.* 140, 1–12. doi:10.1016/j.jaci.2017.05.010.

Canonica, G. W., and Compalati, E. (2009). Minimal persistent inflammation in allergic rhinitis: Implications for current treatment strategies: REVIEW ARTICLE. *Clin. Exp. Immunol.* 158, 260–271. doi:10.1111/j.1365-2249.2009.04017.x.

Chandra, R. K. (1989). Nutritional regulation of immunity and risk of illness. *Indian J. Pediatr.* 56, 607–611. doi:10.1007/BF02722377.

Checkley, W., West, K. P., Wise, R. A., Wu, L., LeClerq, S. C., Khatry, S., et al. (2011). Supplementation with vitamin A early in life and subsequent risk of asthma. *Eur. Respir. J.* 38, 1310–1319. doi:10.1183/09031936.00006911.

Ciencewicki, J., Trivedi, S., and Kleeberger, S. R. (2008). Oxidants and the pathogenesis of lung diseases. *J. Allergy Clin. Immunol.* 122, 456–68; quiz 469–70. doi:10.1016/j.jaci.2008.08.004.

Clark, J., Craig, L., McNeill, G., Smith, N., Norrie, J., and Devereux, G. (2012). A novel dietary intervention to optimize vitamin E intake of pregnant women to 15 mg/Day. *J. Acad. Nutr. Diet.* 112, 297–301. doi:10.1016/j.jada.2011.08.033.

Cohen, H. A., Neuman, I., and Nahum, H. (1997). Blocking effect of vitamin c in exercise-induced asthma. *Arch. Pediatr. Adolesc. Med.* 151, 367–370. doi:10.1001/archpedi.1997.02170410041005.

Cook-Mills, J. M. (2013). Isoforms of Vitamin E Differentially Regulate PKCα and Inflammation: A Review. *J. Clin. Cell. Immunol.* 04. doi:10.4172/2155-9899.1000137.

Cook-Mills, J. M., Abdala-Valencia, H., and Hartert, T. (2013). Two faces of vitamin e in the lung. *Am. J. Respir. Crit. Care Med.* 188, 279–284. doi:10.1164/rccm.201303-0503ED.

Cox, S. E., Arthur, P., Kirkwood, B. R., Yeboah-Antwi, K., and Riley, E. M. (2006). Vitamin A supplementation increases ratios of proinflammatory to anti-inflammatory cytokine responses in pregnancy and lactation. *Clin. Exp. Immunol.* 144, 392–400. doi:10.1111/j.1365-2249.2006.03082.x.

De Luis, D. A., Armentia, A., Aller, R., Asensio, A., Sedano, E., Izaola, O., et al. (2005). Dietary intake in patients with asthma: A case control study. *Nutrition* 21, 320–324. doi:10.1016/j.nut.2004.06.027.

Devereux, G. (2006). The increase in the prevalence of asthma and allergy: Food for thought. *Nat. Rev. Immunol.* 6, 869–874. doi:10.1038/nri1958.

Dunstan, J. A., Breckler, L., Hale, J., Lehmann, H., Franklin, P., Lyons, G., et al. (2007). Supplementation with vitamins C, E, β-carotene and selenium has no effect on anti-oxidant status and immune responses in allergic adults: A randomized controlled trial. *Clin. Exp. Allergy* 37, 180–187. doi:10.1111/j.1365-2222.2007.02657.x.

Fogarty, A., Lewis, S. A., Scrivener, S. L., Antoniak, M., Pacey, S., Pringle, M., et al. (2003). Oral magnesium and vitamin C supplements in asthma: A parallel group randomized placebo-controlled trial. *Clin. Exp. Allergy* 33, 1355–1359. doi:10.1046/j.1365-2222.2003.01777.x.

Fogarty, A., Lewis, S. A., Scrivener, S. L., Antoniak, M., Pacey, S., Pringle, M., et al. (2006). Corticosteroid sparing effects of vitamin C and magnesium in asthma: a randomised trial. *Respir. Med.* 100, 174–9. doi:10.1016/j.rmed.2005.03.038.

Forte, G. C., da Silva, D. T. R., Hennemann, M. L., Sarmento, R. A., Almeida, J. C., and de Tarso Roth Dalcin, P. (2018). Diet effects in the asthma treatment: A systematic review. *Crit. Rev. Food Sci. Nutr.* 58, 1878–1887. doi:10.1080/10408398.2017.1289893.

Ghaffari, J., Farid Hossiani, R., Khalilian, A., Nahanmoghadam, N., Salehifar, E., and Rafatpanah, H. (2014). Vitamin e supplementation, lung functions and clinical manifestations in children with moderate asthma: a randomized double blind placebo- controlled trial. *Iran. J. Allergy. Asthma. Immunol.* 13, 98–103. Available at: http://www.ncbi.nlm.nih.gov/pubmed/24338254 [Accessed February 15, 2020].

Gore, C., and Custovic, A. (2004). Can we prevent allergy? *Allergy* 59, 151–161.

Greenough, A., Shaheen, S. O., Shennan, A., Seed, P. T., and Poston, L. (2010). Respiratory outcomes in early childhood following antenatal vitamin C and E supplementation. *Thorax* 65, 998–1003. doi:10.1136/thx.2010.139915.

Halliwell, B. (2006). Reactive Species and Antioxidants. Redox Biology Is.pdf. *Plant Physology* 141, 312–322. doi:10.1104/pp.106.077073.312.

Halliwell, B. (2012). Free radicals and antioxidants: updating a personal view. *Nutr. Rev.* 70, 257–65. doi:10.1111/j.1753-4887.2012.00476.x.

Halliwell, B., and Gutteridge, J. M. C. (1985). The importance of free radicals and catalytic metal ions in human diseases. *Mol. Aspects Med.* 8, 89–193. doi:10.1016/0098-2997(85)90001-9.

Han, Y. Y., Blatter, J., Brehm, J. M., Forno, E., Litonjua, A. A., and Celedón, J. C. (2013). Diet and asthma: Vitamins and methyl donors. *Lancet Respir. Med.* 1, 813–822. doi:10.1016/S2213-2600(13)70126-7.

Han, Y. Y., Forno, E., Holguin, F., and Celedón, J. C. (2015). Diet and asthma: An update. *Curr. Opin. Allergy Clin. Immunol.* 15, 369–374. doi:10.1097/ACI.0000000000000179.

Hemilä, H. (2013a). Vitamin C and common cold-induced asthma: A systematic review and statistical analysis. *Allergy, Asthma Clin. Immunol.* 9. doi:10.1186/1710-1492-9-46.

Hemilä, H. (2013b). Vitamin C may alleviate exercise-induced bronchoconstriction: A meta-analysis. *BMJ Open* 3. doi:10.1136/bmjopen-2012-002416.

Hemilä, H. (2013c). Vitamin C should be tested against exercise-induced bronchoconstriction. *Am. J. Respir. Crit. Care Med.* 188, 1370. doi:10.1164/rccm.201307-1205LE.

Hernandez, M., Zhou, H., Zhou, B., Robinette, C., Crissman, K., Hatch, G., et al. (2009). Combination treatment with high-dose vitamin C and alpha-tocopherol does not enhance respiratory-tract lining fluid vitamin C levels in asthmatics. *Inhal. Toxicol.* 21, 173–181. doi:10.1080/08958370802161077.

Holgate, S. T. (2004). Cytokine and anti-cytokine therapy for the treatment of asthma and allergic disease. in *Cytokine*, 152–157. doi:10.1016/j.cyto.2004.07.010.

Holt, P. G., and Sly, P. D. (2007). Prevention of allergic respiratory disease in infants: Current aspects and future perspectives. *Curr. Opin. Allergy Clin. Immunol.* 7, 547–555. doi:10.1097/ACI.0b013e3282f14a17.

Kalayci, O., Besler, T., Kilinç, K., Sekerel, B. E., and Saraçlar, Y. Serum levels of antioxidant vitamins (alpha tocopherol, beta carotene, and ascorbic acid) in children with bronchial asthma. *Turk. J. Pediatr.* 42, 17–21. Available at: http://www.ncbi.nlm.nih.gov/pubmed/10731863 [Accessed February 15, 2020].

Kanagaratham, C., Kalivodová, A., Najdekr, L., Friedecký, D., Adam, T., Hajduch, M., et al. (2014). Fenretinide prevents inflammation and airway hyperresponsiveness in a mouse model of allergic asthma. *Am. J. Respir. Cell Mol. Biol.* 51, 783–792. doi:10.1165/rcmb.2014-0121OC.

Kelly, F. J., Dunster, C., and Mudway, I. (2003). Air pollution and the elderly: Oxidant/antioxidant issues worth consideration. in *European Respiratory Journal, Supplement* doi:10.1183/09031936.03.00402903.

Kelly, F. J., Mudway, I., Blomberg, A., Frew, A., and Sandstrom, T. (1999). Altered lung antioxidant status in patients with mild asthma. *Lancet* 354, 482–483. doi:10.1016/S0140-6736(99)01812-7.

Kelly, F. J., Mudway, I., Krishna, M. T., and Holgate, S. T. (1995). The free radical basis of air pullution: focus on ozone. *Respir. Med.* 89, 647–656. doi:10.1016/0954-6111(95)90131-0.

Khan, D. A. (2014). Allergic rhinitis and asthma: Epidemiology and common pathophysiology. in *Allergy and Asthma Proceedings* (OceanSide Publications Inc.), 357–361. doi:10.2500/aap.2014.35.3794.

Kianian, F., Karimian, S. M., Kadkhodaee, M., Takzaree, N., Seifi, B., Adeli, S., et al. (2019). Combination of ascorbic acid and calcitriol attenuates chronic asthma disease by reductions in oxidative stress and inflammation. *Respir. Physiol. Neurobiol.* 270, 103265. doi:10.1016/j.resp.2019.103265.

King, M. R., Ismail, A. S., Davis, L. S., and Karp, D. R. (2006). Oxidative Stress Promotes Polarization of Human T Cell Differentiation Toward a T Helper 2 Phenotype. *J. Immunol.* 176, 2765–2772. doi:10.4049/jimmunol.176.5.2765.

Kongerud, J., Crissman, K., Hatch, G., and Alexis, N. (2003). Ascorbic acid is decreased in induced sputum of mild asthmatics. *Inhal. Toxicol.* 15, 101–9. doi:10.1080/08958370304477.

Kowalski, M. L., and Alam, R. (2001). Signal transduction mechanisms as new targets for allergists. *Allergy* 56, 199–203. doi:10.1034/j.1398-9995.2001.056003199.x.

Kurti, S. P., Murphy, J. D., Ferguson, C. S., Brown, K. R., Smith, J. R., and Harms, C. A. (2016). Improved lung function following dietary antioxidant supplementation in exercise-induced asthmatics. *Respir. Physiol. Neurobiol.* 220, 95–101. doi:10.1016/j.resp.2015.09.012.

Leynaert, B., Neukirch, C., Kony, S., Guénégou, A., Bousquet, J., Aubier, M., et al. (2004). Association between asthma and rhinitis according to atopic sensitization in a population-based study. *J. Allergy Clin. Immunol.* 113, 86–93. doi:10.1016/j.jaci.2003.10.010.

Martindale, S., McNeill, G., Devereux, G., Campbell, D., Russell, G., and Seaton, A. (2005). Antioxidant intake in pregnancy in relation to wheeze and eczema in the first two years of life. *Am. J. Respir. Crit. Care Med.* 171, 121–8. doi:10.1164/rccm.200402-220OC.

Maslova, E., Hansen, S., Strøm, M., Halldorsson, T. I., and Olsen, S. F. (2014). Maternal intake of vitamins A, E and K in pregnancy and child allergic disease: A longitudinal study from the Danish National Birth Cohort. *Br. J. Nutr.* 111, 1096–1108. doi:10.1017/S000711451 3003395.

McEvoy, C. T., Schilling, D., Clay, N., Jackson, K., Go, M. D., Spitale, P., et al. (2014). Vitamin C supplementation for pregnant smoking women and pulmonary function in their newborn infants: A randomized clinical trial. *JAMA - J. Am. Med. Assoc.* 311, 2074–2082. doi:10.1001/ jama.2014.5217.

Milan, S. J., Hart, A., and Wilkinson, M. (2013). Vitamin C for asthma and exercise-induced bronchoconstriction. *Cochrane Database Syst. Rev.* 2013, CD010391. doi:10.1002/14651858.CD010391.pub2.

Misso, N. L. A., Brooks-Wildhaber, J., Ray, S., Vally, H., and Thompson, P. J. (2005). Plasma concentrations of dietary and nondietary antioxidants are low in severe asthma. *Eur. Respir. J.* 26, 257–264. doi:10.1183/09031936.05.00006705.

Montaño Velázquez, B. B., Jáuregui-Renaud, K., Bañuelos Arias, A. D. C., Ayala, J. C., Mogica Martínez, M. D., Campillo Navarrete, R., et al. (2006). Vitamin E effects on nasal symptoms and serum specific IgE levels in patients with perennial allergic rhinitis. *Ann. Allergy, Asthma Immunol.* 96, 45–50. doi:10.1016/s1081-1206(10)61039-3.

Moreno-Macias, H., and Romieu, I. (2014). Effects of antioxidant supplements and nutrients on patients with asthma and allergies. *J. Allergy Clin. Immunol.* 133, 1237–1244. doi:10.1016/j.jaci.2014.03.020.

Murr, C., Schroecksnadel, K., Winkler, C., Ledochowski, M., and Fuchs, D. (2005). Antioxidants may increase the probability of developing allergic diseases and asthma. *Med. Hypotheses* 64, 973–977. doi:10.1016/j.mehy.2004.11.011.

Neurauter G et (2004). Wine and grape juice modulate biochemical pathways in stimulated human peripheral blood mononuclear cells. *Pteridines* 15, 1–6.

Nurmatov, U., Devereux, G., and Sheikh, A. (2011). Nutrients and foods for the primary prevention of asthma and allergy: Systematic review and meta-analysis. *J. Allergy Clin. Immunol.* 127, 724-733.e30. doi:10.1016/j.jaci.2010.11.001.

Nwaru, B. I., Erkkola, M., Ahonen, S., Kaila, M., Kronberg-Kippilä, C., Ilonen, J., et al. (2011). Intake of antioxidants during pregnancy and the risk of allergies and asthma in the offspring. *Eur. J. Clin. Nutr.* 65, 937–943. doi:10.1038/ejcn.2011.67.

Patel, B. D., Welch, A. A., Bingham, S. A., Luben, R. N., Day, N. E., Khaw, K. T., et al. (2006). Dietary antioxidants and asthma in adults. *Thorax* 61, 388–393. doi:10.1136/thx.2004.024935.

Patelarou, E., Giourgouli, G., Lykeridou, A., Vrioni, E., Fotos, N., Siamaga, E., et al. (2011). Association between biomarker-quantified antioxidant status during pregnancy and infancy and allergic disease during early childhood: A systematic review. *Nutr. Rev.* 69, 627–641. doi:10.1111/j.1753-4887.2011.00445.x.

Pathinayake, P. S., Hsu, A. C. Y., Waters, D. W., Hansbro, P. M., Wood, L. G., and Wark, P. A. B. (2018). Understanding the unfolded protein response in the pathogenesis of asthma. *Front. Immunol.* 9. doi:10.3389/fimmu.2018.00175.

Pearson, P. J. K., Lewis, S. A., Britton, J., and Fogarty, A. (2004). Vitamin E supplements in asthma: A parallel group randomised placebo controlled trial. *Thorax* 59, 652–656. doi:10.1136/thx.2004.022616.

Picado, C., Deulofeu, R., Lleonart, R., Agustí, M., Mullol, J., Torra, M., et al. (2001). Dietary micronutrients/antioxidants and their relationship with bronchial asthma severity. *Allergy Eur. J. Allergy Clin. Immunol.* 56, 43–49. doi:10.1034/j.1398-9995.2001.00793.x.

Polk, B. I., and Bacharier, L. B. (2019). Potential Strategies and Targets for the Prevention of Pediatric Asthma. *Immunol. Allergy Clin. North Am.* 39, 151–162. doi:10.1016/j.iac.2018.12.010.

Ram, F. S. F., Rowe, B. H., and Kaur, B. (2004). Vitamin C supplementation for asthma. *Cochrane database Syst. Rev.*, CD000993. doi:10.1002/14651858.CD000993.pub2.

Reifen, R., Berkovich, Z., and Mandelberg, A. (2015). Vitamin A supplementation via aerosol spray in asthmatic children. *Pediatr. Allergy Immunol.* 26, 578–579. doi:10.1111/pai.12443.

Riccioni, G., Mancini, B., Bucciarelli, T., Di Ilio, C., and D'Orazio, N. (2007). Minireview: Antioxidant vitamin supplementation in asthma. *Ann. Clin. Lab. Sci.* 37, 96–101.

Romieu, I., Sienra-Monge, J. J., Ramírez-Aguilar, M., Téllez-Rojo, M. M., Moreno-Macías, H., Reyes-Ruiz, N. I., et al. (2002). Antioxidant supplementation and lung functions among children with asthma exposed to high levels of air pollutants. *Am. J. Respir. Crit. Care Med.* 166, 703–709. doi:10.1164/rccm.2112074.

Romieu, I., and Trenga, C. (2001). Diet and Obstructive Lung Diseases. *Epidemiol. Rev.* 23, 268–287. doi:10.1093/oxfordjournals.epirev.a000806.

Rubin, R. N., Navon, L., and Cassano, P. A. (2004). Relationship of Serum Antioxidants to Asthma Prevalence in Youth. *Am. J. Respir. Crit. Care Med.* 169, 393–398. doi:10.1164/rccm.200301-055oc.

Sackesen, C., Ercan, H., Dizdar, E., Soyer, O., Gumus, P., Tosun, B. N., et al. (2008). A comprehensive evaluation of the enzymatic and nonenzymatic antioxidant systems in childhood asthma. *J. Allergy Clin. Immunol.* 122, 78–85. doi:10.1016/j.jaci.2008.03.035.

Sahiner, U. M., Birben, E., Erzurum, S., Sackesen, C., and Kalayci, Ö. (2018). Oxidative stress in asthma: Part of the puzzle. *Pediatr. Allergy Immunol.* 29, 789–800. doi:10.1111/pai.12965.

Santanam, N., Kavtaradze, N., Murphy, A., Dominguez, C., and Parthasarathy, S. (2013). Antioxidant supplementation reduces endometriosis-related pelvic pain in humans. *Transl. Res.* 161, 189–195. doi:10.1016/j.trsl.2012.05.001.

Seyedrezazadeh, E., Moghaddam, M. P., Ansarin, K., Vafa, M. R., Sharma, S., and Kolahdooz, F. (2014). Fruit and vegetable intake and risk of wheezing and asthma: a systematic review and meta-analysis. *Nutr. Rev.* 72, 411–28. doi:10.1111/nure.12121.

Shaheen, S. O., Newson, R. B., Henderson, A. J., Emmett, P. M., Sherriff, A., Cooke, M., et al. (2004). Umbilical cord trace elements and minerals and risk of early childhood wheezing and eczema. *Eur. Respir. J.* 24, 292–7. doi:10.1183/09031936.04.00117803.

Sienra-Monge, J. J., Ramirez-Aguilar, M., Moreno-Macias, H., Reyes-Ruiz, N. I., Del Río-Navarro, B. E., Ruiz-Navarro, M. X., et al. (2004). Antioxidant supplementation and nasal inflammatory responses among young asthmatics exposed to high levels of ozone. *Clin. Exp. Immunol.* 138, 317–322. doi:10.1111/j.1365-2249.2004.02606.x.

Spergel, J. M. (2010). From atopic dermatitis to asthma: The atopic march. *Ann. Allergy, Asthma Immunol.* 105, 99–106. doi:10.1016/j.anai.2009.10.002.

Tecklenburg, S. L., Mickleborough, T. D., Fly, A. D., Bai, Y., and Stager, J. M. (2007). Ascorbic acid supplementation attenuates exercise-induced bronchoconstriction in patients with asthma. *Respir. Med.* 101, 1770–1778. doi:10.1016/j.rmed.2007.02.014.

Thorburn, A. N., Macia, L., and Mackay, C. R. (2014). Diet, metabolites, and "western-lifestyle" inflammatory diseases. *Immunity* 40, 833–42. doi:10.1016/j.immuni.2014.05.014.

Trenga, C. A., Koenig, J. Q., and Williams, P. V (2001). Dietary antioxidants and ozone-induced bronchial hyperresponsiveness in adults with asthma. *Arch. Environ. Health* 56, 242–9. doi:10.1080/0003989 0109604448.

Vollbracht, C., Raithel, M., Krick, B., Kraft, K., and Hagel, A. F. (2018). Intravenous vitamin C in the treatment of allergies: an interim subgroup analysis of a long-term observational study. *J. Int. Med. Res.* 46, 3640–3655. doi:10.1177/0300060518777044.

Wilkinson, M., Hart, A., Milan, S. J., and Sugumar, K. (2014). Vitamins C and E for asthma and exercise-induced bronchoconstriction. *Cochrane Database Syst. Rev.* 2014. doi:10.1002/14651858.CD010749.pub2.

Wiser, J., Alexis, N. E., Jiang, Q., Wu, W., Robinette, C., Roubey, R., et al. (2008). In vivo gamma-tocopherol supplementation decreases systemic oxidative stress and cytokine responses of human monocytes in normal and asthmatic subjects. *Free Radic. Biol. Med.* 45, 40–9. doi:10.1016/j.freeradbiomed.2008.03.002.

Wood, L. G., Gibson, P. G., and Garg, M. L. (2003). Biomarkers of lipid peroxidation, airway inflammation and asthma. *Eur. Respir. J.* 21, 177–186. doi:10.1183/09031936.03.00017003a.

Wood, L. G., Shivappa, N., Berthon, B. S., Gibson, P. G., and Hebert, J. R. (2015). Dietary inflammatory index is related to asthma risk, lung function and systemic inflammation in asthma. *Clin. Exp. Allergy* 45, 177–183. doi:10.1111/cea.12323.

Wu, H., Zhang, C., Wang, Y., and Li, Y. (2018). Does vitamin E prevent asthma or wheeze in children: A systematic review and meta-analysis. *Paediatr. Respir. Rev.* 27, 60–68. doi:10.1016/j.prrv.2017.08.002.

Yildiz, A., Gullulu, C. B., Ocak, N., Ersoy, A., Sag, S., Oruc, A., et al. (2015). Fluvastatin Decreases Oxidative Stress in Kidney Transplant Patients. *Transplant. Proc.* 47, 2870–2874. doi:10.1016/j.transproceed.2015.10.027.

In: Nutraceuticals: Evidence Based Use
Editor: Raluca Maria Pop
ISBN: 978-1-53618-052-7

Chapter 7

VITAMIN D IN VIRAL INFECTION

Gavriela Feketea[1,*], Octavia Sabin[2] and Ioana Corina Bocşan[2]

[1]Pediatric Allergy Outpatient Clinic, Department of Pediatrics, Karamandaneio Children's Hospital, Patras, Greece

[2]Department of Pharmacology, Toxicology and Clinical Pharmacology, "Iuliu Haţieganu" University of Medicine and Pharmacy, Cluj Napoca, Romania

ABSTRACT

Vitamin D has multiple, important roles in the human organism. It was first recognized for its basic function in bone mineralization and controlling calcium homeostasis. Additionally, through its vitamin D receptor (VDR), it acts as a hormone in the modulation of the innate and adaptive immune responses to various types of viral infection.

Observational studies have demonstrated a significant association between a low vitamin D status and an increased risk of certain viral infections, but the findings of randomized controlled trials have been conflicting. This chapter discusses various aspects of vitamin D interaction

* Corresponding Author's E-mail: gabychri@otenet.gr; Fechetea.Gabriela@umfcluj.ro.

with viral infections, based mainly on *in vivo*, and to a lesser extent on *in vitro* studies, aimed at a better understanding of the clinical benefits of vitamin D.

Keywords: 25-hydroxyvitamin D [25(OH)D, calcidiol], 1,25-dihydroxyvitamin D [1,25(OH)2D, calcitriol], viral infection

INTRODUCTION

Vitamin D is a vitamin that is crucial to the maintenance of bone health, but it also plays an important role as a regulator of both the innate and the adaptive immune response.

It is generally accepted that vitamin D deficiency is a global health problem, with adverse effects, not only on calcium homeostasis and bone health but also in non-skeletal diseases, including viral infections.

VITAMIN D

Sources and Metabolism

Vitamin D in the human organism comes from both internal manufacture, which is dependent on exposure to sunlight, and from food and medicinal supplements. Ultraviolet B radiation converts 7-dehydrocholesterol to pre-vitamin D3 and subsequently to vitamin D3 (Holick 2007). At the equator, during the summer and without sunscreen, human skin will produce approximately 10,000 IU of vitamin D per hour (Weiss and Litonjua 2015). Various foods provide vitamins D2 and D3. Supplements prescribed for treatment purposes in the US contain D2, while those taken for prevention, and all European supplements, contain D3. Vitamin D from all sources is metabolized in the liver to 25-hydroxyvitamin D [25(OH)D, calcidiol] which is further transformed in the kidneys by the enzyme 25-hydroxyvitamin D-1α-hydroxylase [1α(OH)ase, CYP27B1] to its active form, 1,25-dihydroxyvitamin D [1,25(OH)2D, calcitriol]. In addition to the kidney's endocrine production of circulating 1α,25 (OH)2D(3), researchers have detected paracrine production of this steroid hormone in organs other than the kidney (Norman 2008). Synthesis of the biologically active metabolite 1,25(OH)2D takes place intracellularly, where the major circulating metabolite, 25(OH)D is hydroxylated by 1α(OH)ase, CYP27B1 (Rosen 2011) (Figure 1). The effects of 1,25(OH)2D are mediated through specific high-affinity vitamin D receptors (VDR) via upregulating or downregulating target genes (Zdrenghea et al. 2017).

Functions

Vitamin D has multiple, important roles in the human organism. The first function to be recognized and investigated was its role in bone mineralization, controlling calcium homeostasis. Later it was observed that through its receptors, the VDRs, it acts as a hormone, in the modulation of both the innate and the adaptive immune response to viral infections.

Classical (Skeletal) Functions

Vitamin D plays a crucial role in bone mineralization and other related processes, such as calcium and phosphate homeostasis and skeletal growth. It has long been recognized that vitamin D deficiency or insufficiency can result in rickets, skeletal abnormalities and short stature in children, while in adults it is associated with osteomalacia, osteoporosis and an increased risk of fractures (Hossein-nezhad and Holick 2013).

Non-Classical (Non-Skeletal) Functions

In addition to its skeletal effects, vitamin D produces a variety of extraskeletal effects, mainly via its receptors, the VDRs. Many types of cells and various organs in the body are equipped with VDRs and also can produce 1,25(OH)2D, influencing a large number of biological pathways (Rosen, Adams, et al. 2012). In addition to the well-researched functions in calcium homeostasis, vitamin D, via the VDRs, modulates both the innate and the adaptive immune response, and plays a key role in the balance between T-helper1 and T-helper 2 (Th1-Th2) cytokines (Figure 2) (Gruber-Bzura 2018). *In vitro* studies suggest that vitamin D induces a shift in the balance between Th1 and Th2 cytokines towards Th2 dominance (Telcian et al. 2017).

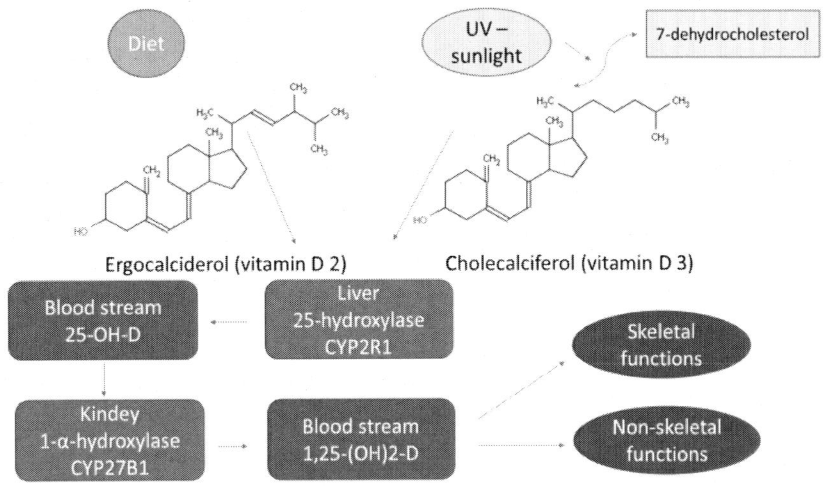

Figure 1. Vitamin D - sources and metabolism. Modified from Holick (Holick 2007).

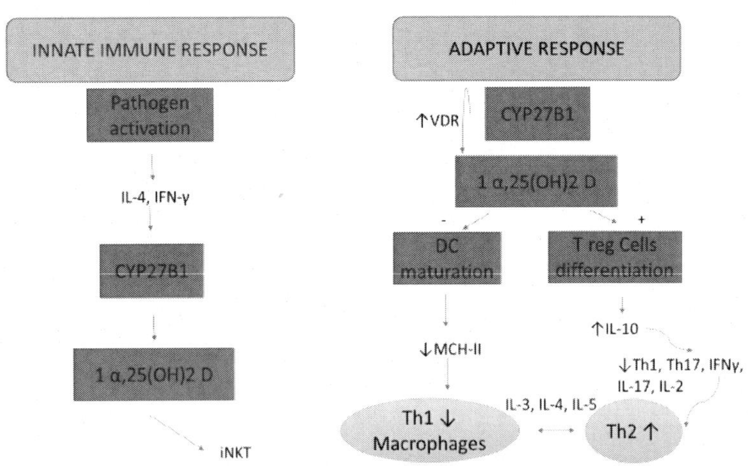

Figure 2. Vitamin D and the immune response (Modified from Gruber-Bzura) (Gruber-Bzura 2018).

Nutrigenomics

Nutrigenomics is the study of how environmental factors, including food intake, influence the expression of the genome, and action of vitamin D3 is an example of nutrigenomics. Its active form, 1α,25OH2D3 binds with a high affinity to the VDR at thousands of loci in the human genome and directly affects the epigenome and transcriptome (Carlberg 2019, Carlberg 2018).

Epidemiological and interventional studies have indicated that individuals differ from each other in terms of both vitamin D levels and their response to vitamin D supplementation. Although it is not yet known which genomic and epigenomic variations determine the vitamin D response index in the general population, it is assumed that this index is an intrinsic property of each individual and does not change during the lifetime (Carlberg 2019).

Requirements and Status

Requirements

Needs for vitamin D intake depend on latitude, season, ethnicity, age, body weight, health status, dress habits and use of sunscreen creams (Medicine 2011). Many experts suggest that, when exposure to sunlight is unable to provide it, both children and adults should take ≥ 800-1,000 IU vitamin D/day from dietary and supplemental sources. In individuals affected by obesity, high doses of vitamin D may be required to normalize serum vitamin D levels, taking into consideration the varying degrees of liposolubility of different supplements (Migliaccio et al. 2019). Sacheck and colleagues showed that children at risk of vitamin D deficiency achieved a higher mean serum level of 25(OH)D after 3 months of daily supplementation with 2,000 IU/day than with 600 and 1,000 IU/day, doses closer to the current recommended daily allowance (Sacheck et al. 2017).

The Endocrine Society Practice Guidelines Committee (Table 1) and the European Academy of Paediatrics (EAP) recommend for prevention in infants (0-1 year) and children (1-18 years) at risk of vitamin D deficiency, dosages of 400-1,000 IU/day, and 600-1,000 IU/day, respectively (Grossman et al. 2017). For adults, the recommended doses are shown in Table 2, based on the guidelines of the Endocrine Society Committee (Rosen, Abrams, et al. 2012).

Table 1. Vitamin D requirements in children and adolescents: Doses recommended by the Endocrine Society Practice Guidelines Committee, for prevention and treatment of vitamin D deficiency, and upper limits for administration without risk of adverse events

Recommended daily dose of vitamin D (IU)	0-6 months	6-12 months	1-3 years	4-8 years	8-18 years
Prevention	400-1,000		600-1,000		
Treatment of deficiency [for at least 6 weeks, to achieve a blood level of 25(OH)D* > 30 ng/ml]	2,000		2,000		
Tolerated upper limit	1,000	1,500	2,500	3,000	4,000

*25(OH)D = 25 hydroxy vitamin D

Table 2. Vitamin D requirements in adults: Doses recommended by the Endocrine Society Practice Guidelines Committee, for prevention and treatment of vitamin D deficiency, and upper limits for administration without risk of adverse events

Recommended daily dose of vitamin D (IU)	Pregnancy	Lactation	18-70 years	>70 years
Prevention	1,500-2,000	1,500-2,000	1,500-2,000	1,500-2,000
Treatment of deficiency [for at least 6 weeks, to achieve a blood level of 25(OH)D* > 30 ng/ml]	10,000	10,000	10,000	10,000
Tolerated upper limit	10,000	10,000	10,000	10,000

*25(OH)D = 25 hydroxy vitamin D

For the treatment of vitamin D deficiency, 2,000 IU/day for at least 6 weeks has been suggested, to achieve a serum level of 25(OH)D > 30 ng/ml, followed by maintenance doses as for prevention. Administration of vitamin D to patients with low baseline 25(OH)D levels should be in doses sufficient to achieve normal levels (i.e., treatment of deficiency, not supplementation). Studies to date have produced conflicting findings of the relative efficacy of the forms of vitamin D, but both vitamin D2 and D3 are recommended (Holick et al. 2011).

Few pediatricians, however, apply these guidelines in practice. Del Giudice and colleagues report that although most primary care providers are aware that vitamin D deficiency is common, fewer than half currently recommend 600 to 1,000 IU supplementation for their pediatric patients (DelGiudice et al. 2018).

Status

The serum or plasma level of total 25(OH)D is currently used as the indicator of vitamin D status. Almost all professional associations define vitamin D status as sufficient when the level of 25(OH)D is at least 30 ng/mL, insufficient at 21-29 ng/mL and deficient at < 20ng/mL. The European Society for Paediatric Gastroenterology, Hepatology and Nutrition (ESPGAHN) characterizes as sufficient concentrations of serum 25(OH)D > 50 nmol/L (20 ng/ml), and as severe deficiency concentrations < 25 nmol/L (10 ng/ml). To convert serum 25(OH) D from ng/ml to nmol/L, multiply by 2.496 (Davidson and Alansari 2017). A large review conducted by the Institute of Medicine (IOM) concluded that a serum level of 25(OH)D serum level equal to or greater than 20 ng/mL is adequate to maintain bone health (Medicine 2011).

Vitamin D deficiency is a health problem worldwide. Even in a sunny country, children and adults living in urban areas are at risk of low vitamin D levels during the winter months in both the northern and the southern hemispheres. The current recommendations for vitamin D screening concern only groups of people at risk for vitamin D deficiency, such as people with increased skin melanin pigmentation, pregnant women, obese children and adults, individuals with an indoor lifestyle, and those wearing sun protection during sun exposure or consuming mostly foods not fortified with vitamin D (Figure 3) (Holick 2017).

Viral Infections

Many studies have been published regarding the relationship between the level of 25(OH)D and various viral infections. Observational and epidemiological studies have clearly revealed an association between vitamin D deficiency and viral infections. The inflammatory processes involved during the clinical course of viral diseases would be expected to reduce the level of 25(OH)D, explaining why low vitamin D status is reported in a wide range of viral disorders (Autier et al. 2014). Trials investigating the influence of vitamin D supplements on the prevention of viral infection and its clinical evolution have had mixed findings, however. In a meta-analysis of data from 16 birth cohorts, Feng showed that increased *in utero* exposure to 25(OH)D is inversely related to the risk of asthma and wheeze in the offspring, but not respiratory tract infections (Feng et al. 2017). Most of the studies, however, used self-reported or parent-reported symptoms as the outcome, and only a few studies identified and specified the exact type of viral infection.

Viral Pathogens Associated with Acute Respiratory Infections

Low serum levels of vitamin D have been reported to be related with an increased frequency of viral respiratory infections (Monlezun et al. 2015, Walker and Modlin 2009, Sabetta et al. 2010, Ginde, Mansbach, and Camargo 2009, Camargo 2018), but only a few reports have identified the exact type of viral infection in vitamin D studies.

The viral pathogens most commonly associated with acute respiratory infections are the influenza virus, rhinoviruses and respiratory syncytial virus (RSV).

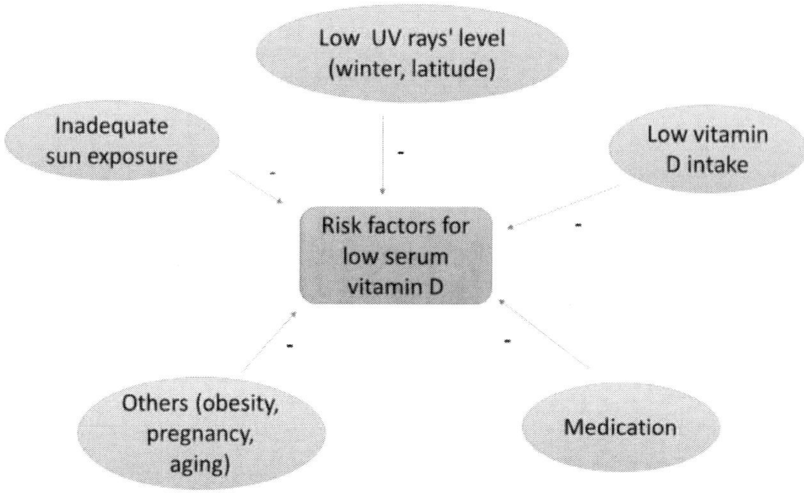

Figure 3. Risk factors for vitamin D deficiency/insufficiency (Modified from Holick (Holick 2017).

Influenza Virus

The influenza viruses (orthomyxoviruses) constitute the genus *Orthomyxovirus*, which consists of three types, A, B, and C. These viruses cause influenza, an acute respiratory disease with prominent systemic symptoms (RB 1996).

Urashima and colleagues studied the effect of 1,200 IU/day of vitamin D supplementation compared with placebo on the incidence of seasonal influenza A and B among children aged 6 to 15 years, and found a reduced incidence of influenza A but not of influenza B infection (Urashima et al. 2010).

A Canadian multisite randomized clinical trial showed that daily high-dose oral vitamin D (2,000 UI) was not superior to a maintenance dose (400 UI) into preventing viral respiratory infections in healthy children aged 1 to 5 years, even in the case of low baseline vitamin D status (Aglipay et al. 2017, Aglipay and Maguire 2017). The number of laboratory-confirmed viral respiratory infections was slightly higher in the high-dose group than the standard-dose group, although not to a statistically significant degree. The incidence of influenza infection in the high-dose group was reduced by 50%, but only a few influenza infections were recorded in comparison with other types of infection (4.4% of all infections in the high-dose group and 8.5% of all infections in the standard-dose group) (Aglipay et al. 2017).

In a recent study in Vietnam of 1,300 healthy children and adolescents between the ages of 3 and 17 years, 650 individuals were randomly assigned to taking 14,000 UI vitamin D weekly and 650 to taking placebo, for 8 months. No significant difference was observed between the vitamin D and placebo groups for influenza confirmed by reverse transcription-polymerase chain reaction (RT-PCR), but non-influenza respiratory viral infections were significantly reduced in the vitamin D group. When considering all respiratory virus infections, including influenza, the effect of vitamin D in reducing infection was also significant (Loeb et al. 2019).

Respiratory Syncytial Virus (RSV)

RSV belongs to the *Paramyxoviridae* family which consists of three genera: *Paramyxovirus*, *Morbillivirus* and *Pneumovirus*, the last of which includes RSV (G 1996).

In a prospective birth cohort study conducted on 156 newborn infants in the Netherlands, the results indicated that vitamin D deficient newborn infants had a greater risk of subsequently developing RSV infection than vitamin D sufficient infants (Belderbos et al. 2011). Beigelman and colleagues found that among infants who required hospital-based care for acute RSV bronchiolitis, vitamin D status at the time of bronchiolitis was not associated with the severity of the bronchiolitis (Beigelman et al. 2015). Conversely, a recent study of 68 patients with RSV bronchiolitis revealed that vitamin D deficiency was significantly associated with admission to the intensive care unit (ICU) and with mechanical ventilation, but not with a length of hospital stay of greater than 3 days (Hurwitz et al. 2017).

Rhinoviruses

The family *Picornaviridae* comprises five genera, including rhinoviruses. Rhinoviruses infect and replicate mainly in the nasopharyngeal epithelium and cause respiratory infections including the common cold. To date, 115 serotypes have been identified (Yin-Murphy M 1996).

A number of *in vitro* studies have investigated the effects of vitamin D and its metabolites on the immune responses of various cells to respiratory viruses. Telcian and colleagues showed that both RSV and rhinoviruses decrease VDR and 25(OH)ase in epithelial cells. Furthermore, vitamin D decreased the replication and release of rhinovirus and increased the activity of rhinovirus induced interferon-stimulated genes, producing the antimicrobial polypeptide cathelicidin, which exerts direct anti-rhinovirus activity (Telcian et al. 2017).

Hepatitis Viruses

Five viruses are responsible for most of the cases of viral hepatitis: hepatitis A virus (HAV), hepatitis B virus (HBV), hepatitis C virus (HCV), hepatitis D virus (HDV) and hepatitis E virus (HEV). Though all hepatitis viruses can cause acute hepatitis, only HBV, HCV and HDV often give rise to chronic hepatitis, and of these, HBV and HCV are responsible for 96% of the mortality from viral hepatitis (WHO 2017).

To date, published studies investigating the possible link between hepatitis viruses and vitamin D concern only HBV and HCV.

Hepatitis B Virus (HBV)

HBV belongs to the *Hepadnaviridae* family, *Orthohepadnavirus* genus. It is a small (3.2 kb size, 42nm diameter) double-stranded DNA virus, which is transmitted predominantly by blood and blood products, injected drug use and unsafe health-care injections (Howard 2002).

In 2015, according to the World Health Organization, the global prevalence of HBV infection in the general population was 3.5%, with the highest rates in the African (6.1%) and Western Pacific regions (6.2%). The global prevalence of HBV infection in human immunodeficiency virus (HIV)-infected patients, according to the same WHO report, was 4% (WHO 2017).

Hoan and colleagues showed that vitamin D deficiency and severe deficiency are more frequent in patients with HBV-related liver disease and that deficiency is significantly associated with end-stage cirrhosis of the liver (Hoan et al. 2016). In a German adult population with chronic HBV infection, a strong association was demonstrated between higher levels of HBV replication and low serum levels of 25(OH)D (Farnik et al. 2013). In that study, only a few patients had greater degrees of liver fibrosis, and thus the authors concluded that HBV-infected patients are at high risk of vitamin D deficiency, even in the absence of significant liver fibrosis (Farnik et al. 2013).

The recent systematic review and meta-analysis of Hu and colleagues included 7 observational studies published between 2012 and 2017, involving a total of 814 patients with HBV infection and 696 healthy control subjects from Asia, the Middle East, and Europe. The authors of the review demonstrated significantly lower levels of vitamin D in patients with chronic hepatitis B than in the healthy control subjects, with an inverse correlation between the serum vitamin D level and the HBV viral load (Hu et al. 2019). Vitamin D deficiency appears to be a risk factor for an adverse clinical outcome, such as cirrhotic complications and hepatocellular carcinoma (HCC), and/or death (Wong et al. 2015).

Hepatitis C Virus (HCV)

HCV belongs to the *Flaviridae* family, *Hepacivirus* genus. It is a small (30-60 nm diameter) single-stranded RNA virus with wide genetic variability, transmitted principally by blood and blood products (Howard 2002). In 2015, the global prevalence of HCV infection in the general population was 1%, significantly lower than that of HBV. The highest prevalence was recorded in the Eastern Mediterranean Region (2.3%) followed by the European Region (1.5%). Among HIV-infected patients, the prevalence of HCV infection was 6.2% (WHO 2017).

Almost 60% patients with HCV infection have been reported to present hypovitaminosis D in different regions over the past decade (Jin, Chen, and Sheng 2018). Although the findings of epidemiological studies suggest a link between vitamin D deficiency or insufficiency and HCV infection, no beneficial role of vitamin D supplementation has yet been demonstrated. Vitamin D synthesized in the skin under the influence of UV rays, and that from dietary sources may be stored in the adipocytes, or it may undergo hepatic 25-hydroxylation, which is mediated by the 25-hydroxylases, isoforms of the P450 cytochrome (CYP2R1, CYP27A1), to produce 25(OH)D or calcidiol (Iruzubieta et al. 2014). Most of the 25(OH)D is then converted to its active form, calcitriol, 1,25(OH)2D, mainly in the kidneys and to a lesser degree in various other tissues, including liver, adipose tissue, macrophages and dendritic cells (Holick 2007).

Vitamin D is a critical regulator of immunity, affecting the natural and acquired immune response, and thus is involved in the regulation of the inflammatory response and fibrinogenesis, especially in liver tissue (Udomsinprasert and Jittikoon 2019). The VDR is expressed in multiple cells of the body in addition to osteoblasts, including cells involved in immune modulation, such as mononuclear cells, macrophages, and activated T and B lymphocytes and dendritic cells, and in most organs in the body including the liver (Marino and Misra 2019). It is widely expressed in the inflammatory cells of patients with chronic hepatic disease and its expression is negatively correlated with the severity of hepatic histology in patients with chronic hepatitis C, suggesting that the vitamin D/VDR system may play a role in the progression of viral chronic liver damage (Barchetta et al. 2012). Arteh and colleagues reported that vitamin D deficiency is very common in patients with chronic liver disease. They found that, although at least 1/3 of them are vitamin D deficient (<30 ng / mL), there was no difference in the prevalence of severe vitamin D deficiency between patients with HCV and non-HCV patients (Arteh, Narra, and Nair 2010). Low vitamin D level has been linked with a low sustained virological response (SVR) to interferon-based therapy in patients with chronic HCV infection (Petta et al. 2010). A meta-analysis of 11 studies including a total of 1,575 patients infected with HCV demonstrated, in addition to a high prevalence of vitamin D deficiency in these individuals, a high SVR in patients with higher serum levels of vitamin D or who were receiving vitamin D supplementation (Villar et al. 2013). Vitamin D related gene polymorphisms (GC, DHCR7, CYP2R1, CYP27B1A, VDR gene) have been shown to explain the interactions between vitamin D deficiency and HCV infection (Jin, Chen, and Sheng 2018).

HUMAN IMMUNODEFICIENCY VIRUS (HIV)

HIV infection continues to be one of the most important health issues worldwide. Global HIV mortality peaked in 2006 with 1.95 million deaths, and the rate has since decreased to 95 million deaths in 2017. HIV belongs to the family *Retroviridae*, *Lentivirus* genus and is a small virus containing 2 single RNA (Volberding, Baker, and Levine 2003). It initially infects human T cells, followed by dendritic cells, monocytes, and macrophages, dramatically compromising the immune function and finally causing the acquired immunodeficiency syndrome (AIDS). HIV can be transmitted through sexual activity, and by intravascular inoculation during the transfusion of infected blood products or by sharing injection equipment for drugs (Azmi, Hassou, and Ennaji 2020)

Patients with HIV infection may have lower levels of vitamin D, although, based on a large number of epidemiological studies, a review conducted by Mansueto and colleagues presented data showing that vitamin D deficiency in HIV-infected persons is similar to that in the general population, ranging from 70.3% to 83.7% (Mansueto et al. 2015). Hidron and colleagues revealed independent risk factors for vitamin D deficiency in HIV-infected patients, including the black race and the winter season, while increasing age and antiretroviral therapy was associated with less vitamin D deficiency (Hidron et al. 2015). In a recent review published by Teymoori-Rad and colleagues, low levels of vitamin D appeared to be associated with HIV infection and viral load mortality, and a variety of other factors, including preterm birth, hematological disorders, subclinical arterial dysfunction, severe diabetes mellitus (DM), severe cardiovascular and renal diseases, mother-to-child transmission and breast-feeding mother-to-child transmission (Teymoori-Rad et al. 2019). Hence, vitamin D supplementation in HIV infected individuals may be expected to have a beneficial effect on immune recovery, and consequently result in a decrease in the incidence and severity of various comorbidities. A review of 29 clinical studies of vitamin D supplementation in HIV infected patients revealed that there was a decrease in inflammation markers associated with bone turnover. Vitamin D supplementation appears to have no impact on cardiovascular disease and is not associated with an increase in the CD4 T cell count (Alvarez, Aguilar-Jimenez, and Rugeles 2019).

Overall, vitamin D deficiency may contribute to the pathogenesis of HIV infection by negative modulation of the innate and adaptive immune responses. Exogenous vitamin D supplementation appears to be beneficial as it may reverse some of the adverse changes in the immune system, especially in individuals with more severe vitamin D deficiency (Jimenez-Sousa et al. 2018).

HERPESVIRUSES

Herpesviruses are classified into three groups: α herpesviruses, namely herpes simplex virus (HSV) types 1 and 2, and varicella-zoster virus (VZV); β herpesviruses, including cytomegalovirus, and human herpesviruses 6 and 7; and γ herpesviruses, including the Epstein-Barr virus (EBV) and human herpesvirus 8 (RJ. 1996).

Epstein Barr Virus (EBV)

Several studies have suggested a link between EBV infection and vitamin D status in individuals with multiple sclerosis (MS), although the results are conflicting. On the one hand, Ramien and colleagues observed that low serum levels of 25(OH)D had no impact on the EBV load or the anti-EBV nuclear antigen status (Ramien et al. 2014). On the other hand,

Perez-Perez and colleagues, in a recent study of 482 patients with MS, found a significantly higher EBV viral load when 25(OH)D levels were low (Perez-Perez et al. 2018). In individuals without comorbidities, Maloney and colleagues found that only one patient of 9 with a positive test for EBV, had a normal 25(OH)D serum level (Maloney, Almarines, and Goolkasian 2014).

Cytomegalovirus (CMV), Herpes Simplex Virus (HSV 1, 2), Varicella-Zoster Virus (VZV)

Several observational studies showed that in certain conditions, including MS, and patients on chronic dialysis, a low serum level of vitamin D was linked with the serum level of viral antibody. Mowry and colleagues reported that in 140 subjects with childhood-onset MS or clinically isolated syndrome vitamin D status was weakly associated with the levels of antibody to CMV, but not to EBV, or HSV 1 or 2 (Mowry et al. 2011). A study on a multicenter hemodialysis cohort revealed significantly lower VZV-IgG in chronic hemodialysis patients with 25(OH)D <30ng/ml (insufficiency or deficiency) than those with sufficient 25(OH)D (Chao et al. 2014).

Other Viruses: Human Papilloma Virus (HPV), Rotavirus

Results from a US nationally representative sample demonstrated that a low serum level of 25(OH)D is associated with the prevalence of cervicovaginal human papillomavirus HPV infection in sexually active women (Shim et al. 2016). To date, there are no human studies regarding vitamin D influence on other rotavirus infections.

In pig and porcine small intestinal epithelial cell line models, Tian and colleagues found that vitamin D3 attenuates rotavirus infection through regulating autophagic maturation and porcine cathelicidin gene expression (Tian et al. 2016).

Conclusion

The discrepancy between the findings of observational and intervention studies suggests that a low serum level of 25(OH)D is a biomarker of certain diseases, rather than a preexistent condition or a predisposing factor. The debate continues whether vitamin D supplementation in individuals with normal serum levels of vitamin D is of any benefit for preventing various viral infections. It appears that vitamin D supplementation may influence the clinical course and severity of certain viral infections in patients with vitamin D deficiency and/or insufficiency.

To date, based on the current published evidence, the clearest available statement supports the use of daily or weekly vitamin D supplements as prophylaxis and influencer of the clinical course of specific viral infections, particularly in individuals with severe vitamin D deficiency.

References

Aglipay, M. & Maguire, J. L. (2017). "Vitamin D Supplementation and Upper Respiratory Tract Infections in Children-Reply." *JAMA*, *318* (21), 2139-2140. doi: 10.1001/jama.2017.15154.

Aglipay, M., Birken, C. S., Parkin, P. C., Loeb, M. B., Thorpe, K., Chen, Y., Laupacis, A., Mamdani, M., Macarthur, C., Hoch, J. S., Mazzulli, T. & Maguire, J. L. (2017). "Effect of High-Dose vs Standard-Dose Wintertime Vitamin D Supplementation on Viral Upper Respiratory Tract Infections in Young Healthy Children." *Jama*, *318* (3), 245-254. doi: 10.1001/jama.2017.8708.

Alvarez, N., Aguilar-Jimenez, W. & Rugeles, M. T. (2019). "The Potential Protective Role of Vitamin D Supplementation on HIV-1 Infection." *Front Immunol*, *10*, 2291. doi: 10.3389/fimmu.2019.02291.

Arteh, J., Narra, S. & Nair, S. (2010). "Prevalence of vitamin D deficiency in chronic liver disease." *Dig Dis Sci*, *55* (9), 2624-8. doi: 10.1007/s10620-009-1069-9.

Autier, P., Boniol, M., Pizot, C. & Mullie, P. (2014). "Vitamin D status and ill health: a systematic review." *Lancet Diabetes Endocrinol*, *2* (1), 76-89. doi: 10.1016/s2213-8587(13)70165-7.

Azmi, Hasna, Najwa Hassou, & Moulay Mustapha Ennaji. (2020). "Chapter 22 - Vitamin D Immunomodulatory Role in Chronic and Acute Viral Diseases." In *Emerging and Reemerging Viral Pathogens*, edited by Moulay Mustapha Ennaji, 489-506. Academic Press.

Barchetta, I., Carotti, S., Labbadia, G., Gentilucci, U. V., Muda, A. O., Angelico, F., Silecchia, G., Leonetti, F., Fraioli, A., Picardi, A., Morini, S. & Cavallo, M. G. (2012). "Liver vitamin D receptor, CYP2R1, and CYP27A1 expression: relationship with liver histology and vitamin D3 levels in patients with nonalcoholic steatohepatitis or hepatitis C virus." *Hepatology*, *56* (6), 2180-7. doi: 10.1002/hep.25930.

Beigelman, Avraham, Mario Castro, Toni L. Schweiger, Brad S. Wilson, Jie Zheng, Huiquing Yin-DeClue, Geneline Sajol, Tusar Giri, Oscar L. Sierra, Megan Isaacson-Schmid, Kaharu Sumino, Kenneth B. Schechtman, & Leonard B. Bacharier. (2015). "Vitamin D Levels Are Unrelated to the Severity of Respiratory Syncytial Virus Bronchiolitis Among Hospitalized Infants." *Journal of the Pediatric Infectious Diseases Society*, *4* (3), 182-188. doi: 10.1093/jpids/piu042.

Belderbos, M. E., Houben, M. L., Wilbrink, B., Lentjes, E., Bloemen, E. M., Kimpen, J. L., Rovers, M. & Bont, L. (2011). "Cord blood vitamin D deficiency is associated with respiratory syncytial virus bronchiolitis." *Pediatrics*, *127* (6), e1513-20. doi: 10.1542/peds.2010-3054.

Camargo, Carlos A. (2018). "Chapter 112 - Vitamin D, Acute Respiratory Infection, and Asthma/Chronic Obstructive Pulmonary Disease." In *Vitamin D (Fourth Edition)*, edited by David Feldman, 1095-1120. Academic Press.

Carlberg, C. (2019). "Nutrigenomics of Vitamin D." *Nutrients*, *11* (3), doi: 10.3390/nu11030676.

Carlberg, Carsten. (2018). "Vitamin D Genomics: From *In Vitro* to *In Vivo*." *Frontiers in Endocrinology*, *9* (250), doi: 10.3389/ fendo.2018.00250.

Chao, C. T., Lee, S. Y., Yang, W. S., Yen, C. J., Chiang, C. K., Huang, J. W. & Hung, K. Y. (2014). "Serum vitamin D levels are positively associated with varicella zoster immunity in chronic dialysis patients." *Sci Rep*, *4*, 7371. doi: 10.1038/srep07371.

Couch, R. B. (1996). Chapter 58: Orthomyxoviruses. In *Medical Microbiology*, edited by Baron S: Galveston (TX): University of Texas Medical Branch at Galveston.

Davidson, B. L. & Alansari, K. (2017). "Vitamin D Supplementation and Upper Respiratory Tract Infections in Children." *Jama*, *318* (21), 2138-2139. doi: 10.1001/jama.2017.15150.

DelGiudice, N. J., Street, N., Torchia, R. J., Sawyer, S. S., Bernard, S. A. & Holick, M. F. (2018). "Vitamin D Prescribing Practices in Primary Care Pediatrics: Underpinnings From the Health Belief Model and Use of Web-Based Delphi Technique for Instrument Validity." *J Pediatr Health Care*, *32* (6), 536-547. doi: 10.1016/j.pedhc.2018.03.003.

Enders, G. (1996). Chapter 59: Paramyxoviruses. In *Medical Microbiology*, edited by Baron S: Galveston (TX): University of Texas Medical Branch at Galveston.

Farnik, H., Bojunga, J., Berger, A., Allwinn, R., Waidmann, O., Kronenberger, B., Keppler, O. T., Zeuzem, S., Sarrazin, C. & Lange, C. M. (2013). "Low vitamin D serum concentration is associated with high levels of hepatitis B virus replication in chronically infected patients." *Hepatology*, *58* (4), 1270-6. doi: 10.1002/hep.26488.

Feng, H., Xun, P., Pike, K., Wills, A. K., Chawes, B. L., Bisgaard, H., Cai, W., Wan, Y. & He, K. (2017). "In utero exposure to 25-hydroxyvitamin D and risk of childhood asthma, wheeze, and respiratory tract infections: A meta-analysis of birth cohort studies." *J Allergy Clin Immunol*, *139* (5), 1508-1517. doi: 10.1016/j.jaci.2016.06.065.

Ginde, A. A., Mansbach, J. M. & Camargo, C. A. Jr. (2009). "Association between serum 25-hydroxyvitamin D level and upper respiratory tract infection in the Third National Health and Nutrition Examination Survey." *Arch Intern Med, 169* (4), 384-90. doi: 10.1001/archinternmed.2008.560.

Grossman, Z., Hadjipanayis, A., Stiris, T., Del Torso, S., Mercier, J. C., Valiulis, A. & Shamir, R. (2017). "Vitamin D in European children-statement from the European Academy of Paediatrics (EAP)." *Eur J Pediatr, 176* (6), 829-831. doi: 10.1007/s00431-017-2903-2.

Gruber-Bzura, B. M. (2018). "Vitamin D and Influenza-Prevention or Therapy?" *Int J Mol Sci, 19* (8), doi: 10.3390/ijms19082419.

Hidron, A. I., Hill, B., Guest, J. L. & Rimland, D. (2015). "Risk factors for vitamin D deficiency among veterans with and without HIV infection." *PLoS One, 10* (4), e0124168. doi: 10.1371/journal.pone.0124168.

Hoan, N. X., Khuyen, N., Binh, M. T., Giang, D. P., Van Tong, H., Hoan, P. Q., Trung, N. T., Anh, D. T., Toan, N. L., Meyer, C. G., Kremsner, P. G., Velavan, T. P. & Song, L. H. (2016). "Association of vitamin D deficiency with hepatitis B virus - related liver diseases." *BMC Infect Dis, 16* (1), 507. doi: 10.1186/s12879-016-1836-0.

Holick, M. F. (2007). "Vitamin D deficiency." *N Engl J Med, 357* (3), 266-81. doi: 10.1056/NEJMra070553.

Holick, M. F. (2017). "The vitamin D deficiency pandemic: Approaches for diagnosis, treatment and prevention." *Rev Endocr Metab Disord, 18* (2), 153-165. doi: 10.1007/s11154-017-9424-1.

Holick, M. F., Binkley, N. C., Bischoff-Ferrari, H. A., Gordon, C. M., Hanley, D. A., Heaney, R. P., Murad, M. H., Weaver, C. M. & Society Endocrine. (2011). "Evaluation, treatment, and prevention of vitamin D deficiency: an Endocrine Society clinical practice guideline." *J Clin Endocrinol Metab, 96* (7), 1911-30. doi: 10.1210/jc.2011-0385.

Hossein-nezhad, A. & Holick, M. F. (2013). "Vitamin D for health: a global perspective." *Mayo Clin Proc, 88* (7), 720-55. doi: 10.1016/j.mayocp.2013.05.011.

Howard, C. R. (2002). "Hepatitis viruses: a pandora's box?" *J Gastroenterol Hepatol, 17* Suppl, S464-7. doi: 10.1046/j.1440-1746.17.s4.11.x.

Hu, Y. C., Wang, W. W., Jiang, W. Y., Li, C. Q., Guo, J. C. & Xun, Y. H. (2019). "Low vitamin D levels are associated with high viral loads in patients with chronic hepatitis B: a systematic review and meta-analysis." *BMC Gastroenterol*, *19* (1), 84. doi: 10.1186/s12876-019-1004-2.

Hurwitz, J. L., Jones, B. G., Penkert, R. R., Gansebom, S., Sun, Y., Tang, L., Bramley, A. M., Jain, S., McCullers, J. A. & Arnold, S. R. (2017). "Low Retinol-Binding Protein and Vitamin D Levels Are Associated with Severe Outcomes in Children Hospitalized with Lower Respiratory Tract Infection and Respiratory Syncytial Virus or Human Metapneumovirus Detection." *J Pediatr*, *187*, 323-327. doi: 10.1016/j.jpeds.2017.04.061.

Iruzubieta, P., Teran, A., Crespo, J. & Fabrega, E. (2014). "Vitamin D deficiency in chronic liver disease." *World J Hepatol*, *6* (12), 901-15. doi: 10.4254/wjh.v6.i12.901.

Jimenez-Sousa, M. A., Martinez, I., Medrano, L. M., Fernandez-Rodriguez, A. & Resino, S. (2018). "Vitamin D in Human Immunodeficiency Virus Infection: Influence on Immunity and Disease." *Front Immunol*, *9*, 458. doi: 10.3389/fimmu.2018.00458.

Jin, C. N., Chen, J. D. & Sheng, J. F. (2018). "Vitamin D deficiency in hepatitis C virus infection: what is old? what is new?" *Eur J Gastroenterol Hepatol*, *30* (7), 741-746. doi: 10.1097/meg.0000000000001134.

Loeb, M., Dang, A. D., Thiem, V. D., Thanabalan, V., Wang, B., Nguyen, N. B., Tran, H. T. M., Luong, T. M., Singh, P., Smieja, M., Maguire, J. & Pullenayegum, E. (2019). "Effect of Vitamin D supplementation to reduce respiratory infections in children and adolescents in Vietnam: A randomized controlled trial." *Influenza Other Respir Viruses*, *13* (2), 176-183. doi: 10.1111/irv.12615.

Maloney, S. R., Almarines, D. & Goolkasian, P. (2014). "Vitamin D levels and monospot tests in military personnel with acute pharyngitis: a retrospective chart review." *PLoS One*, *9* (7), e101180. doi: 10.1371/journal.pone.0101180.

Mansueto, P., Seidita, A., Vitale, G., Gangemi, S., Iaria, C. & Cascio, A. (2015). "Vitamin D Deficiency in HIV Infection: Not Only a Bone Disorder." *Biomed Res Int*, *2015*, 735615. doi: 10.1155/2015/735615.

Marino, R. & Misra, M. (2019). "Extra-Skeletal Effects of Vitamin D." *Nutrients*, *11* (7), doi: 10.3390/nu11071460.

Medicine, Institute of. (2011). *Dietary Reference Intakes for Calcium and Vitamin D*. Edited by A. Catharine Ross, Christine L. Taylor, Ann L. Yaktine and Heather B. Del Valle. Washington, DC: The National Academies Press.

Migliaccio, Silvia, Andrea Di Nisio, Chiara Mele, Lorenzo Scappaticcio, Silvia Savastano, Annamaria Colao, Education Research on behalf of Obesity Programs of nutrition, & Group Assessment. (2019). "Obesity and hypovitaminosis D: causality or casualty?" *International Journal of Obesity Supplements*. doi: 10.1038/s41367-019-0010-8.

Monlezun, D. J., Bittner, E. A., Christopher, K. B., Camargo, C. A. & Quraishi, S. A. (2015). "Vitamin D status and acute respiratory infection: cross sectional results from the United States National Health and Nutrition Examination Survey, 2001-2006." *Nutrients*, *7* (3), 1933-44. doi: 10.3390/nu7031933.

Mowry, E. M., James, J. A., Krupp, L. B. & Waubant, E. (2011). "Vitamin D status and antibody levels to common viruses in pediatric-onset multiple sclerosis." *Mult Scler*, *17* (6), 666-71. doi: 10.1177/1352458510394398.

Norman, A. W. (2008). "From vitamin D to hormone D: fundamentals of the vitamin D endocrine system essential for good health." *Am J Clin Nutr*, *88* (2), 491s-499s. doi: 10.1093/ajcn/88.2.491S.

Perez-Perez, S., Dominguez-Mozo, M. I., Garcia-Martinez, M. A., Aladro, Y., Martinez-Gines, M., Garcia-Dominguez, J. M., Lopez de Silanes, C., Casanova, I., Ortega-Madueno, I., Lopez-Lozano, L., Torrejon, M. J., Arroyo, R. & Alvarez-Lafuente, R. (2018). "Study of the possible link of 25-hydroxyvitamin D with Epstein-Barr virus and human herpesvirus 6 in patients with multiple sclerosis." *Eur J Neurol*, *25* (12), 1446-1453. doi: 10.1111/ene.13749.

Petta, Salvatore, Calogero Cammà, Concetta Scazzone, Claudio Tripodo, Vito Di Marco, Antonino Bono, Daniela Cabibi, Giusalba Licata, Rossana Porcasi, Giulio Marchesini, & Antonio Craxí. (2010). "Low vitamin D serum level is related to severe fibrosis and low responsiveness to interferon-based therapy in genotype 1 chronic hepatitis C." *Hepatology*, *51* (4), 1158-1167. doi: 10.1002/hep.23489.

Ramien, C., Pachnio, A., Sisay, S., Begum, J., Leese, A., Disanto, G., Kuhle, J., Giovannoni, G., Rickinson, A., Ramagopalan, S. V., Moss, P. & Meier, U. C. (2014). "Hypovitaminosis-D and EBV: no interdependence between two MS risk factors in a healthy young UK autumn cohort." *Mult Scler*, *20* (6), 751-3. doi: 10.1177/1352458513509507.

Rosen, C. J. (2011). "Clinical practice. Vitamin D insufficiency." *N Engl J Med*, *364* (3), 248-54. doi: 10.1056/NEJMcp1009570.

Rosen, C. J., Abrams, S. A., Aloia, J. F., Brannon, P. M., Clinton, S. K., Durazo-Arvizu, R. A., Gallagher, J. C., Gallo, R. L., Jones, G., Kovacs, C. S., Manson, J. E., Mayne, S. T., Ross, A. C., Shapses, S. A. & Taylor, C. L. (2012). "IOM committee members respond to Endocrine Society vitamin D guideline." *J Clin Endocrinol Metab*, *97* (4), 1146-52. doi: 10.1210/jc.2011-2218.

Rosen, C. J., Adams, J. S., Bikle, D. D., Black, D. M., Demay, M. B., Manson, J. E., Murad, M. H. & Kovacs, C. S. (2012). "The nonskeletal effects of vitamin D: an Endocrine Society scientific statement." *Endocr Rev*, *33* (3), 456-92. doi: 10.1210/er.2012-1000.

Sabetta, J. R., DePetrillo, P., Cipriani, R. J., Smardin, J., Burns, L. A.& Landry, M. L. (2010). "Serum 25-hydroxyvitamin d and the incidence of acute viral respiratory tract infections in healthy adults." *PLoS One*, *5* (6), e11088. doi: 10.1371/journal.pone.0011088.

Sacheck, J. M., Van Rompay, M. I., Chomitz, V. R., Economos, C. D., Eliasziw, M., Goodman, E., Gordon, C. M. & Holick, M. F. (2017). "Impact of Three Doses of Vitamin D3 on Serum 25(OH)D Deficiency and Insufficiency in At-Risk Schoolchildren." *J Clin Endocrinol Metab*, *102* (12), 4496-4505. doi: 10.1210/jc.2017-01179.

Shim, J., Perez, A., Symanski, E. & Nyitray, A. G. (2016). "Association Between Serum 25-Hydroxyvitamin D Level and Human Papillomavirus Cervicovaginal Infection in Women in the United States." *J Infect Dis*, *213* (12), 1886-92. doi: 10.1093/infdis/jiw065.

Telcian, A. G., Zdrenghea, M. T., Edwards, M. R., Laza-Stanca, V., Mallia, P., Johnston, S. L. & Stanciu, L. A. (2017). "Vitamin D increases the antiviral activity of bronchial epithelial cells in vitro." *Antiviral Res*, *137*, 93-101. doi: 10.1016/j.antiviral.2016.11.004.

Teymoori-Rad, M., F. Shokri, V. Salimi, and S. M. Marashi. (2019). "The interplay between vitamin D and viral infections." *Rev Med Virol*, 29 (2), e2032. doi: 10.1002/rmv.2032.

Tian, G., Liang, X., Chen, D., Mao, X., Yu, J., Zheng, P., He, J., Huang, Z. & Yu, B. (2016). "Vitamin D3 supplementation alleviates rotavirus infection in pigs and IPEC-J2 cells via regulating the autophagy signaling pathway." *J Steroid Biochem Mol Biol*, *163*, 157-63. doi: 10.1016/j.jsbmb.2016.05.004.

Udomsinprasert, W. & Jittikoon, J. (2019). "Vitamin D and liver fibrosis: Molecular mechanisms and clinical studies." *Biomed Pharmacother*, *109*, 1351-1360. doi: 10.1016/j.biopha.2018.10.140.

Urashima, M., Segawa, T., Okazaki, M., Kurihara, M., Wada, Y. & Ida, H. (2010). "Randomized trial of vitamin D supplementation to prevent seasonal influenza A in schoolchildren." *Am J Clin Nutr*, *91* (5), 1255-60. doi: 10.3945/ajcn.2009.29094.

Villar, L. M., Del Campo, J. A., Ranchal, I., Lampe, E. & Romero-Gomez, M. (2013). "Association between vitamin D and hepatitis C virus infection: a meta-analysis." *World J Gastroenterol* 19 (35):5917-24. doi: 10.3748/wjg.v19.i35.5917.

Volberding, Paul A., Kelty R. Baker, & Alexandra M. Levine. (2003). "Human Immunodeficiency Virus Hematology." *Hematology*, *2003* (1), 294-313. doi: 10.1182/asheducation-2003.1.294.

Walker, V. P. & Modlin, R. L. (2009). "The vitamin D connection to pediatric infections and immune function." *Pediatr Res*, *65* (5 Pt 2), 106R-113R. doi: 10.1203/PDR.0b013e31819dba91.

Weiss, S. T. & Litonjua, A. A. (2015). "Vitamin D dosing for infectious and immune disorders." *Thorax*, *70* (10), 919-20. doi: 10.1136/thoraxjnl-2015-207334.

Whitley, R. J. (1996). Chapter 68 Herpesviruses In *Medical Microbiology*, edited by Baron S: Galveston (TX): University of Texas Medical Branch at Galveston.

WHO. (2017). World Health Organization; Geneva: (2017). *Global hepatitis report.* Original edition, http://apps.who.int/iris/bitstream/10665/255016/1/9789241565455-eng.pdf?ua=1.

Wong, G. L., Chan, H. L., Chan, H. Y., Tse, C. H., Chim, A. M., Lo, A. O. & Wong, V. W. (2015). "Adverse effects of vitamin D deficiency on outcomes of patients with chronic hepatitis B." *Clin Gastroenterol Hepatol*, *13* (4), 783-90.e1. doi: 10.1016/j.cgh.2014.09.050.

Yin-Murphy, M. & Almond, J. W. (1996). Chapter 53: Picornaviruses. In *Medical Microbiology*, edited by Baron S: Galveston (TX): University of Texas Medical Branch at Galveston.

Zdrenghea, M. T., Makrinioti, H., Bagacean, C., Bush, A., Johnston, S. L. & Stanciu, L. A. (2017). "Vitamin D modulation of innate immune responses to respiratory viral infections." *Rev Med Virol*, *27* (1), doi: 10.1002/rmv.1909.

In: Nutraceuticals: Evidence Based Use
Editor: Raluca Maria Pop
ISBN: 978-1-53618-052-7

Chapter 8

PEDIATRICS AND NUTRACEUTICALS

Dana Goşa*[*], *Ruxandra Elena Schiotis and Anca Dana Buzoianu

Department of Pharmacology, Toxicology and Clinical Pharmacology, "Iuliu Hatieganu" University of Medicine and Pharmacy, Cluj-Napoca, Romania

ABSTRACT

Nutraceuticals are products designed to prevent or treat certain conditions, either by correcting metabolic disorders or by preventing the progress or recurrence of a pathological situation. Although their use differs from country to country, and the knowledge and confidence of the population in their action is also variable, the nutraceutical market has experienced a spectacular rise in the last period.

Although at present there are not many studes that scientifically substantiate their usefulness, in different situations functional products are sought by young people, adults and the elderly. Concerning nutraceuticals used in pediatrics, their role in prophylaxis, frequent pathologies or as a supplement to drug therapy was commonly investigated. This chapter

* Corresponding Author's E-mail: dgosa@umfcluj.ro.

describes the controversary situations of frequently used nutraceuticals in children like probiotics, vitamin D and omega 3 fatty acids. A comprehensive picture of their use and available evidences is provided.

Keywords: nutraceuticals, children, omega 3 fatty acids, vitamin D, calcium

INTRODUCTION

The nutriceuticals definition is very permissive and includes several different classes like food supplements, herbal supplements, pre- and probiotics, the "fortifying" and "functioning" foods, which all have the same goals (to prevent and to treat different diseases) (DeFelice 1995).

Most of the nutriceuticals used in pediatrics are products having a high concentration in nutrients, most of them extracted from natural food sources. These are perfectly compatible with the human body. Their composition is simple (containing one active compound) or more complex (containing many nutrients) having the main role to complete the supplementary nutrients intake through food. Another important role is to sustain the normal body functions, without associated side effects. The growth of the human body and it's functioning at the cellular level, need a permanent supply of energy and material, usually provided by nutrients. Thus the nutraceuticals used in children are specifically made for their needs. The most common nutraceuticals are the ones used to strengthen the immunity system, to maintain the intellectual performance, to improve memory and attention or for bones, muscles and strong joints. Also, they are mainly produced to prevent or to be used in the treatment of some chronic disease. For example, the nutriceuticals could offer a "plus" in addition to a balanced diet, therefore improving some physiological process or preventing different risk factors for chronic diseases like atherosclerosis, obesity, hypercholesteremia (e.g., supplementation with antioxidants, vitamins (C, D, E), minerals (chromium, magnesium), conjugated linoleic acid, alpha-lipoic acid, omega-3 fatty acids, and dietary fibers) (Pignatelli and Basili 2010).

Usually, the nutriceuticals used in pediatrics can be administrated in children starting 3 years old. A large part of the traditional nutraceuticals like chemicals constituents, probiotics, and nutraceutical enzymes are commercialized as syrups, suspensions, tablets, gelatin or operculated capsules, powders or other medicinal forms which, generally, are not associated with foods. They can be produced in all pharmaceutical forms of the drugs but preponderantly, and unlike drugs, they are of natural origin (e.g., fish oil fatty acids or phytonutrients like polyphenols extracted from plants and fruits). There are also other alternatives like non-traditional nutraceuticals (fortified nutraceuticals or functional foods, and recombinant nutraceuticals), which are addressed especially to kids who don't like or can't swallow tablets (e.g., vitamin D fortified milk, zinc fortified water, vitamins A, C, zinc, iron, and lysine fortified juice) (Angeles-Agdeppa et al. 2011; Khadgawat et al. 2013, Kujinga et al. 2018). The advantage of functional food is that it looks like conventional food, therefore, it can be part of the usual diet. Moreover, it was demonstrated that it has physiological benefits and/or can lower the associated risk of chronic disease, beyond the basic nutritional function (Kim et al. 2010; Kondo et al. 2014).

In the collective imagination, the term "nutriceutical" is very confusing and frequently misinterpreted. This happens because the term is poorly defined and producers and healthcare professionals used it when referred to many different types of products like dietary supplements, vitamins, minerals, purified substances of natural origin, plant extracts, phytonutrients or functional foods (Schmitt and Ferro 2013). Also, the fact that nutriceuticals are situated at the drugs borderline and are framed by more permissive legislation promotes this confusing perception.

These aspects, cumulated with their important health benefits and parents' need for disease prevention in kids allowed the appearance of a diverse and large number of nutraceuticals on the market. Moreover, the advantage of low prices and high compliance from children influence their frequent use. Even though there are not many studies available to demonstrate scientifically their utility in different situations, they continue to be used in all age categories (children, adults, and elders).

To overcome this knowledge gap, researchers are actively involved in a lot of studies to offer valuable information in this regard. Studies related to important aspects like nutraceuticals mechanism of action, pharmacological effects, their potential targets in the fight to improve health, prevent or reduce the risk of pathologies are of major importance. Even if these concerns are known and a lot of efforts are made to overcome them, the nutraceuticals field still needs clear proof about their efficiency and safety use, especially in the pediatric population. Finally, an officially accepted definition and a more clear set of regulations regarding nutriceuticals are needed (Santini et al., 2018). This chapter aims to present the actual situation of probiotics, vitamin D and omega 3 fatty acids, frequently used nutraceuticals in children, regarding their use, benefits and reported side effects as described in the most recent studies.

Common Nutraceuticals Used in Pediatrics

Probiotics

Probiotics, translated from the Greek words "pro" and "bio" as "for life" are microorganisms specifically used for their health beneficial effects (Varzakas et al. 2018). They are naturally found in our bodies, being identified in the human intestinal tract which is populated with approximately 400 – 500 types of bacterial strains (Gorbach 1996).

Through their mechanisms of action, probiotics are linked to various health benefits. Thus, via the intensification of the epithelial barrier function, nutraceuticals create a protective gut filter against microbes. Via an increased adhesion towards the intestinal epithelial cells can help the immunity system. Via the competitive exclusion of pathogens can prevent gastrointestinal disorders and the imbalances who may take to different recurrent vaginal and urinary tract infections, they adjust the intestinal transit and they can limit the irritable colon symptoms (Varzakas et al. 2018). Also, via their support of angiogenesis, food digestion, and fat metabolism (e.g., involvement in the synthesis of vitamins B group (B1, B2, B6, B12), vitamin

K2, butyric acid/butanoic acid and folic acid), probiotics can help the human body to get rid of harmful substance like ammonia, cholesterol or excess of hormones (Holmes et al. 2011; Vernocchi, Del Chierico, and Putignani 2016). Via the production of a large spectrum of substances with an antimicrobial role (like organic acids – acetic and lactic, bacteriocins, H_2O_2, CO_2, lantibiotics, bacteriolysins) probiotics may inhibit cell proliferation, the microorganism pathogenicity or they may even have a bactericidal effect (Russo, Linsalata, and Orlando 2014; Vernocchi, Del Chierico, and Putignani 2016). Also, probiotics promote a faster immune response by the stimulation of immunoglobulin synthesis, which has an important role in the antimicrobial defense (Paineau et al. 2008). Additionally, they have modulator effects of the inflammatory process and T cells activity with an important role in the organism defense against the pathogenic factors, or against those with pathogenic potential (Azad, Sarker, and Wan 2018).

The importance of probiotics was highlighted in 1907 for the first time, by Ellie Metchnikoff, the winner of the Nobel's award and employee of Pasteur Institute of Paris. He noticed that the locals from some rural areas of Bulgaria had a very good health condition and they have a very long-lasting life. He associated this fact with the consumption of fermented lactate which is considered an important source of probiotics (Anukam and Reid 2007). The term "probiotic" appeared for the first time in 1989, and in 1995 the "prebiotic" notion was established. At this time, the probiotics started to be recommended in associations with the antibiotics treatment.

Currently, the probiotics market is very diverse with several probiotics categories based on their ingredients (bacteria or yeast), application (food and beverages, dietary supplements, animal feed), end-use (animal, humans), functions (preventive, therapeutic, regular), and geography (North America, Europe, Asia-Pacific, Latin America, Middle East, and Africa) ("Probiotics Market Size, Share, and Analysis, 2022").

Generally, the probiotic supplements are formulated as powders, tablets or capsules with doses between a few million and a few billions of bacterias. The probiotic foods like yogurt, juices and soya drinks, fermented milk, miso (made from soya), light cheeses, tempeh (fermented food from soya), kefir, sour cabbage, pickles, kimchi (Korean sour cabbage) and in dark chocolate are preferred by consumers.

Usually, the most common bacteria used in the probiotics formulations are *Bacillus* spp., *Bifidobacterium* spp., *Lactobacilli* spp., *Leuconostoc cremoris*, *Streptococcus* spp. alone or in combinations and *Saccharomyces* spp. yeast.

In children, probiotics supplements are suggested as therapy for several health issues. A recent review concerning the use of probiotics in children well established that their prevalence varies from 4% to 51% (Irwin, Davis and Currie 2019). The actual status of probiotic use in pediatrics and critical points were identified. Thus, they pointed out that probiotics consumption is influenced mostly by education and family incomes. The analyzed studies highlighted that the use of probiotics is not well balanced, the reasons for using them being unclear for parents. More than that, most of the parents informed about the probiotic using unofficial sources like different web sites or family members. Also, the strains used in probiotics were unknown (excepting one article), as well as their efficacity (Irwin, Davis and Currie 2019).

Besides this chaotic consumption of probiotics in children, another problem comes from the lack of well-designed studies and/ or good reports, improper data about the used probiotics strains, their efficacity, side effects or the reasons for their use. Therefore, this part will describe the most relevant articles related to the most frequent causes of probiotics use in children.

Probiotics and Gastrointestinal Tract Disorders

It is well known that antibiotics administration changes the microbial balance at the gastrointestinal tract. Diarrhea associated with antibiotics (AAD) appears when the antibiotics disturb the natural balance between "good" and "bad" bacteria from the intestinal system, determining the multiplication of dangerous bacteria beyond their normal number. The AAD symptoms include fragment stools and painful abdominal cramps. In this regard, probiotics can restore intestinal microflora, preventing AAD.

In a randomized, parallel, controlled trial, made on children with ages between (0 – 18 years old) the authors compared the incidence of secondary diarrhea due to antibiotics use. They compared diarrhea episodes in children with probiotics administration versus placebo, active alternative prophylaxis, and children without treatment, respectively. The diarrhea incidence in children's probiotic group was between 9 – 16% as compared with the control group who had 18 – 22% (Johnson et al., 2011). Despite several critical factors identified in the study like strain heterogeneity, dose and time of probiotics administration, and study quality, the results of this study suggested the protective effect of probiotics in acute diarrhea prevention. The effective dose was <5 billion colony-forming units (CFU)/day. Even so, a Grading of Recommendations, Assessment, Development, and Evaluations (GRADE) analysis showed that the general quality of proves for the primary goals (the diarrhea incidence) was decreased because of the higher losses caused by the deficient follow-up children and because of the inaccuracy data too (Johnson et al., 2011).

The evaluation of efficiency and safety administration of the probiotic (all the strain and the doses) used in acute diarrhea prevention in children were also studied (Guo et al., 2019). The results showed a moderate protective effect of probiotics in acute diarrhea in children treatment (NST 9, IC at 95%). Additionally, the proofs suggest that probiotics in high doses (of 5 billion CFU/day) reduced the AAD incidence. The probiotics were generally well tolerated, with rare minor side effects (ex. skin rashes, nausea, flatulence, abdominal bloating, and constipation). There were reported rare serious side effects (in observational studies) in children with disabilities and deficient immune system.

Also, the results showed that probiotics are efficient to moderately reduce diarrhea time episodes with almost a day. About the evaluation of the different probiotics, *Lactobacillus rhamnosus* or *Saccharomyces boulardii* proved to be the most effective in AAD prevention in children with antibiotic treatments.

A multicenter study highlighted the effect of *Bacillus clausii* as an additional treatment for acute community diarrhea (of viral origin or as a result of antibiotics treatments). *Bacillus clausii* is an aerobe bacteria that forms spores being able to inhibit the growth of pathogenic agents at gastrointestinal (GI) tracts level (De Castro et al., 2019). The obtained results indicated a decrease of diarrhea duration, GI symptoms and stools frequency as a result of *Bacillus clausii* administration. This might be an efficient therapeutical option in AAD's treatment in children.

Taking into account that in 2004, the World Health Organization (WHO) recommended Zn supplements associated with Oral Rehydration Therapy (ORT) in all forms of diarrhea treatments and of all ages, the present study included Zn supplementation as well. Even if they showed the positive impact of *Bacillus clausii* and Zn combination in AAD treatment, they recommended caution of Zn supplementation in children with viral diarrhea.

Besides diarrhea, probiotics could be safely used in children for functional constipation treatment as well. The constipation diagnosis in children is done mostly using Rome III criteria, established for functional gastrointestinal disorders like irritable bowel syndrome (2006). According to the European Society for Paediatric Gastroenterology, Hepatology and Nutrition (ESPGHAN) and North American Society For Pediatric Gastroenterology, Hepatology & Nutrition (NASPGHAN) guidelines, the first-line treatment for infant and children's constipation is the improvement of fecal obstruction and use of laxative (polyethylene glycol). Still, in 40% of patients the treatment is still ineffective, and in 30% of pediatric patients, the symptoms of constipation can last till adolescence. Side effects like abdominal pains or defecation difficulties are also encountered (Wenhao et al., 2019).

In the last years, besides standard treatments, alternatives therapies, like probiotics, preponderantly *Lactobacillus* strains, were frequently used. Their administration influence the balance of intestinal flora by the production of lactic acid, aminoacids, and short-chain fatty acids, modifying the fecal pH value and promoting the intestinal peristalsis. Accordingly, symptomatology improvement and side effects reduction are observed (Perceval et al. 2019).

A recent study investigated the *Lactobacillus* benefits in functional constipation treatment, a common chronic idiopathic condition in children, with an incidence ranging from 7% to 30% in most of the countries (Wenhao et al., 2019). They concluded that their utility in functional constipation remains controversial in the lack of the incontestable proves.

Besides AAD and constipation, probiotics are used to prevent other gastrointestinal diseases like acute gastroenteritis, infant colics, and necrotizing enterocolitis (Ong et al. 2019; LaMont 2018; Wang et al. 2019). So far, the use of the probiotic in preventing these diseases did not offer evidence in their usages.

Probiotics and Respiratory Disorders

Probiotics use in acute or chronic respiratory system disease in children was also studied. Marini et al., (2019) study investigated the prophylactic efficacy of *K12 Streptococcus salivarius* probiotic in recurrent episodes of pharyngotonsillitis prevention in children.

Acute pharyngotonsillitis represent one of the most frequent infections of the superior respiratory tract in children. Most of the infections have a viral origin, only 30% of them being caused by bacterial infections. The bacterial infections treatment includes ambulatory antibiotics (penicillins, ampicillin, and, amoxicillin as such or in combination with clavulanic acid). Symptomatic treatment including non-steroidal anti-inflammatory drugs (AINS) like paracetamol or ibuprofen, for fever and pain release, are also used. Lastly, if tonsillectomy is recommended, even if an improvement in the quality of children's lives is observed, there is a lack of studies regarding the effect of tonsillectomy on their general welfare, growing or behavior. Thus, to reduce antibiotic treatment and surgical intervention, Marini et al., (2019) proposed the use of specific oral cavity probiotics treatment (by oral administration). Accordingly, an isolated *Streptococcus salivarius* strain, named Blis K 12® was used. It was demonstrated its capability to destroy *Streptococcus pyogenes*, the bacteria responsible for the majority of tonsillitis pharyngeal infections. Further Blis K 12 genome analysis may demonstrate the mechanism of action against streptococci infection.

The probiotics may be useful in the children's chronic respiratory treatments as well. The study conducted by Moura et al., (2019) evaluated the probiotics use as a supplementary therapy to asthma treatment in children and teenagers from Belo Horizonte, Brazil. Asthma is the most frequent chronic disease in childhood. It is estimated that about 14% of children suffer from asthma (Global Initiative for Asthma, 2018). According to the guidelines of the Global Initiative for Asthma, the purpose of asthma treatment is to maintain optimal disease control (Global Initiative for Asthma, 2018). Thus, absence or minimal symptoms during the day and night, no limitations of physical activity, the use of minimum drugs to improve the symptoms, the maintaining to normal or almost normal the pulmonary function and the absence of health worsening are required. In this regard, the probiotics could offer benefits like strengthening the immune system and better disease control by reduced symptoms and improved pulmonary function (Moura et al. 2019). Also, probiotics increased epithelial barrier, inhibited pathogenic agent adhesion, excluded the pathogenic microorganisms by competition and produced antimicrobial substances. Finally, the results of this study suggested a tendency to improve the clinical profile of asthma in patients who have *L. reuteri* as a supplementary therapy.

Still, the probiotics use in children with asthma is controversial (Lin et a. 2018). There is great variability of the effects of the probiotic stalks. Currently missing links like mechanisms of action, side effects, drug interactions still need to be addressed.

Probiotics and Viral Infections

Smiyan et al., (2019) studied the optimization of rotavirus infection treatment with *Bacillus clausii* in children. All over the world, rotavirus infection (RVI) is ranging from 40 to 60% among among hospitalized children with gastroenteritis. Each year, 2 billion patients are hospitalized with a severe form of RVI. It was found that the acute phase of RVI disease in children is characterized by an increase in serum IgA and IgM concentrations ($p<0,001$), and an increase of sIGA ($p<0,001$) coprofiltrate levels. This study showed that the children group who received adjuvant probiotic treatment additionally to RVI traditional treatment presented a normalization of (A, M, G) immunoglobulin serum levels and coprofiltrate sIgA levels (Smiyan et al. 2019). The authors concluded that probiotics containing *Bacillus clausii* have a positive effect on the immune system in children with rotavirus infection.

Another probiotic containing *Lactobacillus plantarum* IS–10506 was studied for highlighting the effect of lipopolysaccharides (LPS) level in blood and the effect on the immune response in HIV infected children (Athiyyah et al. 2019). The LPS are the main components of the Gram-positive bacteria cell wall. They possess a strong inflammatory action and can be measured in plasma. The LPS has modified levels in gastrointestinal chronic infection in HIV infections (Brenchley et al. 2006).

The enteropathy from HIV may disturb the intestinal barrier, which led to a loss of D4 + T cells and an increase of the intestinal permeability. The probiotic who contains *Lactobacillus plantarum* IS – 10506 may reduce the level of LPS from blood in HIV infected children who followed an ARV therapy for 6 weeks. Thus, probiotic treatment can be used as a supplement in this treatment. The *Lactobacillus plantarum* IS – 10506 treatment affected neither the intestinal mucosa immune response nor the systemic immune response in HIV infected children (Athiyyah et al. 2019).

Concluding, several clinical studies confirm the probiotics efficacy in prophylaxis and treatment of some intestinal diseases like gastrointestinal irritable bowel syndrome infections, diarrhea associated antibiotics (DAA), irritable bowel syndrome, and chronic inflammatory intestinal disease. Still, the probiotics mechanism of action is a very complex one and it's not yet fully elucidated. Hopefully, the next studies will get the answers to these questions and will offer new information about probiotics pharmacokinetics. Also, each stalk must be individually studied to eliminate the variability of the results.

Vitamin D

Vitamin D is a liposoluble vitamin with two major forms: vitamin D2 (ergocalciferol) and D3 (cholecalciferol). Depending on the food source, vitamin D3 is of animal origin (fatty fish, fish oil, liver, butter, egg yolk, dietary supplements) while vitamin D2 is of plant origin (mushrooms, dietary supplements, fortified foods). It is synthesized at skin level from 7 – dehydrocholesterol, an endogenous steroid precursor (provitamin D3), under the ultraviolet ray's actions. It is partially stored in fat and muscle tissues and partially is hydroxylated.

Vitamin D (calciferol) is essential for developing and maintaining bone health. While many people know that calcium is essential in bone growth and development, many don't know that calcium is absorbed in the body only in the presence of vitamin D. Calcium and D vitamin are necessary to have healthy and strong bones (role in osteomalacia and osteoporosis prevention). Many studies showed that low intake of calcium over life (starting from childhood) is associated with a reduced bone mass, and later, with a high percentage of fractures among elder persons (Beto 2015). Also, long-lasting deficiency in vitamin D may be associated with other health issues, including cardiovascular disease, diabetes or autoimmune diseases (Weydert 2014).

Therefore, vitamin D vitamin is not only important in bone health but also in preventing different diseases. By its influence on the innate and adaptive immune system, it can help in preventing infections or to modulate autoimmunity, respectively (Weydert 2014). It is also mentioned that vitamin D can reduce cancer incidence, can prevent viral illnesses and different inflammatory diseases, can treat musculoskeletal pain or can help depression treatment (Weydert 2014).

Accordingly, the main function of vitamin D in the organism is to participate at calcium homeostasis, simulating its active transport in the enterocytes. Lots of cell involved in the immune responses act as D vitamin receptors and have enzymatic means for local synthesis of 1,25 (OH) 2 D3. Thus, the synthesis sites are very important for adjusting and control different immune responses. So, vitamin D hydroxylation has the function to inhibit cell proliferation, promote cell differentiation and to adjust immunologic mechanisms, like activating inadequate control of interleukin-17, who participates in inflammatory disease pathogenicity (Lourenceti et al., 2018).

Vitamin D and Autoimmune Diseases

Recent studies suggest the new role of D vitamin in immunomodulation, it's anti-inflammatory and antimicrobial effects, being proposed as a treatment for autoimmune diseases, like inflammatory intestinal bowel diseases (IBD), multiple sclerosis, systemic lupus erythematosus, rheumatoid arthritis (Agmon – Levin et al., 2013; Ananthakrishnan et al., 2016; Cutolo et al., 2011; Doaa et al., 2018; Kabani et al., 2016; Von Scheven et al., 2011).

Inflammatory intestinal bowel diseases (IBD) such as Crohn disease (CD) and ulcerative colitis (UC) in children is increasing worldwide. It is known that vitamin D level has an important role in IBD. So far, a clear link between vitamin D and IBD couldn't be established, but Kim et al., (2019) study confirmed a high prevalence of vitamin D deficiency in IBD patients. They reported that micronutrient deficiency, including vitamin D, may lead to aggravation of IBD and that its supplementation improves children's symptoms and quality of life. Knowing that vitamin D supplements must be administered in cases of aforementioned diseases that might also affect its metabolism and that currently there are not enough studies to establish the exact link between vitamin D and disease pathogenesis, especially in children, more studies are required.

Another autoimmune inflammatory disease is the juvenile idiopathic arthritis, one of the most common rheumatologic diseases met in childhood. A recent study tried to establish a link between vitamin D and juvenile idiopathic arthritis (Finch et al., 2018). The relationship between vitamin D and disease activity in children with arthritis could not be established since this disease is affected by many factors like medical treatments, way of life, disease duration. Therefore they proposed to investigate vitamin D role in newly diagnosed children to remove as much as possible the influence of other factors. Also, identifying the role of other factors (genetic, environmental) will help to understand vitamin D's role in children's arthritis prevention and evolution or its role in the disease improvement. There are only a few data about the involvement of vitamin D in childhood arthritis.

Vitamin D and Chronic Diseases with Inflammation

It was observed that vitamin D deficit could be associated with a high risk of asthma in children, especially because of its immunomodulatory properties (Bozzetto et al., 2012). A systematic review and meta-analysis of 23 studies that enrolled 13,160 participants observed that vitamin D deficiency and insufficiency were identified in approximately 27% of children with asthma. Also, vitamin D levels were lower in asthmatic children as compared to the nonasthmatic ones. Still, the correlations between vitamin D levels and asthma incidence and control varied a lot among studies, a better approach (well-designed, randomized clinical controlled trials) being required to draw an unequivocal conclusion (Jat and Khairwa 2017).

Vitamin D intake in the pre- or postnatal period was also investigated to see, whether it can be responsible for asthma evolution in children. A meta-analysis conducted by Shen et al., (2018) found no significant statistical association between vitamin D levels in mother's blood or umbilical cord and the vitamin intake in childhood and asthma, at children over or below 5 years old. Even though some studies presented favorable results regarding vitamin D levels and asthma in children under 3 years, but these results were not statistically significant. Therefore, the authors recommend supplementary studies to confirm the results.

Another inflammatory disease where vitamin D has shown its efficacy is atopic dermatitis (AD). Atopic dermatitis is a chronic recurrent inflammatory disease of the skin. Eighty five percent of AD cases are observed in children. Among them, 30% continue to present the disease as adults too. The world incidence is very high continuing to grow, especially in the urban centers.

Atopic dermatitis pathophysiology is very complex with altered skin barrier function and presence of different deficiencies of the immune system. These include high IgE, eosinophil and distinct T-helper cell populations as well as cytokines imbalance. It may occur also as a result of microbial and viral infections with *Staphylococcus aureus* and *Herpes Simplex*, eventually on vitamin D deficiency background (Maliyar et al. 2018). However, the optimal level of 25 (OH) D necessary in preventing the inflammatory skin diseases are not exactly known. A recent study done in both adults and children with atopic dermatitis (Hattangdi-Haridas et al., 2019) revealed that vitamin D supplementation (especially in children) showed clinically relevant improvements when doses between 1500 – 1600 I.U. was administrated for up to 3 months. In case of children, they draw attention that pediatric physicians should consider vitamin D supplementation only after the evaluation of serum level concentration, and to consider also their medical history.

Vitamin D and Viral Infections

The Endocrine Society Task Force clinical practice guideline remains one of the few to contain recommendations addressing D vitamin deficiency in children and young adults, with HIV, although the evidence for these recommendations are limited (Holick et al., 2011).

The supplements with vitamin D could have a positive impact on the musculoskeletal system in HIV infected children and teenagers. The study made by Penner et al., (2018), investigated the impact of vitamin D supplementation on bone system in HIV infected patients with ages between 0 and 25 years old. It is known that during the growth period, a rapid gain in bone growth may occur. This growth starts in childhood and is more pronounced in puberty when the length, width, volume, and strength are forming the new fully mineralized matrix, reaching the maximal bone mass (PBM). A decrease in PBM may be associated with changing vitamin D metabolism in HIV infected young patients (Fakruddin et al., 2003 and Sudjaritruk et al., 2016). This change is caused by the inflammatory and metabolic components of HIV infections as well as of the drug treatment side effects.

Vitamin D and Neurological Disorders

It's a fact that vitamin D has an important role in the central nervous system specifically in the neurotransmission process (Cebeci et al., 2014; Holick et al., 2007 and Hossein et al., 2013). Vitamin D deficiency among children with epilepsy it's often an overlooked problem. The aim of Khalifah et al., (2018) study was to evidentiate the negative impact of vitamin D deficiency among seizures control and of course among bone health in children with epilepsy. Proposing a pragmatic trial design, the authors followed the evolution of epileptic crisis in children with epilepsy receiving two doses of vitamin D (400IU/day and 1000IU/day for 6 months) in combination with antiepileptic drugs monotherapy or polytherapy. The results of the study will probably be used in the current guidelines regarding the supplementation of vitamin D in children with epilepsy, and will probably prove the link between 25 (OH) D vitamin and crisis control.

Autism Spectrum Disorder (ASD) in children is commonly characterized by irritability and hyperactivity. The pharmacological treatments can induce several side effects and, despite limited proves, the caregivers of these children are often using different food supplements like vitamin D or omega 3 fatty acids to improve their symptoms (Mazahery et al., 2019). In a recent randomized controlled trial Mazahery et al., (2019) investigated the effect of vitamin D and omega-3 fatty acids in children with autism focusing on their irritability and hyperactivity. After one year of vitamin D administration (200 IU/day), long-chain polyunsaturated fatty acids (722 mg/day docosahexaenoic acid, omega-3) or both (2000 IU/day vitamin D+docosahexaenoic acid, vitamin D+omega-3) it was found that irritability symptoms in children with autism were reduced. Another study made by Horvath et al., (2017), showed that efficient treatments for the basic symptoms of autism are still missing. Therefore, it was observed a recent increase in alternative methods to control these symptoms, like vitamin D supplementation. These aspects are best reflected in the increased Over the Counter (OTC) vitamin D commercial sales. In this regard, many types of pharmaceutical preparations for vitamin D are available (oil drops, gelatin capsules, jellies, tablets).

Vitamin D and Other Diseases

Knowing that impairment of thiol/disulfide homeostasis, as well as vitamin D deficiency, is responsible for the pathophysiology of many acute and chronic diseases, Mertoglu et al., (2018) examined the relationship between thiol/disulfide homeostasis and vitamin D level supplementation. It was found that in healthy children, a severe deficiency of D vitamin causes thiol/disulfide homeostasis damage, and increases protein oxidation. Supplementary vitamin D administration cannot reverse the changes caused by vitamin D impairment.

Even though vitamin D has an important role in many organisms' functions, it cannot be exactly established which is the beneficial dose and its deficiency values which are influencing human predisposition to diseases. For example, administration in high doses of all vitamin D preparations, can induce toxic effects. Long term administration, in normal responsive persons, may induce hypervitaminosis. Hypervitaminosis may produce calcium mobilization from bones, with consequences like osteoporosis, increased serum calcemia, and increased calcium and phosphate renal elimination. Also, the mobilized calcium could be distributed in tissues, mostly in kidneys, inducing diffuse nephrocalcinosis and/or nephrolithiasis. In children might stop the growth.

On the other hand, vitamin D supplements administration has many indications in its deficiency or in its associated diseases. Some examples could be the treatment and prophylaxis of rickets (in children) and osteomalacia (in adults); hypoparathyroidism associated with calcium oral preparations; osteoporosis associated with oral calcium supplements, to stimulate intestinal absorption of calcium (frequently poor in osteoporosis); osteodystrophy from chronic renal insufficiency (Jacobson et al. 2017).

An important aspect of children is that in their first year of life, when bones are quickly developing, they must assimilate enough vitamin D (Specker et al., 2005). The organism use predominantly endogenous vitamin sources, the sufficient daily intake being of 100 – 200 U.I. / day. The normal serum value of vitamin D3, no matter age or sex is between 20 – 100 ng/mL.

Overall, vitamin D deficiency and rickets remain to be a current problem in the developing countries as well as in the North American and European countries (Kreiter et al., 2002; Hatun et al., 2005; Shaw et al., 2002). Even though among them there are countries that allow a high level of sun exposure, vitamin D deficiency remains an important problem in infants, children, and adolescents. More than that, in Northern Europe, where the population is facing less sunlight exposure, especially in winter (October - April), the question of vitamin D supplementation in children (between 0 –

2 years) remains open. In the UK, in June 2016, the Scientific Advisory Committee on Nutrition (SACN), recommended that the Department of Health (DH) updates its vitamin D guidelines. A study of British researches over the last decades highlighted an increased number of children with rickets disease (Uday et al., 2018). Among included populational ethnicities, a high level of sensibility was observed among the minorities (Asian and black ethnic groups).

Therefore, to efficiently counteract D vitamin deficiency in children, the knowledge about its proper utilization and importance in breastfed children, and even the older ones, must be known and properly disseminated among parents. The results of a study made by Day et al., in 2019 showed that approximately half of the questioned parents reported that they didn't receive any information about D vitamin administration during pregnancy, breastfeeding or for their children. The majority obtained the information individually, by using unofficial websites or from other parents.

Vitamin D or other micronutrients deficiences must be corrected in children by their supplementary administration although, there are few studies with strong evidence that can be used, especially at children. Thus more evidence is needed to support their exact use in children according to the different conditions.

Omega 3 Supplements

The (n-3) polyunsaturated fatty acids (PUFA) regulate many important physiological processes, like blood pressure, platelet function, blood coagulation, plasma TG concentrations, vascular function, cardiac rhythm, heart rate, inflammation, immune function, fatty acid and triglycerides (TG) metabolism, bone turnover, tumor cell growth, visual signaling (via rhodopsin) and brain and central nervous system structural components (Calder 2012). The regulation of these physiological processes can take place because of the n-3 PUFA involvement in the multiple mechanisms related to the connection of the cell membrane, cytosol, and nucleus. (n-3) PUFA can also act via receptors and sensors, therefore they can regulate the signaling processes that influence gene expression associated with various cellular functions (Calder 2012). As consequences, several potential health benefits arise: decrease in blood pressure, decreased in thrombosis risk appearance, decrease in plasma TG concentrations, improve in vascular reactivity, a decrease of cardiac arrhythmias, a decrease of inflammation processes, improve of immune function, a decrease in TG synthesis and storage, maintainance of bone mass, a decrease of tumor cell growth and survival, an optimized visual signaling, and an optimized brain development (Calder 2012). Accordingly, (n-3) PUFA can be proposed to be used in several diseases like hypertension, thrombosis, hypertriglyceridemia, several inflammatory diseases, compromised immunity, osteoporosis, some cancers, poor infant visual development or improvement of childhood cognitive processes especially learning and behavior (Calder 2012). As seen, among PUFA, omega 3 fatty acids are essential fatty acids needed in the organism for its normal functioning, their necessary intake being assured through nutrition or food supplements. Generally, fish oil is considered one of the richest sources of omega-3 fatty (e.g., salmon, tuna, sardine, mackerel, Pacific herring). Omega 3 fatty acids can also be found in vegetable oils (perilla, flaxseed, soybean and canola oil) (Lee 2013). There are many types of omega 3 fatty acids, but the most important, involved in human physiology, are the eicosapentaenoic acid (EPA), the docosahexaenoic acid (DHA) and the α-linolenic acid (ALA).

Omega 3 Supplements and Neurological Disorders

Omega 3 fatty acid intake has many benefits for health, and its role in sustaining cognitive function is known especially by the parents who wish to ensure health develop to their children. In a randomized controlled clinical trial on the effect of early dietary supplementation of PUFA on the mental development in infants was found that the ones feed with DHA rich powdered formula had better outcomes. Better cooperation between eyes and hands, improved concentration and social skills, as well as higher intelligence test scores, were reported (Birch et al. 2000). Even though several studies support the use of powdered formula rich in omega-3 fatty acids, especially to sustain growth and brain development in preterm infants, there is not sufficient evidence, more research being needed in this regard (Gould, Smithers, and Makrides 2013).

Omega 3 fatty acids were studied concerning tolerance and tolerability in children and teenagers with eating disorders (ED) (Woo et al., 2017). Thus, children and adolescents with EDs received omega 3 supplements (300 mg EPA and 200 mg DHA/ day) in addition to standard treatment for 8 weeks. Compared to the control group, during the 8th week of treatment, the authors observed a significant increase in the ideal average body weight percentage. Still, the EDI – 3 (Eating Disorders Inventory - 3), CDI – 2 (Depression Inventory - 2) and MASC (Multidimensional Anxiety Scale for Children) score were not significantly different, concluding that omega 3 supplements can be acceptable and well-tolerated in children with eating disorders.

Therefore intake of food supplements enriched in omega 3 (e.g fish oil – rich source of fatty acids, vitamin D, proteins or mineral supplements) in ED is recommended. Even though this recommendation is given, their long term implications in children and teenagers' nutritional condition is not known, thus this aspect should receive special focus in the next studies.

Omega 3 fatty acids were studied concerning school performance too. Accordingly, two observational studies showed that fish-eating had a positive influence on the school performance of teenagers and their cognitive tests (Åberg et al. 2009; Groot et al., 2012 Winkvist et al., 2017).

Omega 3 fatty acids food supplements administration was associated with improved school results and improved behavior in ADHD children (Milte et al., 2012). Also, a recent meta-analysis that included 534 young people with ADHD showed that omega 3 supplementations significantly improved the relationship with their parents, improved the attention and moderate their hyperactivity (Chang et al., 2018). Another study (Rodrigues et al., 2019) investigated for 6 months the effects of supplements with omega 3 to cognitive and behavioral performance at teenagers and children diagnosed with ADHD. The beneficial effects occur after 3 months of administration. Results were confirmed by improved symptoms as evaluated by parents. The authors noticed that subsequent research is necessary to demonstrate the benefits of omega 3 supplementation and its specific involvement in the ADHD pathogenic mechanism. The effect of omega-3 and methylphenidate was also investigated to see whether their association can reduce attention deficit in ADHD children (Mohammadzadeh et al. 2019). Following 8 weeks, the attention deficit, the ratio of hyperactivity/impulsivity were evaluated based on the Parent's ADHD Rating Scale evaluations (Mohammadzadeh et al. 2019). They showed that there were no statistically significant differences between children who received methylphenidate with omega 3 supplements (180 mg EPA + 120 mg DHA) and the ones treated with methylphenidate and placebo. On contrary, another recent study suggested that omega 3 supplementation (70 mg/day EPA + 250 mg/day DHA) could be a reliable adjuvant in children ADHD therapy and it can improve the effect of methylphenidate when administered in a dose of 1 mg/day (Checa-Ros et al., 2019). The authors have also mentioned that further studies are needed to evaluate their effect over long term administration.

Even though the pharmacological and psychological treatments are considered first-line treatments in ADHD disorder, a lot of non-pharmacological alternatives, like omega 3 dietary supplements constitute attractive options for teenagers. The reason for their use in ADHD disease is based on fatty acids important effects at the central nervous system level.

Knowing that Omega 3 fatty acids have an important role in maintaining and modulating neuronal functions, their anticonvulsant effects, specifically in drug resistant epilepsy patients were studied (Ibrahim et al. 2018). Ibrahim et al. (2018) study demonstrated that EPA (from capsules containing 358 mg EPA +81 mg DHA) and DHA (from capsules containing 417 mg DHA + 50 mg EPA) effectively reduced seizure frequency.

The study conducted by Pourmasoumi and colleagues (2018) investigated the association between omega 3 supplements and epileptic seizures. Among analyzed studies in the aforementioned review, four of them reported a positive association of omega-3 fatty acids administrations and the epileptic seizures. Only one study referred to their administration in epileptic children. Accordingly, Dahlin et al. (2007) investigated the effect of a ketogenic diet and fish oil (1–2 g, 4 times a day). The study was conducted in 25 children (mean age of 6 years) for 12 months. A seizure reduction with more than 50% was already observed after 3 months in 16 children, after 6 months in 15 children and after 12 months in 19 children, respectively. However, serum fatty acids levels couldn't be correlated with the seizure response. Overall, it couldn't be established a significant correlation between omega-3 supplementation and a beneficial effect on epileptic seizure control in epileptic patients (Dahlin et al. 2007; Pourmasoumi et al. 2018). Further studies are necessary to elucidate these controversial results.

Omega 3 Supplements and Asthma

It is thought that omega 3 fatty acids can prevent and adjust respiratory system disorders. Nagakura et al., (2000) investigated the effects of fish oil capsules (approximately 17 + 7 mg/kg body weight of EPA and DHA, respectively) administration in 29 children diagnosed with bronchial asthma. The results were compared against the control group who received olive oil instead of fish oil. The baseline asthma score measurements were significantly higher in the fish oil group as compared with the values from the 6th, 7th, 8th, 9th and 10th months after omega 3 fatty acids supplements administration. Higher baseline values were also registered for the control group towards the values from the 7th and 9th months. Thus the fish oil group had decreased symptom asthma scores. Also, during this study, children did not present significant side-effects. In conclusion, the authors suggested that omega 3 supplements are beneficial for children with bronchial asthma, but they refered to the strictly controlled environment as presented in their study (Nagakura et al. 2000). Therefore, they also recommended the necessity of more studies to show if indeed, fish oil supplements will have a real impact on improving pulmonary function or reducing drug use to control the disease.

Because most of the time, the supplementation of asthmatic children with omega-3 supplements have conflicting results, D'Auria et al., (2014) investigated if genetic polymorphism, as well as gene-nutritional interactions, can explain the existing inconsistent results. They concluded that variants of key genes should be included in the studies. At the moment, they couldn't draw a definitive conclusion because the few studies are performed in this regard (D'Auria et al. 2014).

Until further better evidence, the omega 3 supplements are still included in children's nutrition even in countries that predominantly consume fish products in their normal diets, or use rich PUFA vegetable oils for cooking. In this regard, Sweden, Finland and Japan are good examples of countries that succeed to have a population with a proper intake of essential fatty acids through food. In other countries, where these types of diet are hard to follow, the omega 3 fatty acids nutritional supplements remain a cheap and good alternative to efficiently improve their deficiency. This approach is supported by the National Center for Complementary and Integrative Health (NCCIH) who states that the side effects of these food supplements are usually light. Normally, the side effects can include unpleasant taste, bad breath, bad-smelling, sweat, headache, and gastrointestinal symptoms such as heartburn, nausea, and diarrhea.

CONCLUSION

Probiotics are commonly used in children for both acute or chronic diseases. Probiotics are mostly used to strengthen the immune system, to improve the intellectual performance, memory and attention or for strengthening the bones, muscles and strong joints. In case of children with chronic diseases, parents use nutraceuticals to prevent the development of different risk factors like atherosclerosis, obesity, hypercholesteremia and others. Because most of the time, supplementation with nutraceuticals in children (probiotics, vitamin D, omega 3 fatty acids) have conflicting results, more studies are needed to explain the variabilities between them. Also, informations regarding their complex mechanism of action it's not yet fully elucidated, further studies being needed as well. Even though there are not sufficient evidences for their use in children, nutraceuticals market is continuously growing. They remain a cheap and good alternative to efficiently improve different symtoms like diarrhea in some intestinal disorders or diarrhea associated to antibiotics by probiotics, vitamin D deficiency symptoms, rickets prophylaxy, osteoporosis profilaxy, improved immune system, or omega 3 fatty acids beneficial effects at the central nervous system. In conclusion, nutraceuticals have a lightening future. The market is expanding for consumer expectations satisfactions in the wellness and health field. They are, definitely, the incontestable stars of tomorrow's health.

REFERENCES

Åberg, M. A., Åberg, N., Brisman, J., Sundberg, R., Winkvist, A., Torén, K. (2009). "Fish intake of Swedish male adolescents is a predictor of cognitive performance." *Acta Paediatr.* 98, 555–560.

Agmon-Levin, N., Theodor, E., Segal, R. M., Shoenfeld, Y. (2013). "Vitamin D in systemic and organ-specific autoimmune diseases." *Clin Rev Allergy Immunol.* 45, 256–266.

Ananthakrishnan A. N. (2016). "Vitamin D and inflammatory bowel disease." *Gastroenterol Hepatol* (NY). 12(8), 513–515.

Angeles-Agdeppa, Imelda, Clarita R Magsadia, and Mario V Capanzana. 2011. "Fortified Juice Drink Improved Iron and Zinc Status of Schoolchildren." *Asia Pacific Journal of Clinical Nutrition* 20 (4): 535–43.

Anukam, Kingsley C., and Gregor Reid. (2007). "Probiotics: 100 Years (1907-2007) after Elie Metchnikoff's Observations." *Biology*. 466-474.

D'Auria, Enza, Michele Miraglia Del Giudice, Salvatore Barberi, Marzia Mandelli, Elvira Verduci, Salvatore Leonardi, Enrica Riva, and Marcello Giovannini. 2014. "Omega-3 Fatty Acids and Asthma in Children." *Allergy and Asthma Proceedings* 35 (3): 233–40.

Azad, Md Abul Kalam, Manobendro Sarker, and Dan Wan. (2018). "Immunomodulatory Effects of Probiotics on Cytokine Profiles." *BioMed Research International* 2018: 1–10.

Athiyyah, A. F., Brahmantya, H., Dwiastuti, S., Darma, A., Puspitasari, D., Husada, D., Ranuh, R., Endaryanto, A., Surono, I. Sudarmo, S. M. (2019)."Effect of *Lactobacillus plantarum* IS-10506 on blood lipopolysaccharide level and immune response in HIV-infected children." *Iran J Microbiol.* 11(2), 137-144.

Beto, Judith A. 2015. "The Role of Calcium in Human Aging." *Clinical Nutrition Research* 4 (1): 1.

Birch, Eileen E., Sharon Garfield, Dennis R. Hoffman, Ricardo Uauy, and David G. Birch. 2000. "A Randomized Controlled Trial of Early Dietary Supply of Long-Chain Polyunsaturated Fatty Acids and Mental Development in Term Infants." *Developmental Medicine and Child Neurology* 42 (3): 174–81.

Bozzetto, S., Carraro, S., Giordano, G., Boner, A., Baraldi, E. (2012). "Asthma, allergy and respiratory infections: the vitamin D hypothesis." *Allergy.* 67(1):10–17.

Brenchley, J. M., Price, D. A., Schacker, T. W., Asher, T. E., Silvestri, G., Rao, S., et al. (2006). "Microbial translocation is a cause of systemic immune activation in chronic HIV infection." *Nat Med.* 12:1365–1371.

Calder, Philip C. 2012. "Mechanisms of Action of (n-3) Fatty Acids." *The Journal of Nutrition* 142 (3): 592S-599S.

Dahlin, Maria, Lena Hjelte, Susanne Nilsson, and Per Åmark. 2007. "Plasma Phospholipid Fatty Acids Are Influenced by a Ketogenic Diet Enriched with N-3 Fatty Acids in Children with Epilepsy." *Epilepsy Research* 73 (2): 199–207.

De Castro, J. A., Guno, M. J. V., Perez, M. O. (2019). "*Bacillus clausii* as adjunctive treatment for acute community-acquired diarrhea among Filipino children: a large-scale, multicenter, open-label study (CODDLE)." *Trop. Dis. Travel Med. Vaccines.* 23, 5:14.

DeFelice, Stephen L. 1995. "The Nutraceutical Revolution: Its Impact on Food Industry R&D." *Trends in Food Science and Technology*. Elsevier. https://doi.org/10.1016/S0924-2244(00)88944-X.

Doaa El, Amrousy., Samir, Hassan., Heba, El. Ashry., Mohamed, Yousef., and Hossam, Hodeib. (2018). "Vitamin D supplementation in adolescents with irritable bowel syndrome: Is it useful? A randomized controlled trial." *Saudi J Gastroenterol.* 24(2), 109–114.

Cebeci, A. N., Ekici, B. (2014). "Epilepsy treatment by sacrificing vitamin D." *Expert Rev Neurother.* 14, 481–91.

Chang, J. P. C., Su, K. P., Mondelli, V. (2018). "Pariante CM. Omega-3 polyunsaturated fatty acids in youths with attention deficit hyperactivity disorder: a systematic review and meta-analysis of clinical trials and biological studies." *Neuropsychopharmacology*. 43(3), 534–545.

Checa-Ros, A., Haro-García, A., Seiquer, I., Molina-Carballo, A., Uberos-Fermández, J., Muñoz-Hoyos, A. (2019). "Early monitoring of fatty acid profile in children with attention deficit and/or hyperactivity disorder under treatment with omega-3 polyunsaturated fatty acids." *Minerva Pediatr*. 71(4):313-325.

Cutolo, M., Pizzorni, C., Sulli, A. (2011). "Vitamin D endocrine system involvement in autoimmune rheumatic diseases." *Autoimmun Rev*. 11, 84–87.

Day, R. E., Krishnarao, R., Sahota, P., Christian, M. S. (2019). "We still don't know that our children need vitamin D daily: a study of parents' understanding of vitamin D requirements in children aged 0-2 years." *BMC Public Health*. 19(1), 1119.

Fakruddin, J. M., Laurence, J. (2003). "HIV envelope gp120-mediated regulation of osteoclastogenesis via receptor activator of nuclear factor kappa B ligand (RANKL) secretion and its modulation by certain HIV protease inhibitors through interferon-gamma/RANKL cross-talk." *The Journal of biological chemistry*. 278 (48), 48251–8.

Ibrahim, F. A. S., Ghebremeskel, K., Abdel-Rahman, M. E., Ahmed, A. A. M., Mohmed, I. M., Osman, G., Elseed, M., Hamed, A., Rabinowicz, A. L., Salih, M. A. M., Elbashir, M. I., Daak, A. A. (2018). "The differential effects of eicosapentaenoic acid (EPA) and docosahexaenoic acid (DHA) on seizure frequency in patients with drug-resistant epilepsy - A randomized, double-blind, placebo-controlled trial." *Epilepsy Behav*. 87, 32-38.

Finch, S. L., Rosenberg, A. M., Vatanparast, H. (2018). "Vitamin D and juvenile idiopathic arthritis." *Pediatr Rheumatol Online J*. 16(1), 34.

Global Initiative for Asthma. *Global strategy for asthma management and prevention*. 2018. Available from: www.ginasthma.org.

Gorbach, Sherwood L. 1996. "Chapter 95 Microbiology of the Gastrointestinal Tract Composition and Distribution of the Microflora

Metabolic Activites of the Microflora The Intestinal Microflora and Infection." Baron S, Editor. *Medical Microbiology*. 4th Edition., 1–7.

Gould, Jacqueline F., Lisa G. Smithers, and Maria Makrides. 2013. "The Effect of Maternal Omega-3 (N23) LCPUFA Supplementation during Pregnancy on Early Childhood Cognitive and Visual Development: A Systematic Review and Meta-Analysis of Randomized Controlled Trials1-3." *American Journal of Clinical Nutrition* 97 (3): 531–44.

De Groot, R. H. M., Ouwehand, C., Jolles, J. (2012). "Eating the right amount of fish: inverted U-shape association between fish consumption and cognitive performance and academic achievement in Dutch adolescents." *Prostaglandins Leukot Essent Fatty Acids*. 86, 113–117.

Guo, Q., Goldenberg, J. Z., Humphrey, C., El, Dib, R., Johnston, B. C. (2019). "Probiotiques pour la prévention de la diarrhée associée aux antibiotiques chez l'enfant." *Cochrane Database Syst. Rev.* 30, 4.

Hattangdi-Haridas, S. R., Lanham-New, S. A., Wong, W. H. S., Ho, M. H. K., Darling, A. L. (2019). "Vitamin D Deficiency and Effects of Vitamin D Supplementation on Disease Severity in Patients with Atopic Dermatitis: A Systematic Review and Meta-Analysis in Adults and Children." *Nutrients*.11(8).

Hatun, S., Ozkan, B., Orbak, Z., Doneray, H., Cizmecioglu, F., Toprak, D., et al. (2005). "Vitamin D deficiency in early infancy." *J Nutr*. 135, 279–282.

Holick, M. F., Binkley, N. C., Bischoff-Ferrari, H. A., Gordon, C. M., Hanley, D. A., Heaney, R. P., et al. (2011). "Evaluation, treatment, and prevention of vitamin D deficiency: An endocrine society clinical practice guideline." *Journal of Clinical Endocrinology and Metabolism*. 96(7), 1911–30.

Holmes, Elaine, Jia V. Li, Thanos Athanasiou, Hutan Ashrafian, and Jeremy K. Nicholson. (2011). "Understanding the Role of Gut Microbiome-Host Metabolic Signal Disruption in Health and Disease." *Trends in Microbiology*.7, 349-359.

Horvath, A., Łukasik, J., Szajewska, H. (2017). "ω-3 Fatty Acid Supplementation Does Not Affect Autism Spectrum Disorder in

Children: A Systematic Review and Meta-Analysis." *J Nutr.* 147(3), 367-376.

Hossein-nezhad, A., Holick, M. F. (2013). "Vitamin D for health: a global perspective." *Mayo Clin Proc.* 88, 720–55.

Irwin, N., Davis, D., Currie, M. (2019). „Probiotic supplementation in well children: A scoping review." *J. Child. Health Care.*

Jacobson, D. L., Stephensen, C. B., Miller, T. L., Patel, K., Chen, J. S., Van, Dyke, R. B., et al. (2017). "Associations of low Vitamin D and elevated parathyroid hormone concentrations with bone mineral density in perinatally HIV-infected children." *Journal of Acquired Immune Deficiency Syndromes.* 76(1), 33–42.

Jat, Kana Ram, and Anju Khairwa. (2017). "Vitamin D and Asthma in Children: A Systematic Review and Meta-Analysis of Observational Studies." *Lung India.* 34(4): 355-363.

Johnston, B. C., Goldenberg, J. Z., Vandvik, P. O., Sun, X., Guyatt, G. H. (2011). "Probiotics for the prevention of pediatric antibiotic-associated diarrhea." *Cochrane Database Syst. Rev.* 11.

Kabbani, T. A., Koutroubakis, I. E., Schoen, R. E., Ramos-Rivers, C., Shah, N., Swoger, J, et al. (2016). "Association of vitamin D level with clinical status in inflammatory bowel disease: a 5-year longitudinal study." *Am J Gastroenterol.* 111(5), 712–719.

Khadgawat, R., R. K. Marwaha, M. K. Garg, R. Ramot, A. K. Oberoi, V. Sreenivas, M. Gahlot, N. Mehan, P. Mathur, and N. Gupta. 2013. "Impact of Vitamin D Fortified Milk Supplementation on Vitamin D Status of Healthy School Children Aged 10-14 Years." *Osteoporosis International* 24 (8): 2335–43.

Khalifah, R. A., Hudairi, A., Homyani, D. A., Hamad, M. H., Bashiri, F. A. (2018). "Vitamin D supplementation to prevent vitamin D deficiency for children with epilepsy: Randomized pragmatic trial protocol." *Medicine (Baltimore).* 97(40):e12734.

Kim, Jungmin, Yeonjeong Ko, Yu Kyung Park, Nack In Kim, Woel Kyu Ha, and Yunhi Cho. 2010. "Dietary Effect of Lactoferrin-Enriched Fermented Milk on Skin Surface Lipid and Clinical Improvement of Acne Vulgaris." *Nutrition* 26 (9): 902–9.

Kim, S., Kang, Y., Park, S., Koh, H., Kim, S. (2019). "Association of Vitamin D with Inflammatory Bowel Disease Activity in Pediatric Patients." *J Korean Med Sci*. 19, 34 (32):e204.

Kondo, Tomoo, Mikiya Kishi, Takashi Fushimi, Shinobu Ugajin, and Takayuki Kaga. 2014. "Vinegar Intake Reduces Body Weight, Body Fat Mass, and Serum Triglyceride Levels in Obese Japanese Subjects." *Biosci Biotechnol Biochem*. 73(8):1837-43.

Kreiter, S. R., Schwartz, R. P., Kirkman, H. N., Jr, Charlton, P. A., Calikoğlu, A. S., Davenport, M. L. (2000). "Nutritional rickets in African American breast-fed infants." *J Pediatri*. 137, 153–157.

Kujinga, Prosper, Valeria Galetti, Elizabeth Onyango, Viktor Jakab, Simone Buerkli, Pauline Andang'o, Inge D Brouwer, Michael B Zimmermann, and Diego Moretti. 2018. "Effectiveness of Zinc-Fortified Water on Zinc Intake, Status and Morbidity in Kenyan Pre-School Children: A Randomised Controlled Trial." *Public Health Nutrition* 21 (15): 2855–65.

LaMont, J. Thomas. (2018). "Probiotics for Children with Gastroenteritis." *New England Journal of Medicine*. 379 (21): 2076-2077.

Lee, Ji Hyuk. 2013. "Polyunsaturated Fatty Acids in Children." *Pediatric Gastroenterology, Hepatology and Nutrition*. Korean Society of Pediartic Gastroenterology, Hepatology and Nutrition. 16 (3): 153-161.

Lin, Jilei, Yin Zhang, Chunyan He, and Jihong Dai. (2018). "Probiotics Supplementation in Children with Asthma: A Systematic Review and Meta-Analysis." *Journal of Paediatrics and Child Health*. 54: 953-961.

Lourencetti, M., Abreu, M. M. (2018)."Use of active metabolites of vitamin D orally for the treatment of psoriasis." *Rev Assoc Med Bras*. 64(7):643-648.

Maliyar, Khalad, Cathryn Sibbald, Elena Pope, and R. Gary Sibbald. 2018. "Diagnosis and Management of Atopic Dermatitis." *Advances in Skin & Wound Care*. 31 (12): 538–50.

Marini, G., Sitzia, E., Panatta, M. L., De Vincentiis, G. C. (2019). "Pilot study to explore the prophylactic efficacy of oral probiotic Streptococcus salivarius K12 in preventing recurrent pharyngo-tonsillar episodes in pediatric patients." *Int. J. Gen Med.* 5, 12:213-217.

Mazahery, H., Conlon, C. A., Beck, K. L., Mugridge, O., Kruger, M. C., Stonehouse W., Camargo, C. A. Jr., Meyer, B. J., Jones, B., von Hurst, P. R. (2019). "A randomised controlled trial of vitamin D and omega-3 long chain polyunsaturated fatty acids in the treatment of irritability and hyperactivity among children with autism spectrum disorder." *J Steroid Biochem Mol Biol.* 187:9-16.

Mertoglu, C., Siranli, G., Topal, I., Gok, G., Erel, O. (2018). "Vitamin D supplementation does not improve plasma thiol/disulfide homeostasis." *Pediatr Int*. 60(11), 1008-1013.

Milte, C. M., Parletta, N., Buckley, J. D., Coates, A. M., Young, R. M., Howe, P. R. (2012). "Eicosapentaenoic and docosahexaenoic acids, cognition, and behavior in children with attention-deficit/hyperactivity disorder: a randomized controlled trial." *Nutrition*. 28, 670–677.

Mohammadzadeh, Soleiman, Narmin Baghi, Fayegh Yousefi, and Bahar Yousefzamani. 2019. "Effect of Omega-3 plus Methylphenidate as an Alternative Therapy to Reduce Attention Deficit-Hyperactivity Disorder in Children." *Korean Journal of Pediatrics* 62 (9): 360–66.

Moura, J. C. V., Moura, I. C. G., Gaspar, G. R., Mendes, G. M. S., Faria, B. A. V., Jentzsch, N. S., do Carmo Friche Passos, M., Kurdi, A., Godman, B., Almeida, A. M. (2019). "The use of probiotics as a supplementary therapy in the treatment of patients with asthma: a pilot study and implications." *Clinics (Sao Paulo).*

Nagakura, T., S. Matsuda, K. Shichijyo, H. Sugimoto, and K. Hata. 2000. "Dietary Supplementation with Fish Oil Rich in ω-3 Polyunsaturated Fatty Acids in Children with Bronchial Asthma." *European Respiratory Journal* 16 (5): 861–65.

Ong, Teck Guan, Morris Gordon, Shel S. C. Banks, Megan R. Thomas, and Anthony K. Akobeng. 2019. "Probiotics to Prevent Infantile Colic." *Cochrane Database of Systematic Reviews*. 2019: (3) CD012473.

Paineau, Damien, Didier Carcano, Greg Leyer, Sylviane Darquy, Marie Alexandra Alyanakian, Guy Simoneau, Jean François Bergmann, Dominique Brassart, Francis Bornet, and Arthur C. Ouwehand. (2008). "Effects of Seven Potential Probiotic Strains on Specific Immune

Responses in Healthy Adults: A Double-Blind, Randomized, Controlled Trial." *FEMS Immunology and Medical Microbiology* 53 (1): 107–13.

Penner, J., Ferrand, R. A., Richards, C., Ward, K. A., Burns, J. E., Gregson, C. L. (2018). "The impact of vitamin D supplementation on musculoskeletal health outcomes in children, adolescents, and young adults living with HIV: A systematic review." *PLoS One.* 15, 13(11).

Perceval, C., Szajewska, H., Indrio, F., Weizman, Z., Vandenplas, Y. (2019). "Prophylactic use of probiotics for gastrointestinal disorders in children." *Lancet Child Adolesc Health.* 3 (9), 655-662.

Pignatelli, Pasquale, and Stefania Basili. 2010. "Nutraceuticals in the Early Infancy." *Cardiovascular Therapeutics* 28 (4): 236–45.

Pourmasoumi, Makan, Nooshin Vosoughi, Seyedeh Masoumeh Derakhshandeh-Rishehri, Mostafa Assarroudi, and Motahar Heidari-Beni. 2018. "Association of Omega-3 Fatty Acid and Epileptic Seizure in Epileptic Patients: A Systematic Review." *International Journal of Preventive Medicine*. 9:36.

Prescott, S. L, Bjorksten B. (2007). "Probiotics for the prevention or treatment of allergic diseases." *J Allergy Clin Immunol* 120 (2), 255-62.

"Probiotics Market Size, Share, and Analysis, 2022." n.d. Accessed March 17, 2020. https://www.alliedmarketresearch.com/probiotics-market.

Rodríguez, C., García, T., Areces, D., Fernández, E., García-Noriega, M., Domingo, J. C. (2019). "Supplementation with high-content docosahexaenoic acid triglyceride in attention-deficit hyperactivity disorder: a randomized double-blind placebo-controlled trial." *Neuropsychiatr Dis Treat*. 8, 15:1193-1209.

Russo, Francesco, Michele Linsalata, and Antonella Orlando. 2014. "Probiotics against Neoplastic Transformation of Gastric Mucosa: Effects on Cell Proliferation and Polyamine Metabolism." *World Journal of Gastroenterology*. 20, 13258-13272.

Santini, A., Cammarata, S. M., Capone, G., Ianaro, A., Tenore, G. C., Luca Pani, L. and Ettore Novellino, E., (2018). "Nutraceuticals: opening the debate for a regulatory framework." *Br. J. Clin. Pharmacol.* 84, 659–672.

Schmitt, J., and Ferro A. (2013). "Nutraceuticals: Is There Good Science behind the Hype?" *British Journal of Clinical Pharmacology*. 75, 585-587.

Von Scheven, E., Burnham, J. M. (2011). "Vitamin D supplementation in the pediatric rheumatology clinic." *Curr Rheumatol Rep*. 13, 110–116.

Shaw, N. J., Pal, B. R. (2002). "Vitamin D. Deficiency in UK Asian families: activating a new concern." *Arch Dis Child*. 86:147–149.

Shen, S. Y., Xiao, W. Q., Lu, J. H., Yuan, M. Y., He, J. R., Xia. H. M., Qiu, X., Cheng, K. K., Lam, K. B. H. (2018). "Early life vitamin D status and asthma and wheeze: a systematic review and meta-analysis." *BMC Pulm Med*. 18(1), 120.

Smiyan, O. I., Smiian-Horbunova, K. O., Bynda, T. P., Loboda, A. M., Popov, S. V., Vysotsky, I. Y., Moshchych, O. P., Vasylieva, O. G., Manko, Y. A., Ovsianko, O. L., Kolesnikova, M. V., Dolgova, N. O., Aleksakhina, T. O., Al-Rawashdeh, B. (2019)"Optimization of the treatment of rotavirus infection in children by using bacillus clausii." *Wiad Lek*. 72(7):1320-1323.

Specker, B. L., Valanis, B., Hertzberg, V., Edwards, N., Tsang, R. C. (1985). "Sunshine exposure and serum 25-hydroxyvitamin D concentrations in exclusively breast-fed infants." *J Pediatr*. 107, 372–376.

Sudjaritruk, T., Bunupuradah, T., Aurpibul, L., Kosalaraksa, P., Kurniati, N., Prasitsuebsai, W, et al. (2016). "Hypovitaminosis D and hyperparathyroidism: effects on bone turnover and bone mineral density among perinatally HIV-infected adolescents. *AIDS (London, England)*. 30(7), 1059–67.

Uday, S., Högler, W. (2018). "Prevention of rickets and osteomalacia in the UK: political action overdue." *Arch Dis Child*. 103, 901–906.

Varzakas, Theodoros, Panagiotis Kandylis, Dimitra Dimitrellou, Chryssoula Salamoura, George Zakynthinos, and Charalampos Proestos. 2018. Innovative and Fortified Food: Probiotics, Prebiotics, GMOs, and Superfood. *Preparation and Processing of Religious and Cultural Foods*. Elsevier Ltd.

Vernocchi, Pamela, Federica Del Chierico, and Lorenza Putignani. 2016. "Gut Microbiota Profiling: Metabolomics Based Approach to Unravel

Compounds Affecting Human Health." *Frontiers in Microbiolog.*7, 1144.

Wang, Zheng Li, Li Liu, Xiao Yu Hu, Lu Guo, Qiu Yu Li, Yao An, Ya Jun Jiang, et al. (2019). "Probiotics May Not Prevent the Deterioration of Necrotizing Enterocolitis from Stage i to II/III." *BMC Pediatrics*. 19 (1): 185.

Wenhao, Yang., Tao, He., Weijian, Zhang., Li, Gu., Renyuan, Tu., and Hanmin, Liu., (2019)." Efficacy and safety of lactobacilli in children with functional constipation. Study protocol for a meta-analysis and systematic review." *Medicine* (Baltimore), 98, 20.

Winkvist, A., Åberg, M. A., Åberg. N., Sundberg. R., Torén, K., Brisman, J. (2010). "Fish consumption and school grades in Swedish adolescents: a study of the large general population." *Acta Paediatr*. 99, 72–77.

Woo, J., Couturier, J., Pindiprolu, B., Picard, L., Maertens, C., Leclerc, A., Findlay, S., Johnson, N., Grant, C., Kimber, M. (2017). Acceptability and tolerability of omega-3 fatty acids as adjunctive treatment for children and adolescents with eating disorders. *Eat Disord*. 25(2), 114-121.

In: Nutraceuticals: Evidence Based Use ISBN: 978-1-53618-052-7
Editor: Raluca Maria Pop

Chapter 9

NUTRACEUTICALS IN DERMATOLOGY: FROM ANECDOTIC TO SCIENTIFICALLY PROVEN INDICATIONS

Andreea Nicoleta Boca[1,*] and Roxana Flavia Ilieș[2]
[1]Department of Pharmacology, Toxicology and Clinical Pharmacology, "Iuliu Hatieganu" University of Medicine and Pharmacy, Cluj-Napoca, Romania
[2]Department of Medical Genetics, "Iuliu Hatieganu" University of Medicine and Pharmacy, Cluj-Napoca, Romania

ABSTRACT

Nutraceuticals, often marketed as nutritional supplements, have become immensely popular in recent years. With many supplements being derived from natural extracts, patients' confidence in these substances has soared.

The physicians are generally skeptical regarding nutraceuticals. In light of lax regulations, inflated claims and reports of side effects, this skepticism is well-justified.

[*] Corresponding Author's E-mail: boca.andreea@umfcluj.ro.

The following chapter aims to offer an overview into which supplements are truly proved useful for dermatological purposes, whether as adjuvants in dermatological conditions or as aides in maintaining skin, nail and hair health. Furthermore, dosages, side effects, and specific formulations will be discussed.

Keywords: nutraceuticals, skin disease, hair, vitamins, collagen, probiotics, aging

INTRODUCTION

The term "nutraceuticals," a portmanteau of "nutritional" and "pharmaceutical" was first coined by Stephen DeFelice in 1989 (Brower 1998; Kalra 2003), referring to "food (or a part of it) that provides medical or health benefits, including the prevention and/or treatment of a disease."

A 2017 survey established that 76% of adults in the United States take dietary supplements, with the global market for beauty-related supplements reaching 3.5 billion US dollars in 2016 ("2017 CRN Consumer Survey on Dietary Supplements | Council for Responsible Nutrition" 2019; "Beauty Supplements Market- Sales Data, Size, Trends, Demand" 2019).

However, due to these products being marketed as food supplements, they can bypass efficacy and safety testing, with 2/3rds of supplements on the market lacking label information on safety and efficacy (Katta and Huang 2019). This has led to life-threatening situations, such as one patient ingesting 1000 times the daily allowance for vitamin D (Araki et al. 2011), or 80 patients suffering from lead poisoning due to contamination in a traditional Ayurvedic supplement (Saper et al. 2004, 2008).

These cases highlight the need for more stringent laws regarding food supplements. In the meantime, a solid knowledge regarding supplements, their efficacy, dosage, adverse effects, and interactions is key in controlling the issue.

Different supplement manufacturers use proprietary formulations, combinations, and dosages, as such it is virtually impossible to compare and rank particular products with regards to efficacy and safety. The following chapter will discuss the most popular active ingredients found in nutraceuticals that target skin health-elaborating on their efficacy in pathological as well as physiological states.

Chemical Constituents – The Vitamins

Vitamin D

Vitamin D is a group of fat-soluble secosteroids, with the most important two forms being vitamin D3 (cholecalciferol) and D2 (ergocalciferol). Vitamin D is synthesized at skin level as a response to exposure to UVB radiation but may also be found in foods, such as fatty fish, and supplement form.

Vitamin D is involved in the physiological development of keratinocytes, wound healing and production of skin proteins (Fenner and Silverberg 2019). Lower vitamin D levels were found in pediatric patients suffering from atopic dermatitis, however, the baseline for vitamin D levels was measured in a cohort of healthy adults. As such, comparing vitamin D levels across groups which differ fundamentally from a demographic standpoint does not validate the results of this study. A lower vitamin D serum level was associated with poorer "SCORing Atopic Dermatitis" (SCORAD) scores (Wang et al. 2014; Peroni et al. 2011; Heimbeck, Wjst, and Apfelbacher 2013). However, other studies show no correlation between vitamin D levels and the severity of atopic dermatitis (Samochocki et al. 2013; Robl et al. 2016; Galli et al. 2015) and no benefit found in vitamin D supplementation versus placebo (Bath-Hextall et al. 2012). Thus, there is still insufficient evidence to recommend vitamin D supplementation in atopic dermatitis (Baquerizo Nole, Yim, and Keri 2014).

Vitamin D has been proposed for investigation as a supplement in psoriasis, likely linked to the vitamin's key role in the metabolic syndrome. Topical vitamin D, as the compound calcitriol, is currently used as an important therapeutic option in psoriasis. Regarding oral intake of vitamin

D, there are few studies performed, with positive results, but more research is needed before formulating a solid recommendation (Fu and Vender 2011).

If recommended as a supplement, vitamin D should be adjusted according to laboratory results, due to its sequestration in fatty tissues, leading to a long half-life of the substance.

The threshold for vitamin D supplementation has not been clearly established, with the tolerable upper intake level currently cited at 4000 UI/day (Manson et al. 2011). Vitamin D overdose causes hypercalcemia, manifesting through anorexia, nausea, vomiting, polyuria, polydipsia, weakness, insomnia and culminating with renal failure. Kidney damage following excessive vitamin D consumption can be irreversible.

Recent findings suggest that a significant proportion of the population is vitamin D deficient (Holick 2017); this has prompted self-administration of vitamin D supplements, and several labeling errors in vitamin D supplements have been uncovered, resulting in patients developing toxicity from vitamin D supplementation (Araki et al. 2011).

Vitamin E

The vitamin E group includes 8 fat-soluble compounds, namely tocopherols and tocotrienols. Vitamin E is naturally found in vegetable oils (wheat germ oil, hazelnut oil, canola oil, almond oil, etc) with smaller amounts. It is naturally present in animal food sources, such as eggs, oysters, butter and fish. In supplements, vitamin E is typically found as tocopheryl acetate.

Vitamin E has important antioxidant effects, through reducing the formation of reactive oxygen species, scavenging free radicals and stabilizing cell membranes. Vitamin E boasts photoprotective properties mainly against UVB, increasing MED (Minimal Erythemal Dose) (Souyoul, Saussy, and Lupo 2018). Positive effects of vitamin E supplementation have been noted in telogen effluvium (Shrivastava 2009).

The recommended daily allowance for vitamin E is 15mg/day for adults as established by the National Academy of Medicine in 2015 while the

European Food Safety Authority recommends 11mg/day for women and 13mg/day for men (Manson et al. 2011).

Some trials have pointed out that supplementation with vitamin E increases the risk of prostate cancer among healthy men (Klein et al. 2011). Regarding overdose, vitamin E increases bleeding time, associating symptoms such as nausea, dizziness, headaches, blurred vision, bruising and bleeding (Marosz and Chlubek 2014).

Vitamin C

L-ascorbic acid, known as vitamin C, is a water-soluble compound naturally present in fresh fruit and vegetables, and common addition to food supplements. Vitamin C is not synthesized in the human body. As such, exogenous sources, whether through whole foods or supplements, are necessary.

L-ascorbic acid acts as an antioxidant through the scavenging of free radicals, as well as playing a key role in collagen synthesis and protection against the effects of UVA and UVB radiation (Souyoul, Saussy, and Lupo 2018).

The recommended daily allowance of vitamin C is 75mg/day for women and 90mg/day for men (Souyoul, Saussy, and Lupo 2018).

Due to its hydrophilic nature, excess vitamin C is easily eliminated, however, megadoses may cause diarrhea, nausea, vomiting, heartburn, and insomnia (Marosz and Chlubek 2014).

Folic Acid

Folic acid, known as vitamin B9, is a water-soluble vitamin with significant involvement in protein synthesis. As it cannot be produced endogenously, exogenous sources such as whole foods or supplements are necessary to maintain homeostasis. Symptoms of folate deficiency include fatigue, palpitations, dyspnea and changes in hair and skin color and texture, as well as neuropathy (Koike et al. 2015).

Folic acid has been quoted as a potential supplement in cases of telogen effluvium (Shrivastava 2009) as well as psoriasis (Murzaku, Bronsnick, and Rao 2014).

A Recommended Dietary Allowance (RDA) of 400 μg of folic acid per day is recommended in adults, regardless of gender, with higher requirements for pregnant and lactating women. Consistently increased levels of folic acid have been linked to an increased risk of cognitive impairment in the elderly as well as an increased risk of prostate cancer (Smith 2007).

Biotin

Biotin, also known as vitamin H or B7, is a water-soluble vitamin involved in a plethora of metabolic pathways, of which mentionable: catabolism of fat and carbohydrates, protein synthesis, cell growth. Biotin deficiency manifests through hair thinning, brittle nails and skin rashes ("Biotin Deficiency - an Overview Science Direct Topics" 2019).

Biotin has been proposed as a supplement for telogen effluvium, psoriasis, as well as atopic dermatitis, improving pruritus and xerosis in biotin-deficient pediatric patients suffering from atopic dermatitis (Iikura et al. 1988). Biotin supplementation has yielded significant results when used for brittle nails, triangular worn-down nails, trachonychia or habit tic nail deformation (Möhrenschlager et al. 2000; Gloster and Kindred 2005). It has been used in 2 cases of uncombable hair syndrome with normal serum biotin levels, with marked clinical improvement (Shelley and Shelley 1985; Boccaletti et al. 2007).

The recommended intake for biotin is 30 μg/day (Lipner 2018).

In light of recent events, the FDA has put forward a warning regarding biotin supplementation, as it has been established that biotin interferes with laboratory assays, affecting the accuracy of results for troponin, TSH, NTproBNP, PTH and FT3. As such, patients are recommended to inform their physician regarding any supplements they may be using, and discontinue usage before routine laboratory testing ("The FDA Warns That Biotin May Interfere with Lab Tests: FDA Safety Communication | FDA" 2019).

Nicotinamide

Nicotinamide, a form of the B3 vitamin, is a water-soluble compound, naturally occurring in yeast, meat, milk and green vegetables.

Nicotinamide has been proposed as a preventative agent for non-melanoma skin cancer due to its role in DNA repair. However, the study investigating its efficacy was performed on high-risk individuals, with a personal history of non-melanoma skin cancer, and investigated the effect of nicotinamide supplementation on the occurrence of a second NMSC. At this point, the recommendation can not be broadened to include primary prophylaxis in other categories of patients (Katta and Huang 2019).

Chemical Constituents – The Minerals

Copper

Copper acts as a cofactor in collagen crosslinking and skin pigmentation, as well as promoting keratinocyte and fibroblast proliferation (*Dietary Reference Intakes for Energy, Carbohydrate, Fiber, Fat, Fatty Acids, Cholesterol, Protein, and Amino Acids (Macronutrients)* 2005; Yeowell et al. 1994; Driscoll et al. 2010).

The RDA for copper is 900 µg for adults over 19 years of age (Institute of Medicine 1998); copper occurs naturally in nuts, seeds, seafood, meat, and grains.

Potential indications with regards to skin health are related to its stimulation of collagen synthesis, skin pigmentation, and rejuvenation, wound repair, anti-inflammatory and antibacterial properties (Souyoul, Saussy, and Lupo 2018).

Selenium

Selenium acts as an aid in glutathione peroxidation and thioredoxin reduction and removes damaging lipid hydroperoxides (Wu, Cheng, and Wang 2017; Wintergerst, Maggini, and Hornig 2007; Pinnell 2003). As such, it supports DNA synthesis and repair, guarding against oxidative damage.

In vitro studies have highlighted that selenium protects keratinocyte stem cells (Jobeili et al. 2017). Furthermore, selenium has been studied as a preventative for skin cancer, with no statistically significant effect (Clark et al. 1996).

A recommended dose of 55μg/day in women as well as men can be achieved through whole foods (meat and seafood) or supplementation.

An error in formulation, leading patients to consume 200 times the RDA of selenium, resulted in toxicity symptoms such as hair loss, muscle cramps, diarrhea, joint pain and fatigue (Morris and Crane 2013).

Zinc

Zinc acts as a protective element against lipid peroxidation, UVR-induced cytotoxicity, decreases nitric oxide production and plays an important role in epidermal proliferation and keratinocyte differentiation. Notable effects of use in dermatology are: anti-inflammatory, wound healing, skin cell proliferation and cofactor for cellular activity and defense (Pinnell 2003; Institute of Medicine 1998; Ogawa, Kawamura, and Shimada 2016).

A combined supplement of zinc and phytase was found to increase the duration of the effect of injected botulinum toxin (Koshy et al. 2012). Low serum levels of zinc have been associated with atopic dermatitis (Gray et al.

2019), and there is preliminary evidence of some effectiveness of zinc supplementation in acne vulgaris, hidradenitis suppurativa, atopic dermatitis and diaper dermatitis (Dhaliwal et al. 2019). However, the studies substantiating these claims are of modest quality and should be followed-up by larger-scale studies to formulate a recommendation (Gray et al. 2019).

A daily intake of 11 mg/day for men and 9 mg/day for women is recommended and can be achieved through the consumption of whole foods (non-milled whole grains, red meat, seafood, fortified cereal) or supplements.

Zinc has shown promise as a supplement for telogen effluvium (Kil, Kim, and Kim 2013).

Chemical Constituents – The Polyunsaturated Fatty Acids

Polyunsaturated fatty acids are comprised of two classes: omega 3 and omega 6 fatty acids.

The effects of omega 3 fatty acids such as alpha-linolenic acid (ALA), eicosapentaenoic acid (EPA) and docosahexaenoic acid (DHA) have been studied for a plethora of diseases and tissues. At skin level, omega 3 fatty acids are anti-inflammatory (McCusker and Grant-Kels 2010; Yoshida, Yasutomo, and Watanabe 2016). They lower the levels of proinflammatory cytokines such as prostaglandins, leukotrienes, TNF alpha, gamma interferon, IL-12, and IL-6. Ultraviolet-induced inflammation, as well as Minimal Erythema Dose (MED), is lessened through supplementation with omega 3 fatty acids (McCusker and Grant-Kels 2010).

No RDA has been established for omega 3 fatty acids, however, the appropriate intake range is 12-17 g/day, including 1.1-1.6 g of ALA per day (Souyoul, Saussy, and Lupo 2018).

Omega 3 fatty acids have been studied for their effect in atopic dermatitis, as well as psoriasis, however, there is insufficient evidence to recommend supplementation in these cases (Baquerizo Nole, Yim, and Keri 2014; Murzaku, Bronsnick, and Rao 2014).

Omega 3 fatty acid supplementation may be available through dedicated supplements or through products that contain fish oil.

Regarding Omega 6 PUFAs popular in skin supplements, some notable representatives are evening primrose oil and borage oil. Evening primrose oil has shown potential in several studies, investigating its use in conditions such as atopic dermatitis or psoriasis, however, a recent meta-analysis shows no benefit of prescribing evening primrose oil for these conditions (Fenner and Silverberg 2019).

Evening primrose was previously linked to seizures, an association that has been disproven, however reports of lipoid pneumonia, thrombosis and immunosuppression have been associated with evening primrose supplements: as such, caution and a thorough risk-benefit analysis is needed to recommend this supplement for dermatological patients.

Similar results have been obtained with borage oil supplementation, with reviews unable to arrive at a clear consensus but without being able to exclude any beneficial effects of borage oil in atopic dermatitis (Foster, Hardy, and Alany 2010). Adverse effects noted from the use of borage oil include headaches and minor gastrointestinal upset (Fenner and Silverberg 2019).

Chemical Constituents – The Phytochemicals

Carotenoids

Beta carotene is a red-orange pigment, serving as an inactive precursor to vitamin A, a lipid-soluble vitamin.

Vitamin A reduces free radicals and singlet O_2-induced lipid peroxidation, with photoprotective benefits manifested through an increase in MED (Evans and Johnson 2010).

The recommended daily allowance for trans β carotene is 1800 μg per day for men and 1400 μg/day for women (Juturu, Bowman, and Deshpande 2016).

Long term supplementation with vitamin A or any of its precursors may increase the risk for the development of lung cancer (Grether-Beck et al. 2017) in male smokers, on a 6.1 years median follow-up (Chen et al. 2015). High levels of vitamin A may trigger hair loss (Katta and Huang 2019) and an increase in yellow skin pigmentation may be noted with some formulations (Gollnick et al. 1996).

Lutein and zeaxanthin are xanthophyll carotenoids, boasting antioxidant effects alongside blocking blue wavelengths. Upon supplementation, an increase in MED, skin tone, luminance and color have been noted in the literature (Juturu, Bowman, and Deshpande 2016).

Lutein and zeaxanthin may be administered orally or topically, in a recommended dose of 6-10 mg/day of lutein and 2 mg/day of zeaxanthin.

Probiotics and Prebiotics

Probiotic products contain live microorganisms, which aid in restoring optimal gut flora. Prebiotics are often a complimentary supplement alongside probiotics and are composed of nondigestible foodstuffs, oftentimes oligosaccharides, with a similar scope of promoting optimal development of the gut flora.

Prebiotics and probiotics have been studied extensively for their potential in preventing and managing atopic dermatitis. Currently, prebiotics are recommended for infants and probiotics are recommended prenatally and postnatally as a preventative agent for the development of atopic dermatitis (Bronsnick, Murzaku, and Rao 2014).

Probiotics administered to patients suffering from atopic dermatitis have decreased Th2 response and increased Th1 response (Winkler et al. 2007; Isolauri et al. 2001; De Preter et al. 2004; Schley and Field 2002; Passeron et al. 2006), as well as increased the activity of Treg cells (Kwon et al. 2010). An increase in skin and mucosal barrier function has been observed

(Gueniche et al. 2010; Matsumoto et al. 2007; Philippe, Blum, and Benyacoub 2011), alongside an increase in intestinal microflora diversity (Matsumoto et al. 2007), a decrease in ferment production (De Preter et al. 2004; Schley and Field 2002; Passeron et al. 2006) and S. *aureus* attachment (Prince, McBain, and O'Neill 2012).

Administration of probiotic supplements in patients suffering from acne resulted in a decrease in inflammation, explained by a downregulation of inflammatory cytokines, recruitment of CD8 cells and activation of T regulatory cells. Furthermore, ceramide production was upregulated, with an up to 80% clinical improvement (Muizzuddin et al. 2012; Kang et al. 2009; Hacini-Rachinel et al. 2009).

Regarding wound healing, the administration of probiotic supplements leads to a decrease in bacterial load of the wound, promoting faster healing and decreasing the area of chronic leg ulcers (Maria C. Peral, Huaman Martinez, and Valdez 2009; M. C. Peral et al. 2010).

Probiotics have been commonly used for several years, with some reported side effects such as bloating or diarrhea. However, the available safety studies have been deemed insufficient (Rijkers et al. 2011), with more studies needed to clarify efficacy and safety profiles. Caution must be taken in recommending probiotic supplements to patients with a compromised immune status or other comorbidities. Case reports of sepsis or bowel ischemia associated with probiotic use have emerged in literature (Boyle et al. 2008).

OTHERS

Collagen

Collagen is a fibrillar protein, naturally found within the human body in the skin, joints, and bones (Avila Rodríguez, Rodríguez Barroso, and Sánchez 2018). A loss in thickness and strength of collagen fibers was strongly correlated with skin aging, as such collagen supplements have gained traction as anti-aging supplements.

Collagen in supplement form may be obtained through extraction from plant or animal sources, as well as through recombinant protein production systems. The two most common forms of collagen supplements are based on collagen hydrolysate, studied in doses ranging from 2.5g to 10g/day, and collagen tripeptides, studied in doses of 3g/day (Choi et al. 2019).

Collagen supplements seem to provide a modest but statistically significant increase in skin elasticity and a decrease in transepidermal water loss, however, the evidence regarding skin hydration levels is conflicting (Proksch et al. 2013; Choi et al. 2019; Asserin et al. 2015). Collagen supplements reduced the degree of collagen fragmentation in the dermis while increasing its overall density, as well as promoting glycosaminoglycan synthesis. These supplements show potential as a complementary therapy in wound healing, xerosis, and other dermatological conditions, however, more studies are necessary before establishing their efficacy.

No adverse effects have been reported so far relating to the intake of collagen-based supplements (Proksch et al. 2013).

An Overview of Commercially Available Skincare Nutraceuticals

The following subchapter aims to offer a general overview of nutraceuticals marketed as skin, hair or nails - enhancing supplements. However, as many supplements are advertised for a plethora of conditions, in the next table (Table 1), supplements with more than 4 distinct indications will be classified as "panacea."

Most supplements aimed towards skin, hair and nails' health are complex in their formulation, some containing a multitude of both water and fat-soluble vitamins. This brings forward the difficulty in balancing micronutrient intake from supplements, and the risk of unbalanced vitamin consumption.

Table 1. An overview of supplements targeted towards skin, hair and nails' health, their composition and claimed effects

	Commercial name	Manufacturer	Active ingredients	Marketed for
1	Collanol ®	Vitaslim	470.6 mg Curcuma extract, 80 mg vitamin C, 40 mg type 2 collagen, 1 mg copper, 2 mg manganese, 5.5 mcg D3 vitamin	Regeneration of joints and cartilages, collagen production
2	Qyra Colagen®	Pharmalink	2500 mg Verisol ® collagen, 16mg niacin, 50 mcg biotin, other natural extracts (apple, lemon, Carthamus tinctures)	Antiaging, skin regeneration and health, hair and nail health, collagen production
3	Femovita Beauty®	Naturpharma	1000 mg hydrolyzed collagen, 30 mg hyaluronic acid, 1500 mcg vitamin A, vitamins B, C, D, E, selenium, zinc and other plant extracts in undisclosed concentrations	Hair and skin health, antiaging
4	Kollagen beauty®	Doppelherz	2500 mg collagen peptides, beta carotene equivalent to 160 microg vitamin A, 3 mg vitamin E, 50 microg biotin, 80 mg vitamin C, 5 mg zinc, 0.15 mg copper, acai and fruit extract	Maintains skin hydration, elasticity and firmness
5	Collagen HydraPlump c	Resvitale	30 mg vitamin C, 500 mg Biocell collagen, 300 mg type II collagen, 100 mg chondroitin sulfate, 50 mg hyaluronic acid, 7.5 mg glycosylceramides, 6 mg di-galactosyl diglycerides	Restores skin hydration in 2 weeks, increases skin firmness
6	Beauty3 Collagen Keratin Elastin®	Resvitale	18mg niacinamide, 2 mg vitamin B6, 300 mcg biotin, pantothenic acid, 15 mg zinc, 1.7 mg copper, 1000 mg collagen, 100 mg hyaluronic acid, 500 mg keratin, other herbal extracts	Antiaging, maintaining skin vitality, antioxidant effect
7	Collagen Enhancer®	Resvitale	1000 mg Biocell Collagen, 600 mg hydrolyzed type II collagen, 200 mg chondroitin sulfate, 100 mg hyaluronic acid, other herbal extracts	Aids in skin health, hydration, elasticity, scarring, joint health, muscle tone

	Commercial name	Manufacturer	Active ingredients	Marketed for
8	Ultra Collagen Enhancer®	Resvitale	Resveratrol complex, 1000 mg Biocell Collagen, 600 mg hydrolyzed type II collagen, 200 mg chondroitin sulfate,100 mg hyaluronic acid, other natural extracts	Joint health, muscle tone, antiaging, aids in scarring
9	Beauty Start®	Zenyth	2500 mg bovine peptides, 87.5 mg inulin, 80 mg vitamin C, 80 mg coenzyme q10, 30 mg sodium hyaluronate	Skin, hair, nail health, antiaging
10	S.O.D & Collagen Complex®	Zenyth	636 mg hydrolyzed bovine collagen, melon extract, 4 mg hyaluronic acid	Antiaging, anti-cellulite
11	Hyaluronic Acid with Collagen Complex®	Zenyth	400 mg hydrolyzed bovine collagen, 200 mg type 2 collagen, 100mg sodium hyaluronate	Hair and skin health, joint health, adjuvant in acne treatment
12	Collagen & Hyaluronic acid®	Casa Herba	62.7 mg collagen, 20 mg chondroitin sulfate, 10 mg hyaluronic acid	Hair, skin and nail health, antiaging, joint and bone health
13	Collagen & Hyaluronic acid Forte®	Casa Herba	550 mg collagen, 200 mg vitamin C, 55 mg hyaluronic acid	Antiaging, joint health
14	Women's Evening Primrose Oil 500mg®	GNC	500 mg evening primrose oil	Panacea
15	Women's Evening Primrose Oil 1300mg®	GNC	1300 mg evening primrose oil	Panacea
16	Orissa 1000®	Agetis	1000 mg evening primrose oil, 15 mg alpha-tocopherol	Metabolic support, protection against oxidative stress
17	Beres Beauty Hair, Nails and Skin®	Beres	800 mcg vitamin A, 1.4 mg vitamin B1,1.6 mg vitamin B2, 8 mg vitamin B3, 6 mg vitamin B5, 2 mg vitamin B6, vitamin B12, 10 g vitamin E, 200 mcg folic acid, 150 mcg biotin, 7 mg iron, 15 mg zinc, 700 mcg copper, 150 mcg iodine, 5 mg ubiquinone	Hair, skin, nail health, antiaging
18	Aspasia 40+®	Zdrovit	Collagen, hyaluronic acid, vitamin E, vitamin C, copper, para-aminobenzoic acid, herbal extracts	Antiaging, skin, nail, and hair health

Some supplements have undisclosed amounts of certain components: one example is Aspasia 40+, for which we could not locate an accurate list of components, as the listings available online varied according to the website, and the company does not have their website listing products and ingredients.

With some supplements, the ingredient list mentions various herbal extracts, for instance, apple extract or other botanicals, with no mention in the literature of their effectiveness.

The claims made by supplement manufacturers are heterogeneous, with some simply stating the role and importance of the active ingredients in their product, while others promote their products' effect as guaranteed to appear in a specified, short timeframe.

While some supplements are formulated in accordance to the data available in the literature, some offer only trace amounts: for instance, collagen supplementation has been studied at dosages surpassing 2000 mg, while the first supplement in Table 1 is formulated with 40mg of collagen, yet the mention of collagen appears in its name. This is 50 times less than the minimum amount proven to have an effect in the studies carried out up until now. As such, it is hard to estimate whether this particular dose of collagen would offer any benefit.

General Considerations Regarding Nutraceuticals in Dermatology

The present book chapter has elaborated on some of the most popular active ingredients used in skin, hair and nails supplements. While the science may be clear in some cases, it is important to note that the particular formulation of the supplement or, more so, the combination of actives may yield an important effect through synergy. Due to the heterogeneity of products available and the various methodologies used to study these supplements, elaborating a complete list of supplements and outlining their precise effects is unfeasible.

Mankind has seen food as medicine dating back to Hippocrates. The popularity of natural remedies is on the rise among patients who often resort to self-prescribing supplements in addition to, or replacing the medication prescribed by physicians. This raises a series of concerns, ranging from the purity and real composition of the supplements ingested to the potential interactions between nutraceuticals and prescribed medication. Such is the case of saw palmetto, used as an aid in treating hair loss while increasing bleeding time (Tachjian, Maria, and Jahangir 2010). A mnemonic of points to clarify before recommending a particular supplement was proposed by Katta and Huang: PPIES (Purity, Potency, Interactions, Efficacy and Safety).

Even very common supplementation can be harmful, in the absence of adequate dosing. For example, beta carotene, or vitamins B6 and B12 have been associated with an increase in the risk of developing lung cancer in men; however, this risk was only revealed after long-term surveillance in a clinical trial (Katta and Huang 2019; Grether-Beck et al. 2017).

The lack of toxicology regulations for nutraceuticals, combined with their wide availability, can lead to dire health consequences, such as liver toxicity associated with certain formulations of Chinese herbal medicine (Fenner and Silverberg 2019) or contamination with pathogenic fungi documented in a children's probiotic supplement ("Fatal Gastrointestinal Mucormycosis in an Infant Following Use of Contaminated ABC Dophilus Powder From Solgar Inc. | Fungal Diseases | CDC" 2019).

Essential to consider is also the population subgroup for a particular supplement. For instance, the fact that a vitamin D supplement has offered positive therapeutic effects in a deficient cohort does not allow translation of those particular results for the general population, who may not be deficient for vitamin D.

The rise of nutraceuticals' popularity among patients led to an increase in attention from the scientific community, with subsequent studies aiming to clarify aspects of efficacy and safety. Until these aspects are fully elucidated through short, medium and long-term studies, recommending nutraceuticals for various dermatological conditions remains a gamble.

CONCLUSION

Nutritional supplements are sought after by patients and could play an important role as adjuvant therapy in specific dermatological conditions and general skin management. Yet, contrary to popular opinion, they can be of harm when taken without adequate medical consideration towards dosage, interactions and side effects. Thus, further research and more strict regulations are required for nutraceuticals.

REFERENCES

2017 CRN Consumer Survey on Dietary Supplements | Council for Responsible Nutrition. 2019. Accessed November 16. https://www.crnusa.org/resources/2017-crn-consumer-survey-dietary-supplements.

Araki, Takako, Michael F. Holick, Bianca D. Alfonso, Esti Charlap, Carla M. Romero, Dahlia Rizk, and Lisa G. Newman. 2011. "Vitamin D Intoxication with Severe Hypercalcemia Due to Manufacturing and Labeling Errors of Two Dietary Supplements Made in the United States." In *Journal of Clinical Endocrinology and Metabolism*, 96:3603–8. doi:10.1210/jc.2011-1443.

Asserin, Jérome, Elian Lati, Toshiaki Shioya, and Janne Prawitt. 2015. "The Effect of Oral Collagen Peptide Supplementation on Skin Moisture and the Dermal Collagen Network: Eevidence from an Ex Vivo Model and Randomized, Placebo-Controlled Clinical Trials." *Journal of Cosmetic Dermatology* 14 (4). Blackwell Publishing Ltd: 291–301. doi: 10.1111/jocd.12174.

Avila Rodríguez, María Isabela, Laura G. Rodríguez Barroso, and Mirna Lorena Sánchez. 2018. "Collagen: A Review on Its Sources and Potential Cosmetic Applications." *Journal of Cosmetic Dermatology*. Blackwell Publishing Ltd. doi:10.1111/jocd.12450.

Baquerizo Nole, Katherine L, Elizabeth Yim, and Jonette E Keri. 2014. "Probiotics and Prebiotics in Dermatology." *Journal of the American*

Academy of Dermatology 71 (4): 814–21. doi:10.1016/j.jaad.2014.04.050.

Bath-Hextall, Fiona J., Claire Jenkinson, Rosemary Humphreys, and Hywel C. Williams. 2012. "Dietary Supplements for Established Atopic Eczema." *Cochrane Database of Systematic Reviews*. John Wiley and Sons Ltd. doi:10.1002/14651858.CD005205.pub3.

Beauty Supplements Market-Sales Data, Size, Trends, Demand. 2019. Accessed November 16. https://www.goldsteinresearch.com/report/global-beauty-supplements-market-outlook-2024-global-opportunity-and-demand-analysis-market-forecast-2016-2024.

"Biotin Deficiency - an Overview | ScienceDirect Topics." 2019. Accessed November 27. https://www.sciencedirect.com/topics/biochemistry-genetics-and-molecular-biology/biotin-deficiency.

Boccaletti, Valeria, E. Zendri, G. Giordano, L. Gnetti, and G. De Panfilis. 2007. "Familial Uncombable Hair Syndrome: Ultrastructural Hair Study and Response to Biotin." *Pediatric Dermatology* 24 (3). doi:10.1111/j.1525-1470.2007.00385.x.

Boyle, Robert John, Fiona J. Bath-Hextall, Jo Leonardi-Bee, Dedee F. Murrell, and Mimi L. K. Tang. 2008. "Probiotics for Treating Eczema." *Cochrane Database of Systematic Reviews*. John Wiley and Sons Ltd. doi:10.1002/14651858.CD006135.pub2.

Bronsnick, Tara, Era Caterina Murzaku, and Babar K Rao. 2014. "Diet in Dermatology: Part I. Atopic Dermatitis, Acne, and Nonmelanoma Skin Cancer." *Journal of the American Academy of Dermatology* 71 (6): 1039.e1-1039.e12. doi:10.1016/j.jaad.2014.06.015.

Brower, V. 1998. "Nutraceuticals: Poised for a Healthy Slice of the Healthcare Market?" *Nature Biotechnology*. doi:10.1038/nbt0898-728.

Chen, Andrew C., Andrew J. Martin, Bonita Choy, Pablo Fernández-Peñas, Robyn A. Dalziell, Catriona A. McKenzie, Richard A. Scolyer, et al. 2015. "A Phase 3 Randomized Trial of Nicotinamide for Skin-Cancer Chemoprevention." *New England Journal of Medicine* 373 (17): 1618–26. doi:10.1056/NEJMoa1506197.

Choi, Franchesca D., Calvin T. Sung, Margit L.W. Juhasz, and Natasha Atanaskova Mesinkovsk. 2019. "Oral Collagen Supplementation: A

Systematic Review of Dermatological Applications." *Journal of Drugs in Dermatology : JDD* 18 (1). NLM (Medline): 9–16.

Clark, Larry C., Gerald F. Combs, Bruce W. Turnbull, Elizabeth H. Slate, Dan K. Chalker, James Chow, Loretta S. Davis, et al. 1996. "Effects of Selenium Supplementation for Cancer Prevention in Patients with Carcinoma of the Skin: A Randomized Controlled Trial." *Journal of the American Medical Association* 276 (24): 1957–63. doi:10.1001/jama.276.24.1957.

Dhaliwal, Simran, Mimi Nguyen, Alexandra R. Vaughn, Manisha Notay, Cindy J. Chambers, and Raja K. Sivamani. 2019. "Effects of Zinc Supplementation on Inflammatory Skin Diseases: A Systematic Review of the Clinical Evidence." *American Journal of Clinical Dermatology*, November. doi:10.1007/s40257-019-00484-0.

Dietary Reference Intakes for Energy, Carbohydrate, Fiber, Fat, Fatty Acids, Cholesterol, Protein, and Amino Acids (Macronutrients). 2005. *Dietary Reference Intakes for Energy, Carbohydrate, Fiber, Fat, Fatty Acids, Cholesterol, Protein, and Amino Acids (Macronutrients).* National Academies Press. doi:10.17226/10490.

Driscoll, Marcia S., Eun Kyung M. Kwon, Hadas Skupsky, Soon You Kwon, and Jane M. Grant-Kels. 2010. "Nutrition and the Deleterious Side Effects of Nutritional Supplements." *Clinics in Dermatology* 28 (4). Elsevier Inc.: 371–79. doi:10.1016/j.clindermatol.2010.03.023.

Evans, Julie A., and Elizabeth J. Johnson. 2010. "The Role of Phytonutrients in Skin Health." *Nutrients*. MDPI AG. doi:10.3390/nu2080903.

"Fatal Gastrointestinal Mucormycosis in an Infant Following Use of Contaminated ABC Dophilus Powder from Solgar Inc. | Fungal Diseases | CDC." 2019. Accessed November 16. https://www.cdc.gov/fungal/outbreaks/rhizopus-investigation.html.

Fenner, Justine, and Nanette B. Silverberg. 2019. "Oral Supplements in Atopic Dermatitis." *Clinics in Dermatology* 36 (5): 653–58. Accessed October 30. doi:10.1016/j.clindermatol.2018.05.010.

Foster, Rachel H., Gil Hardy, and Raid G. Alany. 2010. "Borage Oil in the Treatment of Atopic Dermatitis." *Nutrition*. doi:10.1016/j.nut.2009.10.014.

Fu, Lisa Wenyang, and Ronald Vender. 2011. "Systemic Role for Vitamin D in the Treatment of Psoriasis and Metabolic Syndrome." *Dermatology Research and Practice*. Hindawi Publishing Corporation. doi:10.1155/2011/276079.

Galli, E., L. Rocchi, R. Carello, P. G. Giampietro, P. Panei, and P. Meglio. 2015. "Serum Vitamin D Levels and Vitamin D Supplementation Do Not Correlate with the Severity of Chronic Eczema in Children." *European Annals of Allergy and Clinical Immunology* 47 (2): 41–47. http://www.ncbi.nlm.nih.gov/pubmed/25781193.

Gloster, Hugh, and Chesahna Kindred. 2005. "Habit-Tic–like and Median Nail–like Dystrophies Treated with Multivitamins." *Journal of the American Academy of Dermatology* 53 (3): 542–43. doi:10.1016/j.jaad.2005.03.045.

Gollnick, Harald Prof. Dr., W. Hopfenmüller, Christian Hemmes, S. C. Chun, Carrie Schmid, K. Sundermeier, and Hans Konrad Biesalski. 1996. *Systemic Beta Carotene plus Topical UV-Sunscreen Are an Optimal Protection against Harmful Effects of Natural UV-Sunlight : Results of the Berlin-Eilath Study.*

Gray, N. A., A. Dhana, D. J. Stein, and N. P. Khumalo. 2019. "Zinc and Atopic Dermatitis: A Systematic Review and Meta-Analysis." *Journal of the European Academy of Dermatology and Venereology*. Blackwell Publishing Ltd. doi:10.1111/jdv.15524.

Grether-Beck, S., A. Marini, T. Jaenicke, W. Stahl, and J. Krutmann. 2017. "Molecular Evidence That Oral Supplementation with Lycopene or Lutein Protects Human Skin against Ultraviolet Radiation: Results from a Double-Blinded, Placebo-Controlled, Crossover Study." *British Journal of Dermatology* 176 (5). Blackwell Publishing Ltd: 1231–40. doi:10.1111/bjd.15080.

Gueniche, Audrey, Jalil Benyacoub, David Philippe, Philippe Bastien, Nicole Kusy, Lionel Breton, Stephanie Blum, and Isabelle Castiel-Higounenc. 2010. "Lactobacillus Paracasei CNCM I-2116 (ST11) Inhibits Substance P-Induced Skin Inflammation and Accelerates Skin Barrier Function Recovery in Vitro." *European Journal of Dermatology* 20 (6): 731–37. doi:10.1684/ejd.2010.1108.

Hacini-Rachinel, Feriel, Hanane Gheit, Jean Benoit Le Luduec, Fariel Dif, Stéphane Nancey, and Dominique Kaiserlian. 2009. "Oral Probiotic Control Skin Inflammation by Acting on Both Effector and Regulatory T Cells." *PLoS ONE* 4 (3). Public Library of Science. doi:10.1371/journal.pone.0004903.

Heimbeck, I., M. Wjst, and C. J. Apfelbacher. 2013. "Low Vitamin D Serum Level Is Inversely Associated with Eczema in Children and Adolescents in Germany." *Allergy: European Journal of Allergy and Clinical Immunology* 68 (7): 906–10. doi:10.1111/all.12167.

Holick, Michael F. 2017. "The Vitamin D Deficiency Pandemic: Approaches for Diagnosis, Treatment and Prevention." *Reviews in Endocrine and Metabolic Disorders*. Springer New York LLC. doi:10.1007/s11154-017-9424-1.

Iikura, Y., Y. Odajima, T. Nagakura, K. Iinuma, K. Hayakawa, and J. Oizumi. 1988. "Oral Biotin Treatment Is Effective for Atopic Dermatitis in Children with Low Biotinidase Activity." *Acta Paediatrica Scandinavica* 77 (5): 762–63. doi:10.1111/j.1651-2227.1988.tb10748.x.

Institute of Medicine. 1998. *"Dietary Reference Intakes for Thiamin, Riboflavin, Niacin, Vitamin B6, Folate, Vitamin B12, Pantothenic Acid, Biotin, and Choline* Status: Final Book ISBN: 978-0-309-06554-2 Downloads: 6,138 Dietary Reference Intakes for Thiamin, Riboflavin, Niacin, Vitami," 58–86. doi:10.17226/6015.

Isolauri, E., Y. Sütas, P. Kankaanpää, H. Arvilommi, and S. Salminen. 2001. "Probiotics: Effects on Immunity." In *American Journal of Clinical Nutrition*. Vol. 73. doi:10.1093/ajcn/73.2.444s.

Jobeili, Lara, Patricia Rousselle, David Béal, Eric Blouin, Anne Marie Roussel, Odile Damour, and Walid Rachidi. 2017. "Selenium Preserves Keratinocyte Stemness and Delays Senescence by Maintaining Epidermal Adhesion." *Aging* 9 (11). Impact Journals LLC: 2302–15. doi:10.18632/aging.101322.

Juturu, Vijaya, James P. Bowman, and Jayant Deshpande. 2016. "Overall Skin Tone and Skin-Lightening-Improving Effects with Oral Supplementation of Lutein and Zeaxanthin Isomers: A Double-Blind, Placebo-controlled Clinical Trial." *Clinical, Cosmetic and*

Investigational Dermatology 9 (October). Dove Medical Press Ltd.: 325–32. doi:10.2147/CCID.S115519.

Kalra, Ekta K. 2003. "Nutraceutical - Definition and Introduction." *AAPS Journal* 5 (3). doi:10.1208/ps050325.

Kang, Bong Seon, Jae Gu Seo, Gwa Su Lee, Jung Hwa Kim, Sei Yeon Kim, Ye Won Han, Hoon Kang, et al. 2009. "Antimicrobial Activity of Enterocins from Enterococcus Faecalis SL-5 against Propionibacterium Acnes, the Causative Agent in Acne Vulgaris, and Its Therapeutic Effect." *Journal of Microbiology* 47 (1): 101–9. doi:10.1007/s12275-008-0179-y.

Katta, Rajani, and Sophia Huang. 2019. "Skin, Hair and Nail Supplements: An Evidence-Based Approach." *Skin Therapy Letter* 24 (5): 7–13. http://www.ncbi.nlm.nih.gov/pubmed/31584785.

Kil, Min Seong, Chul Woo Kim, and Sang Seok Kim. 2013. "Analysis of Serum Zinc and Copper Concentrations in Hair Loss." *Annals of Dermatology* 25 (4): 405–9. doi:10.5021/ad.2013.25.4.405.

Klein, Eric A., Ian M. Thompson, Catherine M. Tangen, John J. Crowley, Scott Lucia, Phyllis J. Goodman, Lori M. Minasian, et al. 2011. "Vitamin E and the Risk of Prostate Cancer: The Selenium and Vitamin E Cancer Prevention Trial (SELECT)." *JAMA - Journal of the American Medical Association* 306 (14): 1549–56. doi:10.1001/jama. 2011.1437.

Koike, Haruki, Mie Takahashi, Ken Ohyama, Rina Hashimoto, Yuichi Kawagashira, Masahiro Iijima, Masahisa Katsuno, Hiroshi Doi, Fumiaki Tanaka, and Gen Sobue. 2015. "Clinicopathologic Features of Folate-Deficiency Neuropathy." *Neurology* 84 (10). Lippincott Williams and Wilkins: 1026–33. doi:10.1212/WNL.0000000000 001343.

Koshy, John C., Safa E. Sharabi, Evan M. Feldman, Larry H. Hollier, James R. Patrinely, and Charles N S Soparkar. 2012. "Effect of Dietary Zinc and Phytase Supplementation on Botulinum Toxin Treatments." *Journal of Drugs in Dermatology* 11 (4): 507–12.

Kwon, Ho Keun, Choong Gu Lee, Jae Seon So, Chang Suk Chae, Ji Sun Hwang, Anupama Sahoo, Jong Hee Nam, Joon Haeng Rhee, Ki Chul

Hwang, and Sin Hyeog Im. 2010. "Generation of Regulatory Dendritic Cells and CD4+Foxp3 + T Cells by Probiotics Administration Suppresses Immune Disorders." *Proceedings of the National Academy of Sciences of the United States of America* 107 (5): 2159–64. doi:10.1073/pnas.0904055107.

Lipner, Shari R. 2018. "Rethinking Biotin Therapy for Hair, Nail, and Skin Disorders." *Journal of the American Academy of Dermatology* 78 (6): 1236–38. doi:10.1016/j.jaad.2018.02.018.

Manson, Joann E., Glenville Jones, Christopher S. Kovacs, A. Catharine Ross, Richard L. Gallo, J. Christopher Gallagher, Susan T. Mayne, et al. 2011. "The 2011 Dietary Reference Intakes for Calcium and Vitamin D: What Dietetics Practitioners Need to Know**This Article Is a Summary of the Institute of Medicine Report Entitled Dietary Reference Intakes for Calcium and Vitamin D (Available at Http://Www.Io." *Journal of the American Dietetic Association* 111 (4): 524–27. doi:10.1016/j.jada.2011.01.004.

Marosz, Agnieszka, and Dariusz Chlubek. 2014. "The Risk of Abuse of Vitamin Supplements." *Annales Academiae Medicae Stetinensis*.

Matsumoto, M., A. Aranami, A. Ishige, K. Watanabe, and Y. Benno. 2007. "LKM512 Yogurt Consumption Improves the Intestinal Environment and Induces the T-Helper Type 1 Cytokine in Adult Patients with Intractable Atopic Dermatitis." *Clinical and Experimental Allergy* 37 (3): 358–70. doi:10.1111/j.1365-2222.2007.02642.x.

McCusker, Meagen M., and Jane M. Grant-Kels. 2010. "Healing Fats of the Skin: The Structural and Immunologic Roles of the Ω-6 and Ω-3 Fatty Acids." *Clinics in Dermatology* 28 (4). Elsevier Inc.: 440–51. doi:10.1016/j.clindermatol.2010.03.020.

Möhrenschlager, Matthias, T. Schmidt, J. Ring, and D. Abeck. 2000. "Recalcitrant Trachyonychia of Childhood - Response to Daily Oral Biotin Supplementation: Report of Two Cases." *Journal of Dermatological Treatment* 11 (2). Martin Dunitz Ltd: 113–15. doi:10.1080/09546630050517522.

Morris, John Steven, and Stacy B. Crane. 2013. "Selenium Toxicity from a Misformulated Dietary Supplement, Adverse Health Effects, and the

Temporal Response in the Nail Biologic Monitor." *Nutrients* 5 (4). MDPI AG: 1024–57. doi:10.3390/nu5041024.

Muizzuddin, Neelam, Wanda Maher, Michael Sullivan, Steven Schnittger, and Thomas Mammone. 2012. "Physiological Effect of a Probiotic on Skin." *Journal of Cosmetic Science* 63 (6). Society of Cosmetic Chemists: 385–95.

Murzaku, Era Caterina, Tara Bronsnick, and Babar K. Rao. 2014. "Diet in Dermatology: Part II. Melanoma, Chronic Urticaria, and Psoriasis." *Journal of the American Academy of Dermatology*. Mosby Inc. doi:10.1016/j.jaad.2014.06.016.

Ogawa, Youichi, Tatsuyoshi Kawamura, and Shinji Shimada. 2016. "Zinc and Skin Biology." *Archives of Biochemistry and Biophysics* 611 (December). Academic Press Inc.: 113–19. doi:10.1016/j.abb.2016.06.003.

Passeron, T., J. P. Lacour, E. Fontas, and J. P. Ortonne. 2006. "Prebiotics and Synbiotics: Two Promising Approaches for the Treatment of Atopic Dermatitis in Children above 2 Years." *Allergy: European Journal of Allergy and Clinical Immunology* 61 (4): 431–37. doi:10.1111/j.1398-9995.2005.00956.x.

Peral, M. C., M. M. Rachid, N. M. Gobbato, M. A. Huaman Martinez, and J. C. Valdez. 2010. "Interleukin-8 Production by Polymorphonuclear Leukocytes from Patients with Chronic Infected Leg Ulcers Treated with Lactobacillus Plantarum." *Clinical Microbiology and Infection* 16 (3). Blackwell Publishing Ltd: 281–86. doi:10.1111/j.1469-0691.2009.02793.x.

Peral, Maria C., Miguel A. Huaman Martinez, and Juan C. Valdez. 2009. "Bacteriotherapy with Lactobacillus Plantarum in Burns." *International Wound Journal* 6 (1): 73–81. doi:10.1111/j.1742-481X.2008.00577.x.

Peroni, D. G., G. L. Piacentini, E. Cametti, I. Chinellato, and A. L. Boner. 2011. "Correlation between Serum 25-Hydroxyvitamin D Levels and Severity of Atopic Dermatitis in Children." *British Journal of Dermatology* 164 (5): 1078–82. doi:10.1111/j.1365-2133.2010.10147.x.

Philippe, David, Stephanie Blum, and Jalil Benyacoub. 2011. "Oral Lactobacillus Paracasei Improves Skin Barrier Function Recovery and Reduces Local Skin Inflammation." *European Journal of Dermatology*. doi:10.1684/ejd.2010.1242.

Pinnell, Sheldon R. 2003. "Cutaneous Photodamage, Oxidative Stress, and Topical Antioxidant Protection." *Journal of the American Academy of Dermatology* 48 (1). Mosby Inc.: 1–22. doi:10.1067/mjd.2003.16.

Preter, V. De, K. Geboes, K. Verbrugghe, L. De Vuyst, T. Vanhoutte, G. Huys, J. Swings, B. Pot, and K. Verbeke. 2004. "The in Vivo Use of the Stable Isotope-Labelled Biomarkers Lactose-[15 N]Ureide and [2 H 4] Tyrosine to Assess the Effects of pro- and Prebiotics on the Intestinal Flora of Healthy Human Volunteers." *British Journal of Nutrition* 92 (3). Cambridge University Press (CUP): 439–46. doi:10.1079/bjn20041228.

Prince, Tessa, Andrew J. McBain, and Catherine A. O'Neill. 2012. "Lactobacillus Reuteri Protects Epidermal Keratinocytes from Staphylococcus Aureus-Induced Cell Death by Competitive Exclusion." *Applied and Environmental Microbiology* 78 (15): 5119–26. doi:10.1128/AEM.00595-12.

Proksch, E., D. Segger, J. Degwert, M. Schunck, V. Zague, and S. Oesser. 2013. "Oral Supplementation of Specific Collagen Peptides Has Beneficial Effects on Human Skin Physiology: A Double-Blind, Placebo-Controlled Study." *Skin Pharmacology and Physiology* 27 (1): 47–55. doi:10.1159/000351376.

Rijkers, Ger T., Willem M. De Vos, Robert Jan Brummer, Lorenzo Morelli, Gerard Corthier, and Philippe Marteau. 2011. "Health Benefits and Health Claims of Probiotics: Bridging Science and Marketing." *British Journal of Nutrition* 106 (9): 1291–96. doi:10.1017/S000711451100287X.

Robl, Renata, Marjorie Uber, Kerstin Taniguchi Abagge, Monica Nunes Lima, and Vânia Oliveira Carvalho. 2016. "Serum Vitamin D Levels Not Associated with Atopic Dermatitis Severity." *Pediatric Dermatology* 33 (3). Blackwell Publishing Inc.: 283–88. doi:10.1111/pde.12795.

Samochocki, Zbigniew, Jaroslaw Bogaczewicz, Renata Jeziorkowska, Anna Sysa-Jędrzejowska, Olga Glińska, Elizabeth Karczmarewicz, Daniel P. McCauliffe, and Anna Woźniacka. 2013. "Vitamin D Effects in Atopic Dermatitis." *Journal of the American Academy of Dermatology* 69 (2): 238–44. doi:10.1016/j.jaad.2013.03.014.

Saper, Robert B., Stefanos N. Kales, Janet Paquin, Michael J. Burns, David M. Eisenberg, Roger B. Davis, and Russell S. Phillips. 2004. "Heavy Metal Content of Ayurvedic Herbal Medicine Products." *Journal of the American Medical Association* 292 (23): 2868–73. doi:10.1001/jama.292.23.2868.

Saper, Robert B., Russell S. Phillips, Anusha Sehgal, Nadia Khouri, Roger B. Davis, Janet Paquin, Venkatesh Thuppil, and Stefanos N. Kales. 2008. "Lead, Mercury, and Arsenic in US- and Indian-Manufactured Ayurvedic Medicines Sold via the Internet." *JAMA - Journal of the American Medical Association* 300 (8): 915–23. doi:10.1001/jama.300.8.915.

Schley, P. D., and C. J. Field. 2002. "The Immune-Enhancing Effects of Dietary Fibres and Prebiotics." *British Journal of Nutrition* 87 (S2). Cambridge University Press (CUP): S221–30. doi:10.1079/bjn/2002541.

Shelley, W. B., and E. D. Shelley. 1985. "Uncombable Hair Syndrome: Observations on Response to Biotin and Occurrence in Siblings with Ectodermal Dysplasia." *Journal of the American Academy of Dermatology* 13 (1): 97–102. doi:10.1016/s0190-9622(85)70150-8.

Shrivastava, Shyam Behari. 2009. "Diffuse Hair Loss in an Adult Female: Approach to Diagnosis and Management." *Indian Journal of Dermatology, Venereology and Leprology* 75 (1): 20. doi:10.4103/0378-6323.45215.

Smith, A. David. 2007. "Folic Acid Fortification: The Good, the Bad, and the Puzzle of Vitamin B-12." *The American Journal of Clinical Nutrition* 85 (1): 3–5. doi:10.1093/ajcn/85.1.3.

Souyoul, Skylar A., Katharine P. Saussy, and Mary P. Lupo. 2018. "Nutraceuticals: A Review." *Dermatology and Therapy* 8 (1): 5–16. doi:10.1007/s13555-018-0221-x.

Tachjian, Ara, Viqar Maria, and Arshad Jahangir. 2010. "Use of Herbal Products and Potential Interactions in Patients With Cardiovascular Diseases." *Journal of the American College of Cardiology.* doi:10.1016/j.jacc.2009.07.074.

The FDA Warns That Biotin May Interfere with Lab Tests: FDA Safety Communication | FDA. 2019. Accessed November 16. https://www.fda.gov/medical-devices/safety-communications/fda-warns-biotin-may-interfere-lab-tests-fda-safety-communication.

Wang, Susan Shuxin, Kam Lun Hon, Alice Pik shan Kong, Henry Nga hin Pong, Gary Wing kin Wong, and Ting Fan Leung. 2014. "Vitamin D Deficiency Is Associated with Diagnosis and Severity of Childhood Atopic Dermatitis." *Pediatric Allergy and Immunology* 25 (1): 30–35. doi:10.1111/pai.12167.

Winkler, Petra, Darab Ghadimi, Jürgen Schrezenmeir, and Jean-Pierre Kraehenbuhl. 2007. "Molecular and Cellular Basis of Microflora-Host Interactions." *The Journal of Nutrition* 137 (3 Suppl 2): 756S-72S. doi:10.1093/jn/137.3.756S.

Wintergerst, Eva S., Silvia Maggini, and Dietrich H. Hornig. 2007. "Contribution of Selected Vitamins and Trace Elements to Immune Function." *Annals of Nutrition and Metabolism.* doi:10.1159/000107673.

Wu, Xiayu, Jiaoni Cheng, and Xu Wang. 2017. "Dietary Antioxidants: Potential Anticancer Agents." *Nutrition and Cancer.* Routledge. doi:10.1080/01635581.2017.1299872.

Yeowell, Heather N., Melanie K. Marshall, Linda C. Walker, Van Ha, and Sheldon R. Pinnell. 1994. "Regulation of Lysyl Oxidase MRNA in Dermal Fibroblasts from Normal Donors and Patients with Inherited Connective Tissue Disorders." *Archives of Biochemistry and Biophysics* 308 (1): 299–305. doi:10.1006/abbi.1994.1042.

Yoshida, Shinya, Koji Yasutomo, and Toshiyuki Watanabe. 2016. "Treatment with DHA/EPA Ameliorates Atopic Dermatitis-like Skin Disease by Blocking LTB4 Production." *Journal of Medical Investigation* 63 (3–4). University of Tokushima: 187–91. doi:10.2152/jmi.63.187.

In: Nutraceuticals: Evidence Based Use
Editor: Raluca Maria Pop
ISBN: 978-1-53618-052-7

Chapter 10

THE USE OF NUTRACEUTICALS IN NEUROLOGICAL DISEASES

Ştefan Cristian Vesa[1,*], Sabina Istratoaie[1], Silvina Iluţ[2], Ioana-Cristina Bârsan[2], George Săraci[3] and Sonia Irina Vlaicu[4]

[1]Department of Pharmacology, Toxicology and Clinical Pharmacology, "Iuliu Haţieganu" University of Medicine and Pharmacy, Cluj-Napoca, România

[2]Department of Neurology and Pediatric Neurology, Toxicology and Clinical Pharmacology, "Iuliu Haţieganu" University of Medicine and Pharmacy, Cluj-Napoca, România

[3]"Iuliu Haţieganu" University of Medicine and Pharmacy, Cluj-Napoca, România

[4]Department of Internal Medicine, "Iuliu Haţieganu" University of Medicine and Pharmacy of Cluj-Napoca, Romania

[*] Corresponding Author's E-mail: stefanvesa@gmail.com.

ABSTRACT

Neurological diseases, Parkinson's disease, Alzheimer's dementia or multiple sclerosis or cerebral atherosclerosis can benefit from nutraceutical use. As the common pathways involved in the neurodegenerative diseases are oxidative stress, inflammation, and mitotoxicity, some nutraceuticals can help inhibit those processes. Some of them like, coenzyme Q10 or resveratrol, can be used in several neurodegenerative diseases. However, their use must be closely monitored, as they can have drug-to-drug interaction, especially with cardiovascular or neuropsychiatric medication.

Keywords: Parkinson's disease, Alzheimer's dementia, multiple sclerosis, nutraceuticals, anti-inflammatory, antioxidative

INTRODUCTION

Neurological diseases represent a public health challenge especially in low income and developing countries. Many neurological diseases and conditions left untreated or even with treatment will affect an individual's quality of life. During neurological conditions, there is a great probability of the disabilities that will reduce the patient's activities, and, sometimes, will increase the chance of dying.

For several common diseases, like Parkinson's disease, Alzheimer's dementia or multiple sclerosis, the standard treatment will not influence the course of the disease. That is why alternative treatment with specific nutraceuticals has an important role by decreasing the inflammation and by their antioxidative effects.

MECHANISM OF ACTION

The common pathways involved in the neurodegenerative diseases are thought to be oxidative stress, inflammation, mitotoxicity and intracellular Ca^{2+} overload (Glass et al. 2010; Uttara et al. 2009; Federico et al. 2012; Nguyen et al. 2013).

Generally speaking, nutraceuticals might provide neuroprotection through the following mechanisms: reactive oxygen species (ROS) scavenging, antiinflammation and antiapoptosis, transition metals chelation, modulation of cell signaling pathways or regulating mitochondrial homeostasis (Hardy, Hardy, and Ball 2003; Brower 1998; Das et al. 2012).

Alzheimer's Dementia

A few nutraceuticals have been proven efficient against the progression of Alzheimer's dementia (AD). They could help in preventing the accumulation of Aβ peptides by regulating Amyloid Precursor Protein (APP) or by modulation of several signaling pathways (Monroy, Lithgow, and Alavez 2013; Ghosh, Banerjee, and Sil 2015; Ullah et al. 2017).

Curcumin attenuates Aβ peptides aggregation by inhibiting acetylcholinesterase (AChE) activity, improving cognition and behavioral impairment in experimental animals (Kotani et al. 2017). Other studies have reported a potential of curcumin to bind with metal ions such as Zn (II) and Cu (II), another mechanism studied to be neuroprotective (Ghalebani et al. 2012). Also treated mouse primary cortical neurons with curcumin of different concentrations (1–20 μM) showed that 10 μM curcumin can stabilize an unfolded form of APP, which leads to a level of Aβ peptides significantly decreased compared to the control cells (Zhang et al. 2010).

At the same time, *in vivo* studies showed that curcumin can inhibit Tau phosphorylation and attenuate inflammation by downregulating the inflammatory stress molecules such as TNF-α, IL-6, ERK ½ (Xiao et al. 2014). Moreover, curcumin can reduce genomic instability events and reduce oxidative stress (Thomas et al. 2009).

Another studied nutraceutical with anti-neurodegenerative properties is resveratrol. This compound has proven significantly antioxidant and anti-inflammatory effects, but recent evidence highlights its potential role in AD, by promoting the breakdown of APP and facilitating removal of Aβ peptides (Ahmed et al. 2017). The resveratrol role in neuroprotection is also reflected by the activation of sirtuins, which are NAD (+)-dependent histone deacetylases enzymes (Pocernich et al. 2011).

Experimental studies on mice model of AF showed that resveratrol might additionally improve the memory deficit and cognitive impairment (Braidy et al. 2016). Also, its beneficial effect on oxidative stress and inflammation is based on the downregulation of the TNF-α and increased infiltration of macrophage in the adipose tissue (Jeon et al. 2012).

Parkinson's Disease

It is known that Mucuna plant, known as dopa bean, is a leguminous plant growing in tropical areas that contain L dopa and it is preferred over other chemical means of synthesizing L-dopa since it is a natural source (Surwase et al. 2012; Manyam, Dhanasekaran, and Hare 2004). Cilia et al., (2017) studied whether *Mucuna pruriens* (MP), a species of Mucuna plant, may be used as a treatment alternative to levodopa. They demonstrated that the intake of both low-dose and high-dose of MP is superior to placebo and non-inferior to levodopa+dopa-decarboxylase inhibitor (LD+DDCI), in terms of efficacy and safety outcomes. Moreover, while low-dose MP lower dose showed a similar motor response when compared to LD+DDCI, high-dose MP induced a better motor response than LD+DDCI, fewer dyskinesias and longer on duration (Cilia et al. 2017).

Another nutraceutical that may confer neuroprotection in Parkinson's disease is ginsenoside Rb1, the major active pharmacological component of ginseng (Chen et al. 2005). Ginseng is the root of *Panax ginseng* C.A. Meyer (Araliaceae), a herbal medicine widely appreciated and used in the Far East. It has antioxidative properties because it acts by maintain glutathione levels. Its anti-inflammatory effect is reflected by modulation of different inflammatory pathways such as inhibition of TNF-alpha in lipopolysaccharide-stimulated macrophages or suppression of NF-kB activation (Chen et al. 2005; Xu, Chen, and Wong 2009). Zhang et al., (2018) were the first to reveal that treatment with ginsenoside Rb1 improves motor deficit and prevents dopaminergic neurons death in a mouse model of

Parkinson's disease. These effects were explained by the nutraceuticals property to suppress glutamate excitotoxicity in the nigrostriatal and corticonigral systems, by upregulating of glutamate transporter GLT-1 (Zhang et al. 2018).

Epigallocatechin-3-gallate (EGCG), the most studied green tea derived catechin, is a nutraceutical that can cross the blood-brain barrier and could protect dopaminergic neurons loss in Parkinson's disease, due to its numerous neuroprotective mechanisms (Pan, Jankovic, and Le 2003). Studies have shown that neurodegeneration in Parkinson's disease is accelerated by a peripheral immune system activation. Consequently, in peripheral blood of Parkinson's disease mouse model, CD4+ T cells are decreased, while CD8+ T cells are increased, suggesting altered T cell function (Amor et al. 2010; Bas et al. 2001). Another study showed that treatment with EGCG in Parkinson's disease mouse model reversed this dysfunction, exerting the immunoprotective effect of EGCG (Zhou, Zhu, and Liang 2018). Other EGCG properties include free radicals scavenging, iron chelation and antiapoptotic properties. Also, EGCG can increase cytosolic Ca^{2+} levels, by activation of AMPK, which has been demonstrated to be neuroprotective (Spasić, Callaerts, and Norga 2009).

Multiple Sclerosis

The EGCG, the curcumin, and the mustard oil glycosides have been proven efficient in inhibiting the pro-inflammatory signaling through NF-kB or toll-like receptors, thus being associated with reduced demyelination (Schmitz et al. 2015). Another nutraceutical with anti-inflammatory properties is the retinoic acid that acts by regulating T-lymphocyte populations in peripheral blood (Fragoso, Stoney, and McCaffery 2014). These effects might contribute to improved regeneration, cognition and plasticity in Multiple sclerosis (MS) patients.

The isothiocyanate sulforaphane broccoli has both anti-inflammatory and antioxidative properties. Studies on experimental autoimmune encephalomyelitis (EAE) models demonstrated that sulphorane reduced the demyelination and the immune cell infiltration, leading to symptoms less severe (Li et al. 2013; Geisel et al. 2014). It can act through various mechanisms such as inhibition of Th17 response, activation of Nrf2 (Li et al. 2013) or suppression of IL23 and IL12 (Geisel et al. 2014).

CLINICAL USES

Parkinson's Disease

Parkinson's disease is a progressive multifocal neurodegenerative disorder (synucleinopathy), in which several levels are affected, namely: synapses (dopa and non-dopa), the central, peripheral and vegetative nervous systems and extracerebral structures (the gastrointestinal system, heart, skin, adrenal glands, blood cells, etc.). In Parkinson's disease, the loss and degeneration of dopaminergic neurons predominantly from the ventrolateral portion of the black substance (pars compacta), noradrenergic neurons (from locus coeruleus), serotonergic neurons (from the nucleus of the raphe), cholinergic (from the Meynert basal nuclei) of the vagus nucleus and from sympathetic peripheral ganglia.

The nutraceuticals used in Parkinson's disease (Hang, Basil, and Lim 2016) can be grouped into six classes based on their neuroprotective properties:

1) iron chelators (creatine, epigalocatechin-3-galate (EGCG) (Zhou, Zhu, and Liang 2018));
2) modulators of cell signaling (fish oil (Acar et al. 2019);
3) anti-inflammatories (curcuminoids from curry (Wu et al. 2018), baicalin (Halida et al. 2019), oxiresveratrol, Genistein, *Ocimum sanctum* (Siddique et al. 2014), quercetin (Tamtaji et al. 2019), ginkobiloba extract (Kuang et al. 2018));
4) antioxidants (vitamins C and E, coenzyme Q10, *Mucuna pruriens* (Johnson et al. 2018), *Mucuna sanjappae*, curcuminoids from curry, baicalin, resveratrol (Huang et al. 2019);

5) antiapoptotics (vitamin D, ginsenoside, genistein (Siddique et al. 2019) and
6) attenuated mitochondrial dysfunction (vitamin B, coenzyme Q10, *Mucuna pruriens*, *Mucuna sanjappae* (Chao et al. 2012).

Some examples of nutraceuticals and their clinical implications are given below (Nasri et al. 2014):

- Vitamin D promotes the growth of dopaminergic axons, increases glutathione levels, interferes with calcium homeostasis, has an anti-apoptotic, immunomodulatory effect and reduces nitric oxide synthase.
- Vitamins B regulates the level of homocysteine.
- Vitamin C is involved in free radical removal in the cytosol.
- Vitamin E prevents lipid peroxidation in membranes.
- Coenzyme Q10 (CoQ10) maintains the proper electron transfer in the electron transport chain of mitochondria and ATP production; it also has an antioxidant effect by reducing alpha-tocopherol to prevent lipid peroxidation.
- Creatine kinase phosphorylate is used to form phosphocreatine, which maintains ATP levels.
- Fish oil ensures the modulation of dopaminergic neurons in the basal ganglia and has antidepressant effects by increasing serotonergic neurotransmission.
- Mucuna pruriens and Mucuna sanja ppae have significant amounts of CoQ10 and NADH that increase dopamine levels by regulating TH, counteract inhibition of mitochondrial complex activity and protect neurons against DNA damage.
- Epigallocatechin-3-gallate (EGCG), a component of the green tea, activates AMP kinase.

Multiple Sclerosis

Multiple sclerosis is a chronic CNS disorder characterized by episodes of inflammation and focal demyelination with multiple locations, disseminated over time (based on autoimmune mechanisms against myelin protein, mediated by T lymphocytes and triggered by incompletely elucidated exogenous factors, to which a process of degeneration and an abnormal oligodendrocyte function associate) and through a process of axonal degeneration with progressive evolution in a person with genetic susceptibility. The cellular changes in MS consist of an increase in the secretion of prostaglandins by macrophages, of the number of activated T cells and of an *in vitro* secretion of immunoglobulins under T cell lymphocyte dependence. Also a decrease in the suppressive cell function, in the secretion of interferon and in the NK cell function is characteristic.

The use of nutraceuticals results in an imaging improvement (change in the brain volume assessed by MRI) (Marx et al. 2020) and in the clinical manifestations (fatigue, depression, sphincter disorders, and EDSS). The nutraceuticals (Bagur et al. 2017) used in MS, based on their properties, are antioxidants (alpha-lipoic acid, vitamin A, biotin, carnitine, green tea, coenzyme Q10), anti-inflammatory compounds (curcumin, Vitamin D), ginkgobiloba, probiotics, prebiotics (Riccio and Rossano 2015), *Andrographis paniculata*, ginseng, and *lemon verbena*.

Alzheimer's Dementia

Cognitive dysfunction and Alzheimer's dementia is a class of neurodegenerative disorders characterized by persistent and progressive alteration of cognitive functions, with evolution towards disability and premature death.

Nutraceuticals used in dementia (Howes et al. 2019) can be grouped into eight classes based on their properties:

1) neuroprotection (coffee, green tea, cocoa, isoflavones);
2) antioxidants properties (coffee, green tea, cocoa, anthocyanins/ anthocyanidins (Pacheco et al. 2018));
3) lipid peroxidation enhancement/ improvement (epigallocatechin-3-gallate (EGCG));
4) Aβ plaque deposits suppression (caffeoylquinic acids, resveratrol);
5) induced neurogenesis associated with improved levels of hippocampal BDNF mRNA (resveratrol (Shen et al. 2018));
6) mitochondrial dysfunction attenuation and reduced TNF-α levels (citrus flavanones, methoxyflavones and coumarins (Shabani and Mirshekar 2018));
7) anti-inflammatory properties (carotenoids, isoflavones);
8) ACh via nicotinic receptors inhibition (monoterpenes, rosmarinic acid).

Neurodegenerative Disorders

Amyotrophic lateral sclerosis (ALS) and Huntington's disease (HD) are major neurodegenerative diseases. ALS is caused by the loss and degeneration of the peripheral motoneurons (spinal and inferior cranial nerves) with axonal swelling (spheroidal changes) and hyaline inclusion corpuscles (Hirano, Bunina) as well as central motoneurons (pyramidal Betz cells) from the motor cortex. In some cases, the spinal ganglia and the Clarke spine may be involved. HD is an autosomal dominant disorder with mental, cognitive and hyperkinetic changes. The motor disorders that occur are caused by progressive, selective neuronal loss, predominantly in striated bodies, with astrocytic gliosis. This loss is of particular interest to GABA-ergic neurons, as well as to cholinergic interneurons.

Generally, the nutraceuticals used in Neurodegenerative disorders can be grouped into five classes based on their properties, acting on

1) mitochondrial dysfunction (curcumin, α-lipoic acid (Guo et al. 2014), astaxanthin, coenzyme Q10, apigenin),

2) Ca^{2+} overload-mediated neuronal damage (blueberry (Fuentealba et al. 2011)),
3) oxidative stress (resveratrol, L-sulforaphane, curcumin, aged garlic extract), neuroprotective (carnosic acid, rosmarinic acid, resveratrol (Ogle, Speisman, and Ormerod 2013)),
4) lipid peroxidation (eugenol) and
5) inflammation (curcumin, epigallocatechin-3-gallate, mustard oil glycoside, retinoic acid, vitamin D (Gianforcaro and Hamadeh 2014), vitamin E, omega-3 polyunsaturated fatty acid).

Atherosclerosis

Atherosclerosis is incriminated in the case of cerebral ischemic pathology, which is the most important cause of morbidity and long-term disability in Europe. Demographic changes have led to an increase in both incidence and prevalence.

The nutraceuticals used in atherosclerosis can be grouped into four classes based on their properties, they act on:

1) reduction in homocysteine levels (folic acid (Wang et al. 2019));
2) antioxidants (omega-3 LC-PUFA (Khan et al. 2019), anthocyanins (Manolescu et al. 2019), curcumin (Li et al. 2019)), *Lagenaria siceraria*, *Murraya koenigii*, *Trigonella foenum-graecum*),
3) they attenuate the mitochondrial dysfunction (coenzyme Q10, L-carnitine (Endo et al. 2018));
4) they have an anti-inflammatory effect (curcumin, Vitamin E (Cheng et al. 2018), *Allium cepa*, *Allium sativum*, *Beta vulgaris* (Jain et al. 2018)).

DRUG-DRUG INTERACTIONS

The consumption of nutraceutical products for medicinal purposes is common among persons with neurological diseases and it is increasing worldwide (Risvoll et al. 2017). As a consequence, it and can be the cause of numerous drug-nutraceutical product interactions which may lead to unexpected clinical manifestations. These interactions occur more often than thought (Mouly et al. 2017).

An interaction can be defined as a wanted or unwanted chemical, pharmacokinetic, or pharmacodynamic reaction when two or more chemical substances come together in the body (Chan et al. 2016). In other words, it can be defined by any food, herbal medicines, or dietary supplements induced changes in oral bioavailability leading to changes in concentrations of co-administered drugs (Boullata and Hudson 2012). These interactions can have an antagonistic or additive effect and may lead to loss of therapeutic efficacy or increased potential of adverse reactions, even toxic effects of co-administered drugs (Boullata and Hudson 2012; Amadi and Mgbahurike 2017).

It has been shown that the interactions influence the therapeutic outcome by alteration in the bioavailability of co-administered drugs, but food/herb-drug interactions are not as well documented as drug-drug interactions (Amadi and Mgbahurike 2017). Despite the large number of possible interactions between supplements and medication, only a small number of combinations have been examined or reported. Knowledge of the specific effects of nutraceuticals on drug absorption, metabolism, and effect is still incomplete (Boullata 2005). Clinical studies, including the clinical phases of new drug candidate development, robust methodological guidelines for such clinical pharmacology studies are lacking to make definitive clinical and regulatory recommendations (Mouly et al. 2017).

Mechanisms of Interaction

Essentially two types of interactions can be ascribed: pharmacodynamic and pharmacokinetic (Gurley, Rock, and States 2012). Clinically important interactions between an herbal supplement and a drug typically manifest as pharmacokinetic interactions. This is affecting a drug's concentration in the blood and its pharmacologic action. The risk of a pharmacokinetic interaction occurs when the nutraceutical shares the same mechanism of absorption, distribution, metabolism, or excretion (ADME) as the co-administered drug (Asher et al. 2017). Less commonly, interactions may manifest as pharmacodynamic interactions, which involve the clinical effect of a drug or the physiologic effect of a nutrient. These interactions involve transporters and enzymes that are important for drug absorption, distribution, or elimination (Boullata and Hudson 2012).

Knowledge of the mechanisms underlying ADME processes for a drug and an herbal supplement is needed to recognize pharmaco-kinetic herb-drug interactions. Most ADME mechanisms fall under four large gene super-families comprising more than 1,000 proteins: the cytochrome P450 (CYP) drug metabolism enzymes; the uridine diphosphate-glucuronosyltransferase (UGT) conjugating enzymes; the adenosine triphosphate–binding cassette (ABC) drug uptake/efflux transporters; and the organic anion-transporting polypeptide (OATP) drug transporters (Asher et al. 2017).

Factors that may enhance the type and intensity of drug-nutraceutical interactions include drug-nutrient associated factors such as route of administration, nutrient status, and pharmacological/toxicological profile of the drug and patient-related variables (Mouly et al. 2017). Patients at risk for interactions are those with chronic disease, who use multiple medications, particularly medication with a narrow therapeutic index, have genetic variants in drug metabolism, impaired organ function, and are at either end of the age spectrum (Boullata 2005). Age, malnutrition, malabsorption, chronic liver disease, kidney failure, polymedication, long-term drug dosing, and pharmacogenetics are well-known risk factors for drug-drug and drug-food interactions (Mouly et al. 2017).

Even though nutraceuticals are frequently prescribed, knowledge of drug-nutraceutical interactions is still incomplete. It is necessary to mention that there is documented and theoretical evidence in the literature regarding these interactions. Thus, the following interactions of nutraceuticals commonly used in neurological diseases have been described:

Ginkgobiloba (contains flavonoids - quercetin) decreases the bioavailability of amitriptyline, clopidogrel, diazepam, alprazolam, phenytoin, phenobarbital, nifedipine, nicardipine, propranolol, and warfarin by inducing the CYP2C19 gene. In particular, warfarin and phenytoin are of more concern due to their low therapeutic index (Chan, Li, and Perez 2016). It has antiplatelet activity so it increases the risk of bleeding especially in combination with warfarin, aspirin or other anticoagulant/antiplatelet drugs (Gurley et al. 2012; Asher et al. 2017; Abebe 2002). Interacts with acetaminophen. Acetaminophen has an inhibitory effect on thromboxane synthesis potentiating the antiaggregant effect of ginkgo which may induce bleeding (Abebe 2002).

Ginseng. The American ginseng decreases the bioavailability of warfarin having an effect of mild induction of CYP2C9 (Shao and Jia, 2013). The Asian ginseng decreases the bioavailability and the effect of calcium channel blockers, certain antihypertensive, statins, and some antidepressants by inducing CYP3A4 (Asher et al. 2017). Also, the Asian ginseng has antiaggregant activity and increases the risk of bleeding in combination with aspirin and other NSAIDs (Mouly et al. 2017).

Resveratrol increases the bioavailability and the risk of toxicity of 3-hydroxy-3-methylglutaryl-coenzyme A reductase inhibitors (atorvastatin, lovastatin, simvastatin), calcium-channel antagonists (felodipine, nifedipine, nicardipine, nitrendipine, nimodipine, verapamil), antiarrhythmic agents (amiodarone), benzodiazepines (midazolam, triazolam) by inhibiting the activity of CYP3A4 (Detampel et al. 2012).

St. John's worth reduces the bioavailability of midazolam, alprazolam, amitriptyline, phenytoin, carbamazepine, digoxin, talinolol, nifedipine, verapamil, simvastatin, atorvastatin, and warfarin. It is a potent inducer of CYP3A4 and P-glycoprotein. Also, it can induce other cytochrome P450 enzymes depending on the dose, route, and duration of administration (Boullata 2005; Amadi and Mgbahurike 2017; Gurley et al. 2012).

Feverfew because of its constituent parthenolide, has an additive antiplatelet effect in combination with NSAIDs, by blocking the synthesis and/or release of various inflammatory mediators from platelets and polymorphonuclear leucocytes (Abebe 2002).

Curcumin reduces the concentration of many antidepressants and antipsychotic drugs by inducing CYP1A2 metabolization (Asher et al. 2017).

Kava kava increases the bioavailability of acetaminophen by inhibiting CYP2E1 and increases the bioavailability of many NSAIDs, angiotensin receptor blockers, valproic acid, phenytoin, warfarin and clopidogrel due to CYP2C9 inhibition (Asher et al. 2017).

Garlic extract reduces the concentration levels of drugs transported by P-gp: digoxin, rosuvastatin, and verapamil. It has antiplatelet activity, increasing the risk of bleeding in association with oral anticoagulants and antiplatelet drugs (Asher et al. 2017; Mouly et al. 2017).

Green tea extract (contains polyphenols - catechins) decreases the effect of atorvastatin and beta-blockers by inhibiting the uptake transporters OAT1B1 and OAT1B3. Increases simvastatin concentrations, due to P-gp inhibition (Asher et al. 2017).

Grapefruit juice increases the bioavailability of felodipine and nifedipine by inhibiting the cytochrome P450 enzyme system (intestinal CYP3A4) (Amadi and Mgbahurike 2017; Asher et al. 2017).

Melatonin reduces the effect of nifedipine and warfarin (Boullata 2005).

Coenzyme Q reduces the effect of warfarin (Boullata 2005).

Fish oil increases the effect of warfarin and fluoxetine, paroxetine and sertraline (Boullata 2005).

Folic acid reduces the effect of phenytoin (Boullata 2005).

Vitamin K rich foods antagonize the antivitamin K anticoagulants, affecting the anticoagulation stability (Amadi and Mgbahurike 2017).

Vitamin B6, in high doses, reduces the effect of levodopa (Boullata 2005).

Vitamin B9, associated with phenytoin can induce seizures, so the combination should be avoided. The mechanism is still unclear (Mouly et al. 2017).

Vitamin E interacts with oral anticoagulants such as warfarin and aspirin increasing the bleeding risk so the combination should be avoided. The mechanism is unclear (Mouly et al. 2017).

Iron, Zinc, Calcium (contained in enteral nutrition) reduce levodopa and levodopa/carbidopa (Boullata 2005) intestinal absorption. The mechanism is unclear (Mouly et al. 2017).

Zinc alters the effect of immunomodulatory drugs (Boullata 2005).

Conclusion

Nutraceutical uses in neurological diseases are based on their neuroprotective properties (modulation of cell signaling, anti-inflammatory, antioxidative or antiapoptotic effects). Even though nutraceuticals are frequently prescribed, knowledge of their full effectiveness or drug-nutraceutical interactions is still incomplete. Large studies are necessary to elucidate completely their usefulness, but the present evidences show good potential.

References

Abebe, W. 2002. *Herbal Medication: Potential for Adverse Interactions with Analgesic Drugs*, 391–401.

Acar, Nuray, Hande Parlak, Ayse Ozkan, Hakan Soylu, Sema Avcı, Ismail Ustunel, V. Nimet Izgut-Uysal, and Aysel Agar. 2019. "The Effect of Docosahexaenoic Acid on Apelin Distribution of Nervous System in the Experimental Mouse Model of Parkinson's Disease." *Tissue and Cell* 56 (February): 41–51. https://doi.org/10.1016/j.tice.2018.12.002.

Ahmed, Touqeer, Sehrish Javed, Sana Javed, Ameema Tariq, Dunja Šamec, Silvia Tejada, Seyed Fazel Nabavi, Nady Braidy, and Seyed Mohammad Nabavi. 2017. "Resveratrol and Alzheimer's Disease: Mechanistic Insights." *Molecular Neurobiology*. Humana Press Inc. https://doi.org/10.1007/s12035-016-9839-9.

Amadi, Cecilia N, and Amaka A Mgbahurike. 2017. "Selected Food/Herb – Drug Interactions: Mechanisms and Clinical Relevance" 11: 1–11.

Amor, Sandra, Fabiola Puentes, David Baker, and Paul Van Der Valk. 2010. "Inflammation in Neurodegenerative Diseases." *Immunology*. https://doi.org/10.1111/j.1365-2567.2009.03225.x.

Asher, Gary N, Amanda H Corbett, ROYL Hawke, North Carolina, Chapel Hill, and North Carolina. 2017. *Common Herbal Dietary Supplement–Drug Interactions.*

Pocernich, CB, MLB Lange, R Sultana, and DA Butterfield. 2011. "Nutritional Approaches to Modulate Oxidative Stress in Alzheimers Disease." *Current Alzheimer Research* 8 (5): 452–69. https://doi.org/10.2174/156720511796391908.

Bagur, M, José, M Antonia Murcia, Antonia M Jiménez-Monreal, Josep A Tur, M Mar Bibiloni, Gonzalo L Alonso, and Magdalena Martínez-Tomé. 2017. "Influence of Diet in Multiple Sclerosis: A Systematic Review." *Advances in Nutrition: An International Review Journal* 8 (3): 463–72. https://doi.org/10.3945/an.116.014191.

Bas, Jordi, Màtil Calopa, Mariona Mestre, David G. Molleví, Blanca Cutillas, Santiago Ambrosio, and Enric Buendia. 2001. "Lymphocyte Populations in Parkinson's Disease and in Rat Models of Parkinsonism." *Journal of Neuroimmunology* 113 (1): 146–52. https://doi.org/10.1016/S0165-5728(00)00422-7.

Boullata, JI, and LM Hudson. 2012. "Drug–Nutrient Interactions: A Broad View with Implications for Practice." *JAND* 112 (4): 506–17. https://doi.org/10.1016/j.jada.2011.09.002.

Boullata, Joseph. 2005. "Nutrition in Clinical Practice Natural Health Product Interactions with Medication." *Nutr Clin Pract* 20 (1), 33-51. https://doi.org/10.1177/011542650502000133.

Braidy, Nady, Bat-Erdene Jugder, Anne Poljak, Tharusha Jayasena, Hussein Mansour, Seyed Mohammad Nabavi, Perminder Sachdev, and Ross Grant. 2016. "Resveratrol as a Potential Therapeutic Candidate for the Treatment and Management of Alzheimer's Disease." *Current Topics in Medicinal Chemistry* 16 (17): 1951–60. https://doi.org/10.2174/1568026616666160204121431.

Brower, V. 1998. "Nutraceuticals: Poised for a Healthy Slice of the Healthcare Market?" *Nature Biotechnology*. https://doi.org/10.1038/nbt0898-728.

Chan, Noel, Sandy Li, and Evette Perez. 2016. Interactions between Chinese Nutraceuticals and Western Medicines. *Nutraceuticals*. Elsevier Inc. https://doi.org/10.1016/B978-0-12-802147-7.00061-9.

Chen, Xiao Chun, Yi Can Zhou, Ying Chen, Yuan Gui Zhu, Fang Fang, and Li Min Chen. 2005. "Ginsenoside Rg1 Reduces MPTP-Induced Substantia Nigra Neuron Loss by Suppressing Oxidative Stress." *Acta Pharmacologica Sinica* 26 (1): 56–62. https://doi.org/10.1111/j.1745-7254.2005.00019.x.

Cheng, Pengfei, Lihua Wang, Shujun Ning, Zichun Liu, Hao Lin, Shuli Chen, and Jiaying Zhu. 2018. "Vitamin E Intake and Risk of Stroke: A Meta-Analysis." *British Journal of Nutrition* 120 (10): 1181–88. https://doi.org/10.1017/S0007114518002647.

Cilia, Roberto, Janeth Laguna, Erica Cassani, Emanuele Cereda, Nicolò G. Pozzi, Ioannis U. Isaias, Manuela Contin, Michela Barichella, and Gianni Pezzoli. 2017. "Mucuna Pruriens in Parkinson Disease: A Double-Blind, Randomized, Controlled, Crossover Study." *Neurology* 89 (5): 432–38. https://doi.org/10.1212/WNL.0000000000004175.

Das, Lipi, Eshani Bhaumik, Utpal Raychaudhuri, and Runu Chakraborty. 2012. "Role of Nutraceuticals in Human Health." *Journal of Food Science and Technology* 49 (2): 173–83. https://doi.org/10.1007/s13197-011-0269-4.

Detampel, Pascal, Mareike Beck, Stephan Krähenbühl, and Jörg Huwyler. 2012. "Drug Interaction Potential of Resveratrol" *Drug Metab Rev* 44 (May): 253–65. https://doi.org/10.3109/03602532.2012.700715.

Endo, Satoshi, Tadashi Takahashi, Masaya Sato, Yoshinori Noya, and Masayoshi Obana. 2018. "Effects of L-Carnitine Supplementation, Botulinum Neurotoxin Injection, and Rehabilitation for a Chronic Stroke Patient." *Journal of Stroke and Cerebrovascular Diseases* 27 (11): 3342–44. https://doi.org/10.1016/j.jstrokecerebrovasdis.2018.07.033.

Federico, Antonio, Elena Cardaioli, Paola Da Pozzo, Patrizia Formichi, Gian Nicola Gallus, and Elena Radi. 2012. "Mitochondria, Oxidative Stress and Neurodegeneration." *Journal of the Neurological Sciences* 322 (1–2): 254–62. https://doi.org/10.1016/j.jns.2012.05.030.

Fragoso, Yara Dadalti, Patrick N. Stoney, and Peter J. McCaffery. 2014. "The Evidence for a Beneficial Role of Vitamin A in Multiple Sclerosis." *CNS Drugs* 28 (4): 291–99. https://doi.org/10.1007/s40263-014-0148-4.

Fuentealba, Jorge, Andrea J. Dibarrart, Maria Cecilia Fuentes-Fuentes, Francisco Saez-Orellana, Karin Quiñones, Leonardo Guzmán, Claudia Perez, Jose Becerra, and Luis G. Aguayo. 2011. "Synaptic Failure and Adenosine Triphosphate Imbalance Induced by Amyloid-β Aggregates Are Prevented by Blueberry-Enriched Polyphenols Extract." *Journal of Neuroscience Research* 89 (9): 1499–1508. https://doi.org/10.1002/jnr.22679.

Geisel, Julia, Jürgen Brück, Ivana Glocova, Katja Dengler, Tobias Sinnberg, Oliver Rothfuss, Michael Walter, Klaus Schulze-Osthoff, Martin Röcken, and Kamran Ghoreschi. 2014. "Sulforaphane Protects from T Cell–Mediated Autoimmune Disease by Inhibition of IL-23 and IL-12 in Dendritic Cells." *The Journal of Immunology* 192 (8): 3530–39. https://doi.org/10.4049/jimmunol.1300556.

Ghalebani, Leila, Anna Wahlström, Jens Danielsson, Sebastian K.T.S. Wärmländer, and Astrid Gräslund. 2012. "PH-Dependence of the Specific Binding of Cu(II) and Zn(II) Ions to the Amyloid-β Peptide." *Biochemical and Biophysical Research Communications* 421 (3): 554–60. https://doi.org/10.1016/j.bbrc.2012.04.043.

Ghosh, Shatadal, Sharmistha Banerjee, and Parames C. Sil. 2015. "The Beneficial Role of Curcumin on Inflammation, Diabetes and

Neurodegenerative Disease: A Recent Update." *Food and Chemical Toxicology*. Elsevier Ltd. https://doi.org/10.1016/j.fct.2015.05.022.

Gianforcaro, Alexandro, and Mazen J. Hamadeh. 2014. "Vitamin D as a Potential Therapy in Amyotrophic Lateral Sclerosis." *CNS Neuroscience & Therapeutics* 20 (2): 101–11. https://doi.org/10.1111/cns.12204.

Glass, Christopher K., Kaoru Saijo, Beate Winner, Maria Carolina Marchetto, and Fred H. Gage. 2010. "Mechanisms Underlying Inflammation in Neurodegeneration." *Cell*. https://doi.org/10.1016/j.cell.2010.02.016.

Guo, Ying, Desiree Jones, J. Lynn Palmer, Arthur Forman, Shaker R. Dakhil, Maria R. Velasco, Matthias Weiss, et al. 2014. "Oral Alpha-Lipoic Acid to Prevent Chemotherapy-Induced Peripheral Neuropathy: A Randomized, Double-Blind, Placebo-Controlled Trial." *Supportive Care in Cancer: Official Journal of the Multinational Association of Supportive Care in Cancer* 22 (5): 1223–31. https://doi.org/10.1007/s00520-013-2075-1.

Gurley, Bill J, Espero Kim Fifer, Little Rock, and Insect Sciences. 2012. *Pharmacokinetic Herb-Drug Interactions (Part 2): Drug Interactions Involving Popular Botanical Dietary Supplements and Their Clinical Relevance*, no. Part 2: 1490–1514.

Gurley, Bill J, Little Rock, and United States. 2012. "Pharmacokinetic Herb-Drug Interactions (Part 1): Origins, Mechanisms, and the Impact of Botanical Dietary Supplements," *Planta Med* 1: 1478–89.

Halida, Bahetihan, YN Yao, XJ Han, and XL Yang. 2019. "[Effect of Baicalin on the expression of connexin 36 in the cerebral cortex and striatum area of Parkinson's disease model rat]." *Zhonghua Yi Xue Za Zhi* 99 (3): 218–22. https://doi.org/10.3760/cma.j.issn.0376-2491.2019.03.013.

Hang, Liting, Adeline Henry Basil, and Kah-Leong Lim. 2016. "Nutraceuticals in Parkinson's Disease." *NeuroMolecular Medicine* 18 (3): 306–21. https://doi.org/10.1007/s12017-016-8398-6.

Hardy, Gil, Ines Hardy, and Patrick A. Ball. 2003. "Nutraceuticals - A Pharmaceutical Viewpoint: Part II." *Current Opinion in Clinical*

Nutrition and Metabolic Care. https://doi.org/10.1097/00075197-200311000-00010.

Howes, Melanie-Jayne R, Nicolette SL Perry, Carlos Vásquez-Londoño, and Elaine K Perry. 2019. "Role of Phytochemicals as Nutraceuticals for Cognitive Functions Affected in Ageing." *British Journal of Pharmacology*, October, bph.14898. https://doi.org/10.1111/bph.14898.

Huang, Nanqu, Ying Zhang, Mingji Chen, Hai Jin, Jing Nie, Yong Luo, Shaoyu Zhou, Jingshan Shi, and Feng Jin. 2019. "Resveratrol Delays 6-Hydroxydopamine-Induced Apoptosis by Activating the PI3K/Akt Signaling Pathway." *Experimental Gerontology* 124 (September): 110653. https://doi.org/10.1016/j.exger.2019.110653.

Jain, Sajal, Harpal S. Buttar, Meena Chintameneni, and Ginpreet Kaur. 2018. "Prevention of Cardiovascular Diseases with Anti-Inflammatory and Anti- Oxidant Nutraceuticals and Herbal Products: An Overview of Pre-Clinical and Clinical Studies." *Recent Patents on Inflammation & Allergy Drug Discovery* 12 (2): 145–57. https://doi.org/10.2174/1872213X12666180815144803.

Jeon, Byeong Tak, Eun Ae Jeong, Hyun Joo Shin, Younghyurk Lee, Dong Hoon Lee, Hyun Joon Kim, Sang Soo Kang, Gyeong Jae Cho, Wan Sung Choi, and Gu Seob Roh. 2012. "Resveratrol Attenuates Obesity-Associated Peripheral and Central Inflammation and Improves Memory Deficit in Mice Fed a High-Fat Diet." *Diabetes* 61 (6): 1444–54. https://doi.org/10.2337/db11-1498.

Johnson, Shelby, Hyun Park, Nicholas DaSilva, Dhiraj Vattem, Hang Ma, and Navindra Seeram. 2018. "Levodopa-Reduced Mucuna Pruriens Seed Extract Shows Neuroprotective Effects against Parkinson's Disease in Murine Microglia and Human Neuroblastoma Cells, Caenorhabditis Elegans, and Drosophila Melanogaster." *Nutrients* 10 (9): 1139. https://doi.org/10.3390/nu10091139.

Khan, Safi U, Muhammad U Khan, Haris Riaz, Shahul Valavoor, Di Zhao, Lauren Vaughan, Victor Okunrintemi, et al. 2019. "Effects of Nutritional Supplements and Dietary Interventions on Cardiovascular

Outcomes: An Umbrella Review and Evidence Map." *Annals of Internal Medicine* 171 (3): 190. https://doi.org/10.7326/M19-0341.

Kotani, Rina, Yasuomi Urano, Hachiro Sugimoto, and Noriko Noguchi. 2017. "Decrease of Amyloid-β Levels by Curcumin Derivative via Modulation of Amyloid-β Protein Precursor Trafficking." *Journal of Alzheimer's Disease* 56 (2): 529–42. https://doi.org/10.3233/JAD-160794.

Kuang, Shaosong, Lin Yang, Ziliang Rao, Zhiyong Zhong, Jinfeng Li, Haichao Zhong, Lulu Dai, and Xiaojiang Tang. 2018. "Effects of Ginkgo Biloba Extract on A53T α-Synuclein Transgenic Mouse Models of Parkinson's Disease." *Canadian Journal of Neurological Sciences / Journal Canadien Des Sciences* Neurologiques 45 (2): 182–87. https://doi.org/10.1017/cjn.2017.268.

Li, Bin, Wei Cui, Jia Liu, Ru Li, Qian Liu, Xiao Hua Xie, Xiao Li Ge, et al. 2013. "Sulforaphane Ameliorates the Development of Experimental Autoimmune Encephalomyelitis by Antagonizing Oxidative Stress and Th17-Related Inflammation in Mice." *Experimental Neurology* 250 (December): 239–49. https://doi.org/10.1016/j.expneurol.2013.10.002.

Li, Hong, Antoni Sureda, Hari Prasad Devkota, Valeria Pittalà, Davide Barreca, Ana Sanches Silva, Devesh Tewari, Suowen Xu, and Seyed Mohammad Nabavi. 2019. "Curcumin, the Golden Spice in Treating Cardiovascular Diseases." *Biotechnology Advances*, February, S0734975019300102. https://doi.org/10.1016/j.biotechadv.2019.01.010.

Manolescu, Oprea, Mititelu, Ruta, and Farcasanu. 2019. "Dietary Anthocyanins and Stroke: A Review of Pharmacokinetic and Pharmacodynamic Studies." *Nutrients* 11 (7): 1479. https://doi.org/10.3390/nu11071479.

Manyam, Bala V., Muralikrishnan Dhanasekaran, and Theodore A. Hare. 2004. "Neuroprotective Effects of the Antiparkinson DrugMucuna Pruriens." *Phytotherapy Research* 18 (9): 706–12. https://doi.org/10.1002/ptr.1514.

Marx, Wolfgang, Meghan Hockey, Amelia J. McGuinness, Melissa Lane, John Christodoulou, Ingrid van der Mei, Michael Berk, et al. 2020. "The

Effect of Emerging Nutraceutical Interventions for Clinical and Biological Outcomes in Multiple Sclerosis: A Systematic Review." *Multiple Sclerosis and Related Disorders* 37 (January): 101486. https://doi.org/10.1016/j.msard.2019.101486.

Monroy, Adriana, Gordon J Lithgow, and Silvestre Alavez. 2013. "Curcumin and Neurodegenerative Diseases." *BioFactors*. https://doi.org/10.1002/biof.1063.

Mouly, Stéphane, Célia Lloret-linares, Pierre-olivier Sellier, Damien Sene, and J Bergmann. 2017. "Is the Clinical Relevance of Drug-Food and Drug-Herb Interactions Limited to Grapefruit Juice and Saint-John's Wort?" *Pharmacological Research* 118: 82–92. https://doi.org/10.1016/j.phrs.2016.09.038.

Nguyen, Huey T, Darrell R Sawmiller, Olga Markov, and Ming Chen. 2013. "Elevated [Ca2+]i Levels Occur with Decreased Calpain Activity in Aged Fibroblasts and Their Reversal by Energy-Rich Compounds: New Paradigm for Alzheimer's Disease Prevention." *Journal of Alzheimer's Disease* 37 (4): 835–48. https://doi.org/10.3233/JAD-131001.

Ogle, William O, Rachel B Speisman, and Brandi K Ormerod. 2013. "Potential of Treating Age-Related Depression and Cognitive Decline with Nutraceutical Approaches: A Mini-Review." *Gerontology* 59 (1): 23–31. https://doi.org/10.1159/000342208.

Pacheco, Simone Muniz, Mayara Sandrielly Pereira Soares, Jessié Martins Gutierres, Mariana Freire Barbieri Gerzson, Fabiano Barbosa Carvalho, Juliana Hofstatter Azambuja, Maria Rosa Chitolina Schetinger, Francieli Moro Stefanello, and Roselia Maria Spanevello. 2018. "Anthocyanins as a Potential Pharmacological Agent to Manage Memory Deficit, Oxidative Stress and Alterations in Ion Pump Activity Induced by Experimental Sporadic Dementia of Alzheimer's Type." *The Journal of Nutritional Biochemistry* 56: 193–204. https://doi.org/10.1016/j.jnutbio.2018.02.014.

Pan, Tianhong, Joseph Jankovic, and Weidong Le. 2003. "Potential Therapeutic Properties of Green Tea Polyphenols in Parkinson's Disease." *Drugs and Aging*. https://doi.org/10.2165/00002512-200320100-00001.

Riccio, Paolo, and Rocco Rossano. 2015. "Nutrition Facts in Multiple Sclerosis." *ASN Neuro* 7 (1): 175909141456818. https://doi.org/10.1177/1759091414568185.

Risvoll, Hilde, Trude Giverhaug, Kjell H Halvorsen, Marit Waaseth, and Frauke Musial. 2017. "Direct and Indirect Risk Associated with the Use of Dietary Supplements among Persons with Dementia in a Norwegian Memory Clinic," *BMC Complement Altern Med.* 1–12. https://doi.org/10.1186/s12906-017-1765-5.

Schmitz, Katja, Julia Barthelmes, Leonie Stolz, Susanne Beyer, Olaf Diehl, and Irmgard Tegeder. 2015. "'Disease Modifying Nutricals' for Multiple Sclerosis." *Pharmacology and Therapeutics*. Elsevier Inc. https://doi.org/10.1016/j.pharmthera.2014.11.015.

Shabani, Sahreh, and Mohammad Ali Mirshekar. 2018. "Diosmin Is Neuroprotective in a Rat Model of Scopolamine-Induced Cognitive Impairment." *Biomedicine & Pharmacotherapy* 108 (December): 1376–83. https://doi.org/10.1016/j.biopha.2018.09.127.

Shao, Jingwei, and Lee Jia. 2013. "Potential Serious Interactions between Nutraceutical Ginseng and Warfarin in Patients with Ischemic Stroke." *Trends in Pharmacological Sciences* 34 (2): 85–86. https://doi.org/10.1016/j.tips.2012.11.008.

Shen, Jun, Linling Xu, Chujie Qu, Huimin Sun, and Junjian Zhang. 2018. "Resveratrol Prevents Cognitive Deficits Induced by Chronic Unpredictable Mild Stress: Sirt1/MiR-134 Signalling Pathway Regulates CREB/BDNF Expression in Hippocampus in Vivo and in Vitro." *Behavioural Brain Research* 349 (September): 1–7. https://doi.org/10.1016/j.bbr.2018.04.050.

Siddique, Yasir Hasan, Falaq Naz, Smita Jyoti, Fahad Ali, and Rahul. 2019. "Effect of Genistein on the Transgenic Drosophila Model of Parkinson's Disease." *Journal of Dietary Supplements* 16 (5): 550–63. https://doi.org/10.1080/19390211.2018.1472706.

Siddique, Yasir Hasan, Mohammad Faisal, Falaq Naz, Smita Jyoti, and Rahul. 2014. "Role of Ocimum Sanctum Leaf Extract on Dietary Supplementation in the Transgenic Drosophila Model of Parkinson's

Disease." *Chinese Journal of Natural Medicines* 12 (10): 777–81. https://doi.org/10.1016/S1875-5364(14)60118-7.

Spasić, Miloš R, Patrick Callaerts, and Koenraad K Norga. 2009. "AMP-Activated Protein Kinase (AMPK) Molecular Crossroad for Metabolic Control and Survival of Neurons." *The Neuroscientist* 15 (4): 309–16. https://doi.org/10.1177/1073858408327805.

Surwase, Shripad N, Sushama A Patil, Shekhar B Jadhav, and Jyoti P Jadhav. 2012. "Optimization of <scp>l</Scp> -DOPA Production by Brevundimonas Sp. SGJ Using Response Surface Methodology." *Microbial Biotechnology* 5 (6): 731–37. https://doi.org/10.1111/j.1751-7915.2012.00363.x.

Tamtaji, Omid Reza, Tooba Hadinezhad, Maryam Fallah, Arash Rezaei Shahmirzadi, Mohsen Taghizadeh, Mohammad Behnam, and Zatollah Asemi. 2019. "The Therapeutic Potential of Quercetin in Parkinson's Disease: Insights into Its Molecular and Cellular Regulation." *Current Drug Targets* 20 (November). https://doi.org/10.2174/1389450120666191112155654.

Thomas, Philip, Yan Jiang Wang, Jin Hua Zhong, Shantha Kosaraju, Nathan J O'Callaghan, Xin Fu Zhou, and Michael Fenech. 2009. "Grape Seed Polyphenols and Curcumin Reduce Genomic Instability Events in a Transgenic Mouse Model for Alzheimer's Disease." *Mutation Research - Fundamental and Molecular Mechanisms of Mutagenesis* 661 (1–2): 25–34. https://doi.org/10.1016/j.mrfmmm.2008.10.016.

Ullah, Faheem, Andy Liang, Alejandra Rangel, Erika Gyengesi, Garry Niedermayer, and Gerald Münch. 2017. "High Bioavailability Curcumin: An Anti-Inflammatory and Neurosupportive Bioactive Nutrient for Neurodegenerative Diseases Characterized by Chronic Neuroinflammation." *Archives of Toxicology*. Springer Verlag. https://doi.org/10.1007/s00204-017-1939-4.

Uttara, Bayani, Ajay Singh, Paolo Zamboni, and R Mahajan. 2009. "Oxidative Stress and Neurodegenerative Diseases: A Review of Upstream and Downstream Antioxidant Therapeutic Options." *Current Neuropharmacology* 7 (1): 65–74. https://doi.org/10.2174/157015909787602823.

Wang, Yuan, Yang Jin, Yao Wang, Li Li, Yanhong Liao, Yun Zhang, and Dan Yu. 2019. "The Effect of Folic Acid in Patients with Cardiovascular Disease: A Systematic Review and Meta-Analysis." *Medicine* 98 (37): e17095. https://doi.org/10.1097/MD.0000000000017095.

Wu, You, Shunli Liang, Bin Xu, Rongbo Zhang, and Linsheng Xu. 2018. "[Protective effect of curcumin on dopamine neurons in Parkinson's disease and its mechanism]." Zhejiang Da Xue Xue Bao. Yi Xue Ban = Journal of Zhejiang University. *Medical Sciences* 47 (5): 480–86.

Xiao, Zijian, Aiwu Zhang, Jianwen Lin, Zhenyang Zheng, Xiaolei Shi, Wei Di, Weiwei Qi, Yingting Zhu, Guijuan Zhou, and Yannan Fang. 2014. "Telomerase: A Target for Therapeutic Effects of Curcumin and a Curcumin Derivative in Aβ1-42 Insult in Vitro." *PLoS ONE* 9 (7). https://doi.org/10.1371/journal.pone.0101251.

Xu, Li, Wen-Fang Chen, and Man-Sau Wong. 2009. "Ginsenoside Rg1 Protects Dopaminergic Neurons in a Rat Model of Parkinson's Disease through the IGF-I Receptor Signalling Pathway." *British Journal of Pharmacology* 158 (3): 738–48. https://doi.org/10.1111/j.1476-5381.2009.00361.x.

Zhang, Can, Andrew Browne, Daniel Child, and Rudolph E. Tanzi. 2010. "Curcumin Decreases Amyloid-β Peptide Levels by Attenuating the Maturation of Amyloid-β Precursor Protein." *Journal of Biological Chemistry* 285 (37): 28472–80. https://doi.org/10.1074/jbc.M110.133520.

Zhang, Yun Long, Yan Liu, Xin Pan Kang, Chun Yan Dou, Ren Gong Zhuo, Shu Qiong Huang, Li Peng, and Lei Wen. 2018. "Ginsenoside Rb1 Confers Neuroprotection via Promotion of Glutamate Transporters in a Mouse Model of Parkinson's Disease." *Neuropharmacology* 131 (March): 223–37. https://doi.org/10.1016/j.neuropharm.2017.12.012.

Zhou, Tingting, Mengru Zhu, and Zhanhua Liang. 2018. "(-)-Epigallocatechin-3-Gallate Modulates Peripheral Immunity in the MPTP-Induced Mouse Model of Parkinson's Disease." *Molecular Medicine Reports* 17 (4): 4883–88. https://doi.org/10.3892/mmr.2018.8470.

Zhou, Tingting, Mengru Zhu, and Zhanhua Liang. 2018. "(-)-Epigallocatechin-3-Gallate Modulates Peripheral Immunity in the MPTP-Induced Mouse Model of Parkinson's Disease." *Molecular Medicine Reports* January. https://doi.org/10.3892/mmr.2018.8470.

In: Nutraceuticals: Evidence Based Use ISBN: 978-1-53618-052-7
Editor: Raluca Maria Pop

Chapter 11

NUTRACEUTICALS IN PSYCHIATRY

Sebastian-Mihai Armean**, Krisztina-Agota Armean and Anca-Dana Buzoianu
Department of Pharmacology, Toxicology and Clinical Pharmacology,
University of Medicine and Pharmacy "Iuliu Haţieganu,"
Cluj-Napoca, Romania

ABSTRACT

Nutrition and dietary components can influence brain functioning. Modern life comes with stress and altered sleep. Illness may arise from this. Nutraceuticals appear as a feasible way to balance this effect or to prevent it. We shall summarize the most important and up-to-date findings regarding our present-day knowledge in the matter, present the safety concerns and clinical uses in psychiatry alongside their evidence, and future perspectives. Nutraceuticals may represent a valuable asset in psychiatry, but future research is required to establish their full therapeutic potential and to translate preclinical findings into clinical practice.

* Corresponding Author's E-mail: sebastian.armean@umfcluj.ro.

Keywords: nutraceuticals, psychiatry, PUFA, psychobiotic, psychiatric disorders

INTRODUCTION

It has been postulated that nutrition and dietary components can influence brain functioning (Meeusen and Decroix 2018). Research supports the existence of interactions between the cognitive-part and the metabolic-part of the brain (Berthoud 2007).

We may have the impression that our present-day life is subjected to speed, information and, unfortunately, stress. Having that much on our to-do-list, it seems that our 24/7 may just not be enough. To be as effective as possible, to achieve as many goals as possible, to live our lives to the fullest, we may tend to neglect our sleep – for sleeping is not productive.

However, sleeping is a physiological process (Ji, Grandner, and Liu 2017), a universal need. As surprisingly as it may be, we spend about one-third of our lifetime sleeping. Noorwali et al. (2019) provided evidence that the recommended amount of sleeping (~8h daily) is associated with higher intakes of fruits and vegetables in a sub-group of adults in the United Kingdom compared to people who sleep shorter or longer amounts of time, who consume fewer fruits and vegetables. The authors suggest that fruits and vegetable consumption may play a part in sleep regulation via the gut-brain axis, because of their content in polyphenol. Another explanation would be that they augment the mitochondrial function and energetic metabolism (Noorwali, Hardie, and Cade 2019). Would it be possible to use *polyphenol-rich dietary supplements* (Kumar and Khanum 2012) to *enhance* our sleep, by maximizing the beneficial effects and by reducing the duration?

We shall begin our journey on the fascinating path of novel targets with therapeutic potential with the statement of Bickford et al. (2017) who indicate that microglia plays an important role when it comes to adapting to the microenvironment because microglia assesses it continually and by doing this, it can respond to numerous stimuli. This response is possible through a rapid movement between the activation states (Bickford, Flowers, and Grimmig 2017).

From Theory to Practice: What Do We Know So Far?

As we grow old, our brain grows old too, and in time, the microglia cannot further respond to the environment in the same way it did when we were young (Gemma et al. 2002). The authors conclude that the process of aging leads to an alteration of the microglial function, which on the one hand, reduces the brain's resiliency, and on the other hand increases the vulnerability to develop neurodegenerative diseases, such as Parkinson's disease or Alzheimer's disease. After thoroughly presenting the mechanisms involved in these pathophysiological processes, the authors offer insights and personal studies on potential therapeutic effects in humans for polyphenols as a promise for neuroprotection. The authors suggest that polyphenol-rich dietary supplements may change the environment where the illnesses are unfolding and rebalance the resilience. Normally, we lose resilience as we grow old. The authors conducted their studies in rats (Bickford et al. 2015) and human stem cells (Bickford et al. 2006). But, as the authors concluded that researchers need to be aware that not all the phenomena that happens in mice will similarly occur in humans. There are notable differences between the immune systems of the two, mice and men, and we should not jump to conclusions (Nava Catorce and Gevorkian 2016).

Martínez-Cengotitabengoa et al. (2017) provide a synthesis of the mechanisms by which nutrition affects mental health and enumerate the following as being the most important ones: the high metabolic rate of the human brain; the necessity of an adequate supply of (micro-) nutrients; the immune system function modulation via eating habits; the cofactors and phytochemicals that ensure a good antioxidant defense system; the plasticity and neuronal maintenance as roles of the neurotrophic factors (Martínez-Cengotitabengoa and González-Pinto 2017).

Polyunsaturated Fatty Acids (PUFAs)

The ω-3 Long Chain (LC)-PUFAs administration in healthy humans may have an association with the augmentation of the functional activation in the prefrontal cortex of children. When aging, it seems that LC-PUFAs administration in healthy humans may have an association with the integrity of the white matter and a greater volume of the gray matter. These indications were drawn from naturalistic studies and clinical trials. But most of the naturalistic studies were cross-sectional or did not find cognitive effects. From this perspective, it is important to be aware that so far, we are unable to estimate the size of the beneficial effects, allegedly there exists some. The supplementation with ω-3 LC-PUFA in Attention-Deficit/Hyperactivity Disorder (ADHD), Schizophrenia, Major Depressive Disorder or Alzheimer's Disease (AD) is thought to have beneficial effects, but the evidence is scarce and limited (Bos et al. 2016).

There is clinical evidence about ω-3 PUFAs' preventive benefits on mood and anxiety disorders but more prospective clinical trials in high-risk populations are needed (Su, Matsuoka, and Pae 2015).

Agostoni et al. (2017) addressed the efficacy of ω-3 PUFAs in Autism Spectrum Disorder (ASD), ADHD and Psychosis. The findings suggest that there is a need to conduct prospective interventional clinical studies with larger amounts of participants. Although the results seem to highlight the beneficial role of PUFAs, the intervention should consist of administering ω-3 PUFAs in therapeutic dosage right at the start of the disease (Agostoni et al. 2017). Similar conclusions were drawn by other authors (Bozzatello et al. 2016; Sinn, Milte, and Howe 2010).

Psychobiotics

O'Hagan et al. (2017) showed that long-term multi-species *Lactobacillus* and *Bifidobacterium* dietary supplement enhances memory and changes regional brain metabolites in middle-aged rats (O'Hagan et al. 2017).

De Lorenzo et al. (2017) assessed a 3-week intake of selected probiotics – psychobiotic oral suspension (POS) – using a randomized, double-blinded controlled design, in 48 women normal weight obese (NWO), pre-obese–obese (PreOB/OB) compared to normal weight lean women (NWL). The conclusions were that there may be a possibility for a therapeutic potential as strategies used for preventing and/or treating some specific eating behavior disorders and anxiety-like conditions. Other therapeutic potentials were expressed in terms of avoiding the progression of psychiatric symptoms and the prevention or limitation of a global functional impairment in patients. Another issue was the potential improvement in the patients' quality of life (De Lorenzo et al. 2017).

Kim et al. (2018) doubt the therapeutic potential of probiotics suggesting the discussion to be premature when there is still a need for additional research to differentiate the clinical uses of probiotics and prebiotics, as well as for the fecal microbiota transplantation (Kim and Shin 2018), supported by similar findings (Keunen et al. 2015; Hasan Mohajeri et al. 2018; Mayer et al. 2014; Pusceddu et al. 2016; Romijn and Rucklidge 2015).

Kelly et al. (2017) found that *L. rhamnosus* was not superior over placebo in an 8 week, randomized, placebo-controlled, cross-over design, in twenty-nine healthy male volunteers (Kelly, Allen, et al. 2017). In opposition, Bagga et al. (2018) showed that there exists evidence on multidimensional levels that the administration of a multi-strain probiotic and the subsequent change in the gut microbiota composition associated with that administration has a significant impact on behavioral tests, as well as fMRI tests in healthy volunteers. This impact is interrelated and is notably observed in certain brain areas that are involved in the processes of emotional decision-making and emotional memory (Bagga et al. 2018), consistent with other findings (Kim et al. 2018).

If we are to design and implement effective therapies that consist of probiotics and prebiotics, we must first understand how the development of the CNS is modulated via the gut's microbiota. It is mandatory to discover the molecular mechanisms of this modulation. Cerdó et al. (2017) advocate for the necessity to conduct *in vivo* double-blind placebo studies by using the power of high-throughput multi-omic technologies (Cerdó et al. 2017), similar to another author (Sasmita 2019), while Sarkar et al. (2016) insist on the role of moderators and on the drug interactions as important parts of the knowledge gap (Sarkar et al. 2016).

Creatine

Although creatine has been studied for the potential of having a positive effect on brain functions, as Rae et al. (2015), together with Rawson et al. (2018) have demonstrated, the evidence so far for a positive effect of creatine on brain function in healthy individuals is only moderate (Rae and Bröer 2015; Rawson, Miles, and Larson-Meyer 2018). Furthermore, Dolan et al. (2019) discuss the evidence that indicates that augmented levels of brain creatine may be beneficial for mild traumatic brain injury, but as the authors agree, up to date, there is only limited data in humans to fully support this hypothesis (Dolan, Gualano, and Rawson 2019).

Plant-Derived Alkaloids

Hussain et al. (2018) reviewed the potential efficacy and potential therapeutic effects against several neurodegenerative disorders. The authors agree that safety studies are required, as well as to design clinical trials (Hussain et al. 2018).

Safety Concerns and Potential Side Effects: How Safe Are They?

Nutraceuticals are *generally regarded as safe* (GRAS). However, although probiotics are considered safe, Rao et al. (2018) publish a paper on June, 19th with research conducted in a cohort of patients without the short bowel and describe a syndrome of brain fogginess (BF), gas and bloating.

They postulated a possible relation to the use of probiotic and small intestinal bacterial overgrowth (SIBO) as well as D-lactic acidosis. 30 patients presented BF and 8 did not. It turned out that the patients suffering from BF presented an increase prevalence of SIBO and D-lactic acidosis and that the symptoms had improved when antibiotics were administered and probiotics were stopped. The authors recommend clinicians to recognize this condition and treat it accordingly (Rao et al. 2018). According to the cited literature, BF consists of a constellation of symptoms and includes mental confusion, judgment impairment, short-term memory loss or poorer functioning, and concentration difficulties. These symptoms appear to be transient and disabling repeatedly. Similar manifestations, alongside slurred speech and gait disturbances, have been previously reported, since 1979, in patients with short bowel syndrome (Oh et al. 1979). It was not a surprise that comments to the article rose quickly, on September, 20th (Sachdeva et al. 2018), and on September 24th (Quigley, Pot, and Sanders 2018), when subsequent papers expressed their contradicting opinions to the authors.'

Long-term glutamine (GLN) supplementation's potential side effects are thoroughly analyzed. The most discussed ones are the accumulation of GLN precursors because of the altered endogenous synthesis of GLN, the increased production of GLN metabolites because of the increased breakdown of GLN, the altered transport of amino acid through the cellular membrane, the concentrations modification in body fluids for the amino acid, the immune system alteration, the tumor growth effect, and last but not least the withdrawal effect of GLN supplementation (Holecek 2013).

Use of Nutraceuticals in Different Psychiatry Diseases

Brain Development, Aging, Memory and Cognitive Impairment

Docosahexaenoic acid (DHA) is highly concentrated in the human brain (Calder 2016; Sun et al. 2018). The author provides a thorough analysis of DHA biosynthesis, dietary sources of DHA, DHA status and mechanism of action and emphasized the need for pregnant and lactating women and infants to have a sufficient intake of preformed DHA to support the brain development and the brain function. Moreover, an animal model of dietary DHA administration, formulated as lysophosphatidylcholine, has been shown to increase brain DHA by >2-fold in healthy adult mice and to improve memory in adult mice. The findings suggest potential cognitive benefits for adult mammals due to the increases in brain DHA and BDNF, but confirmation in humans is necessary (Sugasini et al. 2017). But according to Meldrum et al. (2016), so far we cannot conclude for sure that a supplementation regime with DHA improves the development of the CNS in healthy on-term delivered babies and children, regardless the supplementation occurs in pregnancy, during lactation or starting only in infancy, as there is a lack of definite evidence (Meldrum and Simmer 2016).

Derbyshire (2018) accounts that even there are numerous studies in older individuals that indicate that ω fatty acids can be a protective measure against the neurodegeneration, there is still a need for more research in the years before the actual diagnosis of these conditions (Derbyshire 2018).

Hu et al. (2013) conclude that a decreased risk of AD development seems to be associated with increased consumption of fruits and vegetables, legumes, nuts, and cereals increased consumption of fish and decreased consumption of meats, as well as refrain from high-fat dairy and sweets, or sodium intake and refined grains. Unfortunately, there are very few studies conducted in this area. (Hu et al. 2013).

When discussing the dietary interventions in children and adolescents with obesity or overweight we cannot definitively conclude, based on the current levels of evidence, the effectiveness of the impact on cognition and academic outcomes of such interventions (Martin et al. 2018).

Nyaradi et al. (2013) consider that a balanced and healthy diet can enhance, in a synergistic manner, the cognitive development in children. Being so, the public health sector should focus mainly on measures to promote an overall diet quality, both in children and adolescents, in opposition to only isolated micronutrient intake or dietary components intake (Nyaradi et al. 2013).

The conclusions postulated by Meeusen et al. (2018) show that only carbohydrates and caffeine influence some aspects of cognition. Most of the herbal products alleged to have any influence on the brain functions whatsoever present "anecdotal" proof (Meeusen and Decroix 2018).

Ginkgo biloba, curcumin, and huperzine A for AD require further evaluation (Kelley and Knopman 2008). Other potential therapeutic candidates are represented by ascorbic acid (Harrison, Bowman, and Polidori 2014) and *Centella asiatica* (Gray et al. 2018).

For a more incisive point of view, we will quote on Hellmuth et al. (2019): "with neurodegenerative disease, the most common example of pseudomedicine is the promotion of dietary supplements to improve cognition and brain health. [...] No known dietary supplement prevents cognitive decline or dementia, yet supplements advertised as such are widely available and appear to gain legitimacy when sold by major [...] retailers" (Hellmuth, Rabinovici, and Miller 2019).

Anxiety

Liu et al. (2018) conducted a meta-analysis on published RCTs on probiotics efficacy in anxiety. They concluded that the evidence is insufficient to postulate the anxiolytic efficacy of probiotics (B. Liu et al. 2018). For a positive result, see the paper of Colica et al. (2017), but with the limitation that the prospective intervention study design did not account for a placebo-arm (Colica et al. 2017).

Tran et al. (2019) conducted the first study to examine the effect of the colony-forming unit and species count on probiotics' efficacy in a randomized placebo-controlled trial of healthy young adults and suggested that probiotics may have the therapeutic potential to treat anxiety (Tran et al. 2019).

Other potential therapeutic agents are represented by black cohosh, chamomile, chasteberry, lavender, passionflower, and saffron (Yeung et al. 2018). As per usual, additional research needs to be conducted to establish the efficacy of one or all of the above mentioned therapeutic agents.

Stress-Induced Neurovegetative Disorders

Chaborski et al. (2015) did not find any improvement compared to the placebo group when having a daily supplementation of a specific amino acid composition with micronutrients in psychologically stressed patients (Chaborski et al. 2015).

Ait-Belgnaoui et al. (2014) provided that a probiotic formulation (*Lactobacillus helveticus* R0052 and *Bifidobacterium longum* R0175) administered for 2 weeks as a treatment in mice had the positive outcome of preventing the neurogenesis reduction due to the inducement of stress in the hippocampal area (Ait-Belgnaoui et al. 2014). Nishida et al. (2019) showed beneficial effects of *Lactobacillus gasseri* in a randomized, double-blind, placebo-controlled study in young adults exposed to chronic stress (Nishida et al. 2019), similar to previous findings (Allen et al. 2016).

Dietary supplementation of phosphatidylserine augments the supply to the human brain and safely attenuates the increase in cortisol secretion (Glade and Smith 2015).

Amborst et al. (2018) tested a specific amino acid composition with micronutrients and their pilot-study showed promising results (Armborst et al. 2018). The results are difficult to interpret because of the complex formulae that have 4 compounds under the amino acid section, 9 compounds under the vitamin section, and 5 more compounds under the mineral section.

Obsessive Compulsive and Related Disorders

Oliver et al. (2015) conducted a systematic review of clinical trials and case reports about N-acetyl cysteine (NAC) and its potential benefits. The authors conclude that even if NAC may appear to be effective in most of the published literature, the data is far from being conclusive. Moreover, the topic should be addressed rigorously in further research (Oliver et al. 2015).

Depressive Disorders

Martínez-Cengotitabengoa et al. (2017) provided a thorough analysis of the studied substances about depressive disorders. Listed in alphabetical order, these substances are S-adenosyl methionine, vitamin B group, magnesium, ω-3 fatty acids – eicosapentaenoic acid (EPA and DHA), probiotics, tryptophan and zinc (Martínez-Cengotitabengoa and González-Pinto 2017).

The authors show some evidence available at present to support the recommendation that a part of the nutritional supplements investigated may be useful in the depressive symptoms' preventions in healthy individuals. Other uses may account for mild depressive symptoms monotherapy. When talking about major depressive disorder treated with antidepressants, these substances may account for associated adjuvant treatment. Their findings report that the experience is greatest with ω-3 fatty acids, but even so, there is still scarce evidence. This having said, both preclinical and clinical studies are needed.

For probiotics, things are quite clear from the authors' perspective: we need more future studies to support their efficacy. Another problem is the determination of the adequate species and strains composition in the supplement that we want to use, as well as the timeframe for the probiotic treatment.

These findings are supported by other authors, such as Pirbaglou et al. (2016) (Pirbaglou et al. 2016), who agree that probiotics may suggest a useful and novel approach in the therapy of neuropathological disorders. For psychiatric disorders, they may be used as adjunct therapies in conditions such as depression. However, to draw clear conclusions, we need well-designed clinical trials (Vlainic et al. 2016). The challenge to indicate the underlying mechanisms which contribute to clinical efficacy must be noted because of the great complexity of the gut microbiota, as well as of the heterogeneity of depression (Yong et al. 2020). These ideas are similar to a narrative review that evaluated systematic reviews (Nadeem et al. 2019).

On the other hand, Vaghef-Mehrabany et al. (2019) state that present evidence is not enough to agree or disagree with the probiotics' alleged antidepressant effect, because the probiotics may intervene in the depression in a strain-specific manner. Furthermore, when studying the antidepressant effect of prebiotics and synbiotics we are left with the same conclusion – we only have inconclusive results (Vaghef-Mehrabany et al. 2019).

Other potential therapeutic agents include soybean isoflavones (Messina and Gleason 2016), black cohosh, chamomile, chasteberry, lavender, passionflower, and saffron (Yeung et al. 2018), bioactive substances in selected species of Genera *Hemerocallis* and *Gladiolus* (Matraszek-Gawron et al. 2019).

Schizophrenia and Psychotic Disorders

Balanzá Martínez (2017) provided a thorough analysis of the studied substances concerning Schizophrenia and Psychotic Disorders. Listed in alphabetical order, these substances are vitamin A, vitamin B group, vitamin C, vitamin D, vitamin E, PUFAs, and probiotics (Balanzá Martínez 2017).

The author can only hope that one-day specific nutrients, such as ω-3, vitamin D and vitamin B group, could be useful as complementary strategies in the management of schizophrenia, even though clinical trials so far are still limited and have inconsistent results. The author states that even if we may develop interventions with formulae that combine several nutrients, thought to have a greater potential than those based on a single nutrient, we must be aware that these nutrients are not going to improve all the psychotic symptoms for all the patients suffering from psychotic disorders, but very likely in some specific subgroups, due to the marked heterogeneity of these disorders.

Attention Deficit-Hyperactivity Disorder

Mohammadzadeh et al. (2019) studied the effect of PUFAs in ADHD, in a double-blinded clinical trial, and found no difference between the methylphenidate with ω-3 group and methylphenidate with the placebo group, concluding that even if a specific dose of ω-3 was administered for 8 weeks, this treatment has not proven any effect on ADHD, whatsoever (Mohammadzadeh et al. 2019).

Pelsser et al. (2017) show that it is unlikely to provide a concrete contribution to the ADHD treatment when supplementing with PUFA (Pelsser et al. 2017).

Autism Spectrum Disorders

El-Ansary et al. (2020) discuss the existing evidence that nutritional factors that limit PUFAs may be involved in autism spectrum disorders development (El-Ansary et al. 2020), as shown in previous studies (Agostoni et al. 2017), (Jory 2016). Insufficient consumption of fish, meat, and nuts was suggested as an explanation for the deficiencies of docosahexaenoic acid (DHA), arachidonic acid (AA), and linoleic acid (LA). However, the limited data in humans, to date, cannot allow a definitive conclusion about ASD and diet supplementation.

The study of Adams et al. (2011), concerning a vitamin/mineral supplement on children and adults with autism, consisted of a randomized, double-blind, placebo-controlled three-month vitamin/mineral treatment study and involved 141 children and adults suffering from autism, suggest, as the authors state "that a vitamin/mineral supplement is a reasonable adjunct therapy for most children and adults with autism" (Adams et al. 2011). The results are difficult to interpret because of the complex formulae of the vitamin/mineral supplement that has 20 compounds under the vitamin section and another 14 compounds under the mineral section, and it is hard to attribute the potential therapeutic effect to one or the other, or all of the compounds. The limitations to the study provided by the authors present some of the aspects in question.

Lofthouse et al. (2012) reviewed the existing data concerning melatonin, B6 and magnesium, methyl B12, multivitamin/mineral supplements, folic acid, Ω-3 fatty acids (EPA and DHA), L-carnosine, ascorbic acid, probiotics, and GI medication, cyproheptadine, iron supplementation. Based on their analysis, if the authors were to recommend any supplements for those individuals who suffer from a limited diet and/or face poor appetite, their recommendations would only be melatonin and multivitamin/mineral (Lofthouse et al. 2012). For an almost similar opinion, Anagnostou et al. (2011) encourage the use of melatonin when indicated, as well-tolerated and effective (Anagnostou and Hansen 2011).

Kelly et al. (2017) suggest that psychobiotics can improve certain defects by showing pre-clinical benefits. They agree that translating these findings to humans is "challenging" (Kelly, Minuto, et al. 2017). Fattorusso et al. (2019) exert that there is a promise for the treatment of bowel dysfunctions and the treatment of neurobehavioral symptoms if we are to administer some specific probiotics – mostly a mixture of *Bifidobacteria*, *Streptococci,* and *Lactobacilli*. However, we are still in the land of promises, because the clinical trials to date are heterogeneous and limited (Fattorusso et al. 2019), with similar findings from other authors (Y. W. Liu et al. 2019), (Yang, Tian, and Yang 2018).

Trudeau et al. (2019) show that even if a variety of supplements and dietary interventions are prevalently used, we cannot support dietary therapy or nutritional supplements use based on the current evidence, which is little. (Trudeau et al. 2019) and Navarro et al. (2016) conclude that regarding behavior in children suffering from ASD the evidence does not confirm the benefit of probiotics (Navarro, Liu, and Rhoads 2016).

PERSPECTIVES: WHAT LIES AHEAD?

Jacka (2017) describes the emerging field of Nutritional Psychiatry and advocates to collaborate with key allies in the public health domain to be capable to advocate for a change in policy to improve the food environment at a populational level. Other aspects would be to develop and build knowledge and increase capacity within clinical settings to promote the benefits of dietary improvement for patients suffering from mental disorders (Jacka 2017).

Firth et al. (2019) supplied a meta-review of meta-analyses and concluded that: ω-3, EPA in particular, and as high-EPA formulas, is the nutritional intervention with the strongest evidentiary support, especially when given adjunctively to antidepressants, whereas N-acetylcysteine (at doses of 2,000 mg/day or higher) is potentially effective for reducing depressive symptoms and improving functional recovery and significant reductions in total symptoms of schizophrenia as an adjunctive treatment (Firth et al. 2019).

Conclusion

As we have seen, there is a tremendous interest in this rising field of medicine, that combines psychiatry and nutrition with the upmost promising research techniques. However, so far, medical research has not been able to provide definite answers to all the questions concerning this matter.

Although many signs of progress have been made, many are still needed.

The most important thing would be that there are still numerous opportunities for improving our present-day medical knowledge for the benefit of our patients.

Probably, the most important challenge would be to fully decipher the underlying mechanism of actions of the studied substances and to fully translate preclinical findings into clinical evidence. This requirement is needed to practice evidence-based medicine.

References

Adams, James B., Tapan Audhya, Sharon McDonough-Means, Robert A. Rubin, David Quig, Elizabeth Geis, Eva Gehn, et al. 2011. "Effect of a Vitamin/Mineral Supplement on Children and Adults with Autism." *BMC Pediatrics* 11 (1): 111. https://doi.org/10.1186/1471-2431-11-111.

Agostoni, Carlo, Maria Nobile, Valentina Ciappolino, Giuseppe Delvecchio, Alessandra Tesei, Stefano Turolo, Alessandro Crippa, Alessandra Mazzocchi, Carlo A. Altamura, and Paolo Brambilla. 2017. "The Role of Ω-3 Fatty Acids in Developmental Psychopathology: A Systematic Review on Early Psychosis, Autism, and ADHD." *International Journal of Molecular Sciences* 18 (12): 2608. https://doi.org/ 10.3390/ ijms18122608.

Ait-Belgnaoui, A., A. Colom, V. Braniste, L. Ramalho, A. Marrot, C. Cartier, E. Houdeau, V. Theodorou, and T. Tompkins. 2014. "Probiotic Gut Effect Prevents the Chronic Psychological Stress-Induced Brain Activity Abnormality in Mice." *Neurogastroenterology and Motility* 26 (4): 510–20. https://doi.org/10.1111/nmo.12295.

Allen, A. P., W. Hutch, Y. E. Borre, P. J. Kennedy, A. Temko, G. Boylan, E. Murphy, J. F. Cryan, T. G. Dinan, and G. Clarke. 2016. "Bifidobacterium Longum 1714 as a Translational Psychobiotic: Modulation of Stress, Electrophysiology and Neurocognition in Healthy Volunteers." *Translational Psychiatry* 6 (11). https://doi.org/ 10.1038/tp.2016.191.

Anagnostou, Evdokia, and Robin Hansen. 2011. "Medical Treatment Overview: Traditional and Novel Psycho-Pharmacological and Complementary and Alternative Medications." *Current Opinion in Pediatrics* 23 (6): 621–27. https://doi.org/10.1097/MOP.0b013e32 834cba3e.

Armborst, Deborah, Christine Metzner, Birgit Alteheld, Norman Bitterlich, Daniela Rösler, and Roswitha Siener. 2018. "Impact of a Specific Amino Acid Composition with Micronutrients on Well-Being in Subjects with Chronic Psychological Stress and Exhaustion Conditions: A Pilot Study." *Nutrients* 10 (5): 1–23. https://doi.org/ 10.3390/nu10050551.

Bagga, Deepika, Johanna Louise Reichert, Karl Koschutnig, Christoph Stefan Aigner, Peter Holzer, Kaisa Koskinen, Christine Moissl-Eichinger, and Veronika Schöpf. 2018. "Probiotics Drive Gut Microbiome Triggering Emotional Brain Signatures." *Gut Microbes* 9 (6): 486–96. https://doi.org/10.1080/19490976.2018.1460015.

Balanzá Martínez, Vicent. 2017. "Nutritional Supplements in Psychotic Disorders." *Actas Espanolas de Psiquiatria* 45 (Supplement): 16–25. https://www.actaspsiquiatria.es/repositorio//suplements/19/ENG/19-ENG-648645.pdf. [*Spanish Acts of Psychiatry*]

Berthoud, Hans Rudolf. 2007. "Interactions between the 'Cognitive' and 'Metabolic' Brain in the Control of Food Intake." *Physiology and Behavior* 91 (5): 486–98. https://doi.org/10.1016/j.physbeh.2006.12.016.

Bickford, Paula C., Antwoine Flowers, and Bethany Grimmig. 2017. "Aging Leads to Altered Microglial Function That Reduces Brain Resiliency Increasing Vulnerability to Neurodegenerative Diseases." *Experimental Gerontology* 94 (1): 4–8. https://doi.org/10.1016/j.exger. 2017.01.027.

Bickford, Paula C, Yuji Kaneko, Bethany Grimmig, Colleen Pappas, Brent Small, Cyndy D Sanberg, Paul R Sanberg, Jun Tan, and R Douglas Shytle. 2015. "Nutraceutical Intervention Reverses the Negative Effects of Blood from Aged Rats on Stem Cells." *Age (Dordrecht, Netherlands)* 37 (5): 103. https://doi.org/10.1007/s11357-015-9840-7.

Bickford, Paula C, Jun Tan, R Douglas Shytle, Cyndy D Sanberg, Nagwa El-Badri, and Paul R Sanberg. 2006. "Nutraceuticals Synergistically Promote Proliferation of Human Stem Cells." *Stem Cells and Development* 15 (1): 118–23. https://doi.org/10.1089/scd.2006.15.118.

Bos, Dienke J., Simone J. T. van Montfort, Bob Oranje, Sarah Durston, and Paul A. M. Smeets. 2016. "Effects of Ω-3 Polyunsaturated Fatty Acids on Human Brain Morphology and Function: What Is the Evidence?" *European Neuropsychopharmacology* 26 (3): 546–61. https://doi.org/10.1016/j.euroneuro.2015.12.031.

Bozzatello, Paola, Elena Brignolo, Elisa De Grandi, and Silvio Bellino. 2016. "Supplementation with Ω-3 Fatty Acids in Psychiatric Disorders: A Review of Literature Data." *Journal of Clinical Medicine* 5 (8): 67. https://doi.org/10.3390/jcm5080067.

Calder, Philip C. 2016. "Docosahexaenoic Acid." *Annals of Nutrition and Metabolism* 69 (1): 8–21. https://doi.org/10.1159/000448262.

Cerdó, Tomás, Alicia Ruíz, Antonio Suárez, and Cristina Campoy. 2017. "Probiotic, Prebiotic, and Brain Development." *Nutrients* 9 (11): 1–19. https://doi.org/10.3390/nu9111247.

Chaborski, Katrin, Norman Bitterlich, Birgit Alteheld, Elke Parsi, and Christine Metzner. 2015. "Placebo-Controlled Dietary Intervention of Stress-Induced Neurovegetative Disorders with a Specific Amino Acid Composition: A Pilot-Study." *Nutrition Journal* 14 (1): 1–8. https://doi.org/10.1186/s12937-015-0030-3.

Colica, Carmela, Ennio Avolio, Patrizio Bollero, Renata Costa De Miranda, Simona Ferraro, Paola Sinibaldi Salimei, Antonino De Lorenzo, and Laura Di Renzo. 2017. "Evidences of a New Psychobiotic Formulation on Body Composition and Anxiety." *Mediators of Inflammation* 2017. https://doi.org/10.1155/2017/ 5650627.

Derbyshire, Emma. 2018. "Brain Health across the Lifespan: A Systematic Review on the Role of Ω-3 Fatty Acid Supplements." *Nutrients* 10 (8). https://doi.org/10.3390/nu10081094.

Dolan, Eimear, Bruno Gualano, and Eric S. Rawson. 2019. "Beyond Muscle: The Effects of Creatine Supplementation on Brain Creatine, Cognitive Processing, and Traumatic Brain Injury." *European Journal of Sport Science* 19 (1): 1–14. https://doi.org/10.1080/17461391.2018.1500644.

El-Ansary, Afaf, Salvatore Chirumbolo, Ramesa Shafi Bhat, Maryam Dadar, Eiman M. Ibrahim, and Geir Bjørklund. 2020. "The Role of Lipidomics in Autism Spectrum Disorder." *Molecular Diagnosis and Therapy* 24 (1): 31–48. https://doi.org/10.1007/s40291-019-00430-0.

Fattorusso, Antonella, Lorenza Di Genova, Giovanni Battista Dell'isola, Elisabetta Mencaroni, and Susanna Esposito. 2019. "Autism Spectrum Disorders and the Gut Microbiota." *Nutrients* 11 (3). https://doi.org/10.3390/nu11030521.

Firth, Joseph, Scott B. Teasdale, Kelly Allott, Dan Siskind, Wolfgang Marx, Jack Cotter, Nicola Veronese, et al. 2019. "The Efficacy and Safety of Nutrient Supplements in the Treatment of Mental Disorders: A Meta-Review of Meta-Analyses of Randomized Controlled Trials." *World Psychiatry* 18 (3): 308–24. https://doi.org/10.1002/wps.20672.

Gemma, Carmelina, Michael H Mesches, Boris Sepesi, Kevin Choo, Douglas B Holmes, and Paula C Bickford. 2002. "Diets Enriched in Foods with High Antioxidant Activity Reverse Age-Induced Decreases in Cerebellar Beta-Adrenergic Function and Increases in Proinflammatory Cytokines." *The Journal of Neuroscience : The Official Journal of the Society for Neuroscience* 22 (14): 6114–20. https://doi.org/20026622.

Glade, Michael J., and Kyl Smith. 2015. "Phosphatidylserine and the Human Brain." *Nutrition* 31 (6): 781–86. https://doi.org/10.1016/j.nut. 2014.10.014.

Gray, Nora E, Armando Alcazar Magana, Parnian Lak, Kirsten M Wright, Joseph Quinn, Jan F Stevens, Claudia S Maier, and Amala Soumyanath. 2018. "Centella Asiatica: Phytochemistry and Mechanisms of Neuroprotection and Cognitive Enhancement." *Phytochemistry Reviews* 17 (1): 161–94. https://doi.org/10.1007/ s11101-017-9528-y.

Harrison, Fiona E., Gene L. Bowman, and Maria Cristina Polidori. 2014. "Ascorbic Acid and the Brain: Rationale for the Use against Cognitive Decline." *Nutrients* 6 (4): 1752–81. https://doi.org/10.3390/nu6041 752.

Hasan Mohajeri, M., Giorgio La Fata, Robert E. Steinert, and Peter Weber. 2018. "Relationship between the Gut Microbiome and Brain Function." *Nutrition Reviews* 76 (7): 481–96. https://doi.org/10.1093/ nutrit/ nuy009.

Hellmuth, Joanna, Gil D. Rabinovici, and Bruce L. Miller. 2019. "The Rise of Pseudomedicine for Dementia and Brain Health." *JAMA* 321 (6): 543. https://doi.org/10.1001/jama.2018.21560.

Holecek, Milan. 2013. "Side Effects of Long-Term Glutamine Supplementation." *Journal of Parenteral and Enteral Nutrition* 37 (5): 607–16. https://doi.org/10.1177/0148607112460682.

Hu, Nan, Jin Tai Yu, Lin Tan, Ying Li Wang, Lei Sun, and Lan Tan. 2013. "Nutrition and the Risk of Alzheimer's Disease." *BioMed Research International* 2013. https://doi.org/10.1155/2013/524820.

Hussain, Ghulam, Azhar Rasul, Haseeb Anwar, Nimra Aziz, Aroona Razzaq, Wei Wei, Muhammad Ali, Jiang Li, and Xiaomeng Li. 2018. "Role of Plant Derived Alkaloids and Their Mechanism in

Neurodegenerative Disorders." *International Journal of Biological Sciences* 14 (3): 341–57. https://doi.org/10.7150/ijbs.23247.

Jacka, Felice N. 2017. "Nutritional Psychiatry: Where to Next?" *EBioMedicine* 17: 24–29. https://doi.org/10.1016/j.ebiom.2017.02.020.

Ji, Xiaopeng, Michael A Grandner, and Jianghong Liu. 2017. "The Relationship between Micronutrient Status and Sleep Patterns: A Systematic Review." *Public Health Nutrition* 20 (4): 687–701. https://doi.org/10.1017/S1368980016002603.

Jory, Joan. 2016. "Abnormal Fatty Acids in Canadian Children with Autism." *Nutrition* 32 (4): 474–77. https://doi.org/10.1016/j.nut.2015.10.019.

Kelley, Brendan J., and David S. Knopman. 2008. "Alternative Medicine and Alzheimer Disease." *Neurologist* 14 (5): 299–306. https://doi.org/10.1097/NRL.0b013e318172cf4d.

Kelly, John R., Andrew P. Allen, Andriy Temko, William Hutch, Paul J. Kennedy, Niloufar Farid, Eileen Murphy, et al. 2017. "Lost in Translation? The Potential Psychobiotic Lactobacillus Rhamnosus (JB-1) Fails to Modulate Stress or Cognitive Performance in Healthy Male Subjects." *Brain, Behavior, and Immunity* 61: 50–59. https://doi.org/10.1016/j.bbi.2016.11.018.

Kelly, John R., Chiara Minuto, John F. Cryan, Gerard Clarke, and Timothy G. Dinan. 2017. "Cross Talk: The Microbiota and Neurodevelopmental Disorders." *Frontiers in Neuroscience* 11 (SEP): 1–31. https://doi.org/10.3389/fnins.2017.00490.

Keunen, Kristin, Ruurd M. Van Elburg, Frank Van Bel, and Manon J.N.L. Benders. 2015. "Impact of Nutrition on Brain Development and Its Neuroprotective Implications Following Preterm Birth." *Pediatric Research* 77 (1): 148–55. https://doi.org/10.1038/pr.2014.171.

Kim, Namhee, Misun Yun, Young Joon Oh, and Hak Jong Choi. 2018. "Mind-Altering with the Gut: Modulation of the Gut-Brain Axis with Probiotics." *Journal of Microbiology* 56 (3): 172–82. https://doi.org/10.1007/s12275-018-8032-4.

Kim, Yong-Ku, and Cheolmin Shin. 2018. "The Microbiota-Gut-Brain Axis in Neuropsychiatric Disorders: Pathophysiological Mechanisms and

Novel Treatments." *Current Neuropharmacology* 15: 559–73. https://doi.org/10.2174/1570159x15666170915141036.

Kumar, G. Phani, and Farhath Khanum. 2012. "Neuroprotective Potential of Phytochemicals." *Pharmacognosy Reviews* 6 (12): 81–90. https://doi.org/10.4103/0973-7847.99898.

Liu, Bangshan, Yunan He, Mi Wang, Jin Liu, Yumeng Ju, Yan Zhang, Tiebang Liu, Lingjiang Li, and Qi Li. 2018. "Efficacy of Probiotics on Anxiety—A Meta-Analysis of Randomized Controlled Trials." *Depression and Anxiety* 35 (10): 935–45. https://doi.org/10.1002/da.22811.

Liu, Yen Wenn, Min Tze Liong, Yu Chu Ella Chung, Hui Yi Huang, Wu Shun Peng, Yun Fang Cheng, Yu Siou Lin, Yu Yu Wu, and Ying Chieh Tsai. 2019. "Effects of Lactobacillus Plantarum PS128 on Children with Autism Spectrum Disorder in Taiwan: A Randomized, Double-Blind, Placebo-Controlled Trial." *Nutrients* 11 (4): 1–16. https://doi.org/10.3390/nu11040820.

Lofthouse, Nicholas, Robert Hendren, Elizabeth Hurt, L. Eugene Arnold, and Eric Butter. 2012. "A Review of Complementary and Alternative Treatments for Autism Spectrum Disorders." *Autism Research and Treatment* 2012 (3): 1–21. https://doi.org/10.1155/2012/870391.

Lorenzo, Antonino De, Micaela Costacurta, Giuseppe Merra, Paola Gualtieri, Giorgia Cioccoloni, Massimiliano Marchetti, Dimitrios Varvaras, Raffaella Docimo, and Laura Renzo. 2017. "Can Psychobiotics Intake Modulate Psychological Profile and Body Composition of Women Affected by Normal Weight Obese Syndrome and Obesity? A Double Blind Randomized Clinical Trial." *Journal of Translational Medicine* 15 (1): 1–12. https://doi.org/10.1186/s12967-017-1236-2.

Martin, A, Booth Jn, Y Laird, J Sproule, Reilly Jj, Saunders Dh, A Martin, et al. 2018. "Adolescents with Obesity or Overweight (Review)." *Cochrane Database of Systematic Reviews Physical*, no. 3. https://doi.org/10.1002/14651858.CD009728.pub4.www.cochranelibrary.com.

Martínez-Cengotitabengoa, Mónica, and Ana González-Pinto. 2017. "Nutritional Supplements in Depressive Disorders." *Actas Espanolas de Psiquiatria* 45: 8–15.

Matraszek-Gawron, Renata, Mirosława Chwil, Paulina Terlecka, and Michał M. Skoczylas. 2019. "Recent Studies on Anti-Depressant Bioactive Substances in Selected Species from the Genera Hemerocallis and Gladiolus: A Systematic Review." *Pharmaceuticals* 12 (4). https://doi.org/10.3390/ph12040172.

Mayer, Emeran A., Rob Knight, Sarkis K. Mazmanian, John F. Cryan, and Kirsten Tillisch. 2014. "Gut Microbes and the Brain: Paradigm Shift in Neuroscience." *Journal of Neuroscience* 34 (46): 15490–96. https://doi.org/10.1523/JNEUROSCI.3299-14.2014.

Meeusen, Romain, and Lieselot Decroix. 2018. "Nutritional Supplements and the Brain." *International Journal of Sport Nutrition and Exercise Metabolism* 28 (2): 200–211. https://doi.org/10.1123/ijsnem.2017-0314.

Meldrum, Suzanne, and Karen Simmer. 2016. "Docosahexaenoic Acid and Neurodevelopmental Outcomes of Term Infants." *Annals of Nutrition and Metabolism* 69 (1): 23–28. https://doi.org/10.1159/000448271.

Messina, Mark, and Carey Gleason. 2016. "Evaluation of the Potential Antidepressant Effects of Soybean Isoflavones." *Menopause* 23 (12): 1348–60. https://doi.org/10.1097/GME.0000000000000709.

Mohammadzadeh, Soleiman, Narmin Baghi, Fayegh Yousefi, and Bahar Yousefzamani. 2019. "Effect of Ω-3 plus Methylphenidate as an Alternative Therapy to Reduce Attention Deficit-Hyperactivity Disorder in Children." *Korean Journal of Pediatrics* 62 (9): 360–66. https://doi.org/10.3345/kjp.2018.06982.

Nadeem, Ibrahim, Mohammed Z. Rahman, Yasser Ad-Dab'bagh, and Mahmood Akhtar. 2019. "Effect of Probiotic Interventions on Depressive Symptoms: A Narrative Review Evaluating Systematic Reviews." *Psychiatry and Clinical Neurosciences* 73 (4): 154–62. https://doi.org/10.1111/pcn.12804.

Nava Catorce, Miryam, and Goar Gevorkian. 2016. "LPS-Induced Murine Neuroinflammation Model: Main Features and Suitability for Pre-

Clinical Assessment of Nutraceuticals." *Current Neuropharmacology* 14 (2): 155–64. https://doi.org/10.2174/1570159x14666151204122017.

Navarro, Fernando, Yuying Liu, and Jon Marc Rhoads. 2016. "Can Probiotics Benefit Children with Autism Spectrum Disorders?" *World Journal of Gastroenterology* 22 (46): 10093–102. https://doi.org/10.3748/wjg.v22.i46.10093.

Nishida, Kensei, Daisuke Sawada, Yuki Kuwano, Hiroki Tanaka, and Kazuhito Rokutan. 2019. "Health Benefits of Lactobacillus Gasseri Cp2305 Tablets in Young Adults Exposed to Chronic Stress: A Randomized, Double-Blind, Placebo-Controlled Study." *Nutrients* 11 (8). https://doi.org/10.3390/nu11081859.

Noorwali, Essra, Laura Hardie, and Janet Cade. 2019. "Bridging the Reciprocal Gap between Sleep and Fruit and Vegetable Consumption: A Review of the Evidence, Potential Mechanisms, Implications, and Directions for Future Work." *Nutrients* 11 (6). https://doi.org/10.3390/nu11061382.

Nyaradi, Anett, Jianghong Li, Siobhan Hickling, Jonathan Foster, and Wendy H. Oddy. 2013. "The Role of Nutrition in Children's Neurocognitive Development, from Pregnancy through Childhood." *Frontiers in Human Neuroscience* 7 (MAR): 1–16. https://doi.org/10.3389/fnhum.2013.00097.

O'Hagan, Caroline, Jia V. Li, Julian R. Marchesi, Sue Plummer, Iveta Garaiova, and Mark A. Good. 2017. "Long-Term Multi-Species Lactobacillus and Bifidobacterium Dietary Supplement Enhances Memory and Changes Regional Brain Metabolites in Middle-Aged Rats." *Neurobiology of Learning and Memory* 144 (October): 36–47. https://doi.org/10.1016/j.nlm.2017.05.015.

Oh, Man S., Kenneth R. Phelps, Morris Traube, Jose L. Barbosa-Saldivar, Carlton Boxhill, and Hugh J. Carroll. 1979. "D-Lactic Acidosis in a Man with the Short-Bowel Syndrome." *New England Journal of Medicine* 301 (5): 249–52. https://doi.org/10.1056/NEJM197908023010505.

Oliver, Georgina, Olivia Dean, David Camfield, Scott Blair-West, Chee Ng, Michael Berk, and Jerome Sarris. 2015. "N-Acetyl Cysteine in the

Treatment of Obsessive Compulsive and Related Disorders: A Systematic Review." *Clinical Psychopharmacology and Neuroscience* 13 (1): 12–24. https://doi.org/10.9758/cpn.2015.13.1.12.

Pelsser, Lidy M., Klaas Frankena, Jan Toorman, and Rob Rodrigues Pereira. 2017. "Diet and ADHD, Reviewing the Evidence: A Systematic Review of Meta-Analyses of Double-Blind Placebo-Controlled Trials Evaluating the Efficacy of Diet Interventions on the Behavior of Children with ADHD." *PLoS ONE* 12 (1): 1–25. https://doi.org/10.1371/journal.pone.0169277.

Pirbaglou, Meysam, Joel Katz, Russell J. de Souza, Jennifer C. Stearns, Mehras Motamed, and Paul Ritvo. 2016. "Probiotic Supplementation Can Positively Affect Anxiety and Depressive Symptoms: A Systematic Review of Randomized Controlled Trials." *Nutrition Research* 36 (9): 889–98. https://doi.org/10.1016/j.nutres.2016.06.009.

Pusceddu, Matteo M., Philip Kelly, Catherine Stanton, John F. Cryan, and Timothy G. Dinan. 2016. "N-3 Polyunsaturated Fatty Acids through the Lifespan: Implication for Psychopathology." *International Journal of Neuropsychopharmacology* 19 (12): 1–23. https://doi.org/10.1093/ijnp/pyw078.

Quigley, Eamonn M.M., Bruno Pot, and Mary Ellen Sanders. 2018. "'Brain Fogginess' and D-Lactic Acidosis: Probiotics Are Not the Cause." *Clinical and Translational Gastroenterology* 9 (9): 0–1. https://doi.org/10.1038/s41424-018-0057-9.

Rae, Caroline D., and Stefan Bröer. 2015. "Creatine as a Booster for Human Brain Function. How Might It Work?" *Neurochemistry International* 89 (October): 249–59. https://doi.org/10.1016/j.neuint. 2015.08.010.

Rao, Satish S.C., Abdul Rehman, Siegfried Yu, and Nicole Martinez De Andino. 2018. "Brain Fogginess, Gas and Bloating: A Link between SIBO, Probiotics and Metabolic Acidosis Article." *Clinical and Translational Gastroenterology* 9 (6). https://doi.org/10.1038/s41424-018-0030-7.

Rawson, Eric S., Mary P. Miles, and D. Enette Larson-Meyer. 2018. "Dietary Supplements for Health, Adaptation, and Recovery in Athletes." *International Journal of Sport Nutrition and Exercise Metabolism* 28 (2): 188–99. https://doi.org/10.1123/ijsnem.2017-0340.

Romijn, Amy R., and Julia J. Rucklidge. 2015. "Systematic Review of Evidence to Support the Theory of Psychobiotics." *Nutrition Reviews* 73 (10): 675–93. https://doi.org/10.1093/nutrit/nuv025.

Sachdeva, Sanjeev, A. S. Puri, Ajay Kumar, Ashok Dalal, and Amol Sonyabapu Dahale. 2018. "Brain Fogginess and SIBO: A Link or Just a Mirage?" *Clinical and Translational Gastroenterology* 9 (9): 9–10. https://doi.org/10.1038/s41424-018-0055-y.

Sarkar, Amar, Soili M. Lehto, Siobhán Harty, Timothy G. Dinan, John F. Cryan, and Philip W. J. Burnet. 2016. "Psychobiotics and the Manipulation of Bacteria–Gut–Brain Signals." *Trends in Neurosciences* 39 (11): 763–81. https://doi.org/10.1016/j.tins.2016. 09.002.

Sasmita, Andrew Octavian. 2019. "Modification of the Gut Microbiome to Combat Neurodegeneration." *Reviews in the Neurosciences*. https://doi.org/10.1515/revneuro-2019-0005.

Sinn, Natalie, Catherine Milte, and Peter R. C. Howe. 2010. "Oiling the Brain: A Review of Randomized Controlled Trials of Ω-3 Fatty Acids in Psychopathology across the Lifespan." *Nutrients* 2 (2): 128–70. https://doi.org/10.3390/nu2020128.

Su, Kuan Pin, Yutaka Matsuoka, and Chi Un Pae. 2015. "Ω-3 Polyunsaturated Fatty Acids in Prevention of Mood and Anxiety Disorders." *Clinical Psychopharmacology and Neuroscience* 13 (2): 129–37. https://doi.org/10.9758/cpn.2015.13.2.129.

Sugasini, Dhavamani, Riya Thomas, Poorna C. R. Yalagala, Leon M. Tai, and Papasani V. Subbaiah. 2017. "Dietary Docosahexaenoic Acid (DHA) as Lysophosphatidylcholine, but Not as Free Acid, Enriches Brain DHA and Improves Memory in Adult Mice." *Scientific Reports* 7 (1): 1–11. https://doi.org/10.1038/s41598-017-11766-0.

Sun, Grace Y., Agnes Simonyi, Kevin L. Fritsche, Dennis Y. Chuang, Mark Hannink, Zezong Gu, C. Michael Greenlief, Jeffrey K. Yao, James C. Lee, and David Q. Beversdorf. 2018. "Docosahexaenoic Acid (DHA): An Essential Nutrient and a Nutraceutical for Brain Health and Diseases." *Prostaglandins Leukotrienes and Essential Fatty Acids* 136: 3–13. https://doi.org/10.1016/j.plefa.2017.03.006.

Tran, Nhan, Masha Zhebrak, Christine Yacoub, Joseph Pelletier, and Darby Hawley. 2019. "The Gut-Brain Relationship: Investigating the Effect of Multispecies Probiotics on Anxiety in a Randomized Placebo-Controlled Trial of Healthy Young Adults." *Journal of Affective Disorders* 252 (March): 271–77. https://doi.org/10.1016/j.jad.2019.04.043.

Trudeau, Madden, Parnell, Gibbard, and Shearer. 2019. "Dietary and Supplement-Based Complementary and Alternative Medicine Use in Pediatric Autism Spectrum Disorder." *Nutrients* 11 (8): 1783. https://doi.org/10.3390/nu11081783.

Vaghef-Mehrabany, E., V. Maleki, Maryam Behrooz, Fatemeh Ranjbar, and Mehrangiz Ebrahimi-Mameghani. 2019. "Can Psychobiotics 'Mood' Ify Gut? An Update Systematic Review of Randomized Controlled Trials in Healthy and Clinical Subjects, on Anti-Depressant Effects of Probiotics, Prebiotics, and Synbiotics." *Clinical Nutrition*, no. xxxx. https://doi.org/10.1016/j.clnu.2019.06.004.

Vlainic, Josipa, Jelena Suran, Toni Vlainic, and Antonella Letizia Vukorep. 2016. "Probiotics as an Adjuvant Therapy in Major Depressive Disorder." *Current Neuropharmacology* 14 (8): 952–58. https://doi.org/10.2174/1570159X14666160526120928.

Yang, Yongshou, Jinhu Tian, and Bo Yang. 2018. "Targeting Gut Microbiome: A Novel and Potential Therapy for Autism." *Life Sciences* 194 (December 2017): 111–19. https://doi.org/10.1016/j.lfs.2017.12.027.

Yeung, K. Simon, Marisol Hernandez, Jun J. Mao, Ingrid Haviland, and Jyothirmai Gubili. 2018. "Herbal Medicine for Depression and Anxiety: A Systematic Review with Assessment of Potential Psycho-Oncologic Relevance." *Phytotherapy Research* 32 (5): 865–91. https://doi.org/10.1002/ptr.6033.

Yong, Shin Jie, Tommy Tong, Jactty Chew, and Wei Ling Lim. 2020. "Antidepressive Mechanisms of Probiotics and Their Therapeutic Potential." *Frontiers in Neuroscience* 13 (January). https://doi.org/10.3389/fnins.2019.01361.

ABOUT THE EDITOR

Raluca Maria Pop - MD, PhD, Researcher, has a master degree in food quality management, one master degree in clinical pharmacology and one PhD in biotechnology. In 2008 she started her work as a research assistant at the Department of Chemistry and Biochemistry at the University of Agricultural Sciences and Veterinary Medicine Cluj-Napoca, Romania. During this period her expertise in the experimental work included several fields like Spectroscopy, Chromatography, Mass-Spectrometry, Chemometrics, Plant metabolomics, and Quality and authenticity of plant biomarkers determination. Further, her work as researcher at Research Centre for Applied Biotechnology in Diagnosis and Molecular Therapy, Cluj-Napoca was focused on the molecular pathogenesis and progression of different diseases, to discovery their molecular biomarkers, to diagnose or establish the prognosis of the diseases and treatment response by using both targeted and untargeted metabolomics. In the last years, her work as researcher at the Department of Pharmacology, Toxicology and Clinical Pharmacology, University of Medicine and Pharmacy "Iuliu Hatieganu" Cluj-Napoca enlarged her research experience. She started the work on pharmacogenetics, determining different gene polymorphisms to evaluate drug efficiency and safety. She was also focused on studying the antioxidant, anti-inflammatory and cardio protective effects of bioactive compounds in plants and their use in experimental animal models.

INDEX

#

A

B

C

D

E

F

G

H

P

Q

R

U

V

W

Y

Z

α

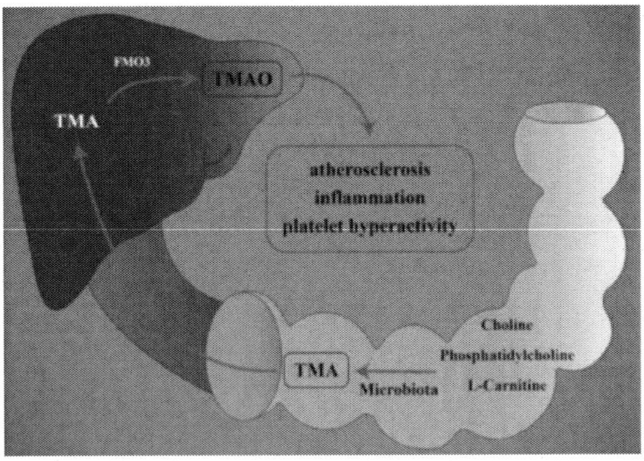

Endogenous antioxidants		**Exogenous antioxidants**
Enzymatic	Non-enzymatic	
Catalase	Glutathione	Vitamin A
Superoxide dismutase	Coenzyme Q10	Vitamin E
Glutathione peroxidase	Bilirubin	Vitamin C
Peroxiredoxin	Uric acid	Polyphenols
Thioredoxin		Selenium
		Phytochemicals

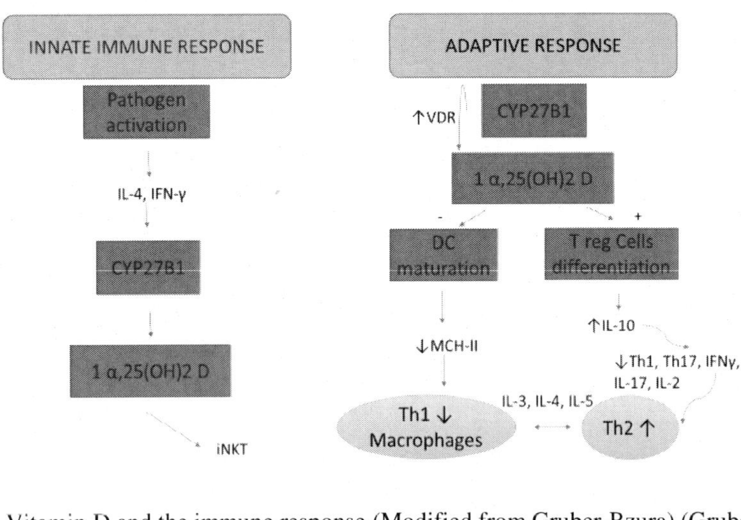

